T0134008

Network Based High Speed
Product Development
Second Edition

RIVER PUBLISHERS SERIES IN MULTI BUSINESS MODEL INNOVATION, TECHNOLOGIES AND SUSTAINABLE BUSINESS

Series Editors

PETER LINDGREN
Aarhus University
Denmark

ANNABETH AAGAARD
Aarhus University
Denmark

Indexing: All books published in this series are submitted to Thomson Reuters Book Citation Index (BkCI), CrossRef and to Google Scholar.

The River Publishers Series in Multi Business Model Innovation, Technologies and Sustainable Business includes the theory and use of multi business model innovation, technologies and sustainability involving typologies, ontologies, innovation methods and tools for multi business models, and sustainable business and sustainable innovation. The series cover cross technology business modeling, cross functional business models, network based business modeling, Green Business Models, Social Business Models, Global Business Models, Multi Business Model Innovation, interdisciplinary business model innovation. Strategic Business Model Innovation, Business Model Innovation Leadership and Management, technologies and software for supporting multi business modeling, Multi business modeling and strategic multi business modeling in different physical, digital and virtual worlds and sensing business models. Furthermore the series includes sustainable business models, sustainable & social innovation, CSR & sustainability in businesses and social entrepreneurship.

Key topics of the book series include:

- Multi business models
- Network based business models
- Open and closed business models
- Multi Business Model eco systems
- Global Business Models
- Multi Business model Innovation Leadership and Management
- Multi Business Model innovation models, methods and tools
- Sensing Multi Business Models
- Sustainable business models
- Sustainability & CSR in businesses
- Sustainable & social innovation
- Social entrepreneurship and -intrapreneurship

For a list of other books in this series, visit www.riverpublishers.com

Network Based High Speed
Product Development
Second Edition

Peter Lindgren

Aarhus University
Denmark

River Publishers

Published, sold and distributed by:
River Publishers
Alsbjergvej 10
9260 Gistrup
Denmark

River Publishers
Lange Geer 44
2611 PW Delft
The Netherlands

Tel.: +45369953197
www.riverpublishers.com

ISBN: 978-87-93519-27-5 (Hardback)
    978-87-93519-05-3 (Ebook)

Layout: Dorte Haslev
English, Grammar and Drawing: Anne Birgitte Kanstrup Lindgren
Print First Edition: Buchs Grafiske
Publisher Second and Revised Edition: River Publishers

Contents

PART II: Theoretical Foundation

2   Project Method   45

3   Concepts of Product Development   81

PART III: Analysis Model

PART IV: Empirical Results

PART V: Comparing Theoretical Framework
and Empirical Results

PART VI: Conclusion

Preface

This book is the second edition of original my PhD project handed in and defended in 2003 at Aalborg University, Denmark. The PhD was original printed in 11 pieces but several ask me to produce some more. Eventually I got the opportunity at my research stay in 2016–2017 at University Tor Vergata, Italy at the CTIF Global Capsule–(http://www.ctifglobalcapsule. com/) research center at Villa Mondracone.

At this stay I got the opportunity to reflect on my research work in 2000–2003 and beyond. It also laid the ground to commence my new book–to be published in 2017.

I revised very little of my original work in the PhD project. Of course much have change since I handed in and defended the PhD project in 2003. All though many of the findings are still actual – also in 2016 – I change some words e.g. business to the word business – because this was more in line with my research interest now – multi business model and technology innovation.

Actually the PhD project on Network Based High Speed Product development laid the ground for my interest in Business models and later Multi Business Model Innovation.

I also stressed in this book that research and findings were all related to data and situation as up to the year 2003. All cases mentioned in the book can be found at my MBIT LAb in a physical case book, which I will be happy to show and discuss.

Due to some important reflections and comments from the PhD committee and research colleges I change one model related to classification of the task of innovation. I took in the network as equal to technology and market to classify whether a innovation task was incremental or radical.

This book and the research behind was very much a product development project based on networks. It had a core, many stage and gates, many internal and external functions and many network partners who were involved in different stage- and gates. At the very beginning I decided on an overall product architecture and found that I couldn't use a traditional stage-gate model. The project was far too uncertain and dynamic. Uncertain because the evolutions in "customer needs" and the "technologies" required to meet these needs were difficult to predict – as nobody at the beginning had enter the field of Network based high speed product development. Dynamic because the evolution on "the market" occurred rapidly.

After reading "the field of product development" I decided on an explorative architecture with a flexible design. The "product" was kept "digital and floating" from the very beginning and gave me the possibility to change major functions in "the product" right up to the "market introduction".

Now "the product" is introduced to the market but far from finished. The development of "the product" is an ongoing process where "customers", "technology", "networks" and even my own "competences" I hope keeps on developing as if it was a never ending process of product and process development.

The very big question is how the product and the process will look like some 100 years from now.

I chose a focus of studying high speed network based product development but turned out to develop I presume a product of right speed network based product development. I could have delivered "the product" 1 year before – facing short term success criteria but the performance would not have meet the needs and wants of "the market" if I had done so. The costs had for sure been too high.

I learned during the process the importance of following long-term success criteria – continuous improvement, continuous innovation and learning.

Thank you Kim Bohn, Poul K. Hansen, John Johansen, Jens Riis, Poul Dreisler, Harry Boer, Mariano Corso, Roberto Verganti and Agnar Gudmundson. Also a big thanks to Junko Nakajima and her team at River Publishers for working together with me on this book. Thanks to River Publishers for letting me publish this work through Your Business.

I also learned the importance of having a strong internal organisation.

Thank you – Anne-Birgitte, Frederikke, Amalie and Thorbjørn and Lupo

Professor Multi Business Model Innovation and Technology Peter Lindgren

Professor PhD Peter Lindgren, Peter Lindgren holds a full Professorship in Multi business model and Technology innovation at Aarhus University, Denmark – Business development and technology innovation and has researched and worked with network based high speed innovation since 2000. He has been head of Studies for Master in Engineering – Business Development and Technology at Aarhus University from 2014–2016. He has been researcher at Politechnico di Milano in Italy (2002/03), Stanford University, USA (2010/11), University Tor Vergata, Italy (2016/2017) and has in the time period 2007–2011 been the founder and Center Manager of International Center for Innovation www.ici.aau.dk at Aalborg University, founder of the MBIT research group and lab – http://btech.au.dk/forskning/mbit/ – and is cofounder of CTIF Global Capsule – www.ctifglobalcapsule.com. He works today as researcher in many different multi business model and technology innovations projects and knowledge networks among others E100 – http://www.entovation.com/kleadmap/, Stanford University project Peace Innovation Lab http://captology.stanford.edu/projects/peace-innovation.html, The Nordic Women in business project – www.womeninbusiness.dk/, The Center for TeleInFrastruktur (CTIF) at Aalborg University www.ctif.aau.dk, EU FP7 project about "multi business model innovation in the clouds" – www.Neffics.eu, EU Kask project – www.Biogas2020.se. He is author to several articles and books about business model innovation in networks and

Emerging Business Models. He has an entrepreneurial and interdisciplinary approach to research.

His research interests are multi business model and technology innovation in interdisciplinary networks, multi business model typologies, sensing and persuasive business models.

Contact detail:

**Professor Multi Business Model Innovation and Technology
Peter Lindgren**
Aarhus University
School of Business and Social Sciences
Birk Centerpark 15, Office: MBIT LAB
DK - 7400 Herning
Denmark
T: +45 29442211/ +45 23425504
e-mail: peterli@btech.au.dk
Pure.: http://pure.au.dk/portal/en/persons/peter-lindgren(244bfceb-2a4c-4ef0-8c5a-34ad8238b5eb)/publications.html
W: http://www.riverpublishers.com/journal.php?j=JMBMIT/1/2

List of Figures

List of Tables

PART I

Problem Statement

The problem statement defines the contents of the book and gives a statement of the problem – network based high speed product development. In the problem statement a discussion of network based high speed product development is carried out, and a systematic and precise description of the problems and issues pertaining to network based high speed product development is generated. The problem statement also delimitates the focus of this book on network based high speed product development. This includes the main questions and the main hypotheses for network based high speed product development.

This part is completed with an overview of the structure of the book.

1

Project Introduction

The chapter will describe the studies which the book project has finally featured and how the analysis and thesis forming part of the book have been arranged. The chapter will describe and show how the thorough study of the subject – **High Speed Product Development Processes and Models Based on Networks (NB HS PD)** – together with the explorative analysis of the product development system of generic industrial case businesses including related models and processes have come up. The aim was to describe and analyse the current context in 2000–2003 of network based high speed product development and make an identification and introduction of the NB HS PD problem – **what is the problem?**, an identification and introduction of **why NB HS PD is a problem?** The aim was further to describe **when NB HS PD is a problem** and give an introduction to the theoretical model apparatus for high speed product development processes and models, Further the chapter gives an introduction to **when NB HS PD models and processes can be used?** with preference.

The chapter gives an introduction to how the theoretical framework outlined above was subsequently empirically tested in case Businesses, focus groups and survey with an explorative focus. The chapter introduce shortly the empirical part and study of Danish and foreign international case businesses which had established network based product development processes and models and which employ and try to employ HS PD based on networks. This chapter explains how the study contributed towards an explanation to processes and models for product development carried out at high speed and deeply rooted in networks.

On the basis of the above the chapter finally will describe how the final part of the book is concluded.

1.1 Context for NB HS PD in 2000–2003

In the first decade of the 21st century product development in networks was predicted to be of ever-increasing importance to businesses of all sizes because of changes in markets, in technology, in networks, and in the competences of Businesses (Nonaka & Takeuchi, 1995) (Sanchez, 1996) (Coldmann & Price, 1998) (Child and Faulkner, 1998). The growth in new products' share of businesses' total turnover and earnings were increasing at an unprecedented speed.

The entrepreneurial innovations and technological improvements had resulted in the increasingly fast development of new goods and services. Businesses and industries in different countries became more and more linked and interdependent in networks with respect to materials, business operations and particularly product development to match the wants and needs of the global market environment to high speed product development.

Businesses were therefore encountering increasingly dynamic market fragmentation, shrinking time in market, increasing product variety, demands of production to customer specifications, reduced product lifetimes, and globalisation of production. In the years up to the 2000–2003, many industrial businesses had seen the necessity of applying network based high speed product development in order to compete on the global market.

Previously, Businesses "simply" had to match their product development competences to the market and the technology, but now a match to the network component on the "field of product development" was vital. Networks were vital because the competition in the 21st century was not business against business, but network against network. Networks were vital because an increasing part of product development was carried out in all types of networks containing physical, ICT, dynamic, and virtual networks (Goldman & Price, 1998).

Speed and pressure on time in product development seemed to continue to increase because customer demands for new products seemed to continue to increase. However, a Business seldom possessed all needed competences, and managers therefore saw product development based on networks as an important solution to meet the strong competition of the future global markets and the strong demand for innovation and innovativeness (Grunert & Harmsen, 1997) (Boer 2001).

The evolution of market demands and focus (required) on competencies of businesses could be characterised as a development from a focus on efficiency, to a focus on quality and flexibility, to a focus on speed and innovativeness. This is shown in Figure 1.1.

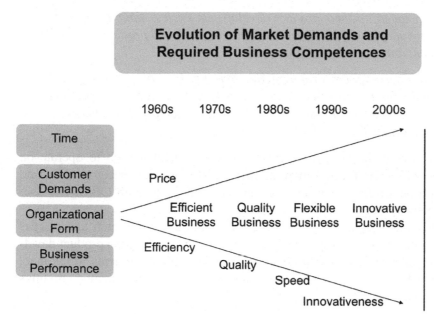

Figure 1.1 Evolution of market demands (model adapted from Harry Boer, 2001).

Source: Harry Boer, 2001 (adapted from Bolwijn and Kumpe, 1998).

Speed, uniqueness, innovativeness and the innovative business was said to be a must for all businesses who wants to join the global market (Baker & Hart, 1999) (Sanchez, 2000) (Albaum, 1994) (MacCormack, Verganti & Iansiti, 2001) but researchers claimed that businesses have to do even better for the future. Researchers claimed that in future, businesses had to show efficiency, quality, flexibility, speed and innovativeness (Boer, 2001) at the same time.

A constant high introduction ratio and "high speed to market" of new products, based on new available and stable technologies was an important competitive parameter (Baker & Hart, 1999) and

> *"Dozens of large businesses had fallen victim to competitors with faster, more flexible new-product development programmes."* (Kotler, 1996) (Case No. 1 Zara)

A major task, a major problem and a strong strategic demand therefore seemed to be to develop, implement high speed product development based on network cooperation.

> *"To speed up their product development cycles, many businesses were adopting a faster, more agile, team-oriented product development approach."* (Goldman, 1998)

> *"Businesses were dropping the sequential product development method in favour of the faster, more flexible simultaneous product development approach. to save time and increase effectiveness."* (Kotler, 1996)

As a result of the above, industrial businesses who wished to operate on the global market faced a more intensive and much fiercer competition and demand on speed in new product development from both OEM, end-user customers as well as from competitors.

This was why it was interesting and important to research and discuss product development and especially to understand high speed product development of individualised products in fragile market segments. (Goldman, 1998) (Child & Faulkner, 1998) (Sanchez, 1996). Consequently, findings and learning on:

1. Enablers
2. Management tools
3. Technological tools
4. Product development models
5. Product development processes
6. Network tools

to speed new product development were central. Likewise, it was important to understand how and when to speed new product development.

1.2 The Need to Improve Speed and Join Network to Gain HS PD – Why?

The book has initially put two main questions to NB HS NPD:

• Why do SMEs need to improve speed in NPD?

and

• Why do SMEs need to join networks to speed NPD?

These two questions are of main importance.

1.2.1 The Need to Improve Speed within New Product Development at the Idea and Concept Stage – Why?

The need to improve speed within NB NPD particularly in the idea and concept phase was of special interest because many SMEs had a very good performance on the lower part of the product development model and process prototyping, tests and marked introduction (Cooper, 1995) (Wheelwright & Clark, 1992) (Baker & Hart, 1999). Firstly, due to long and intense research (Sanchez, 2001) and to strong investment, the lower part of the NPD process was well investigated and defined. Secondly, due to clever and hard work by engineers and production managers this had been systemized and formalized. Yet, there was still a long way to go before the optimal model and process for each NPD project were defined.

The scenario was different when it came to the upper part of the NPD model and process. Many businesses had seen especially the idea stage as a "black box" without structure, filled with creativeness, costs and a lot of trials and errors (Verworn, 2002). Often management at SMEs have realised that trying to formalise, systematize and structure these phases of the PD model and process turned out to be a failure and very difficult to manage.

On the other hand the demand from market, customers, and pressure from competitors to introduce new products at increasingly higher speed had coerced the management to focus on this particular area of the PD model and process.

This was why many SMEs today had realized the importance of generating and collecting new ideas and quickly conceptualize these ideas into new product concepts to the lower part of the PD process.

It was the hypothesis of our research project that the ability to handle speed in this upper part of the NPD model would give the involved SME a major competitive advantage which would give them the opportunity to:

- "harvest the markets first"
- "introduce the products when customers were ready" to buy and consume
- in a long term perspective with continuously high speed PD bring the SME into a strong strategic competitive position where NB HS PD was a strategic competitive weapon.

1.2.2 The Need to Do Join Networks to Do High Speed Product Development – Why?

Networking has always played a major and central role in economy. Networks, knowing about networks, and how to do networking has always been central in business (Håkansson & Johanson, 1992) (Child and Faulkner, 1998) (Kotler, 2000) (Albaum, 1994) (Hollensen, 2003). So why were networks and PD suddenly important, and what was new about joining networks to achieve high speed product development in 2000–2003?

The new issue about networking was that there were an increasing understanding in industries and individual SMEs that the individual SME faced a global competition and world economy (Goldman, 1998) (Child & Faulkner, 1998) (Kræmmergaard, 2002) in which different competences were necessary to develop the customers' increasing demands for new products.

The individual SME could choose to develop and hold all necessary competences inside the Business and inside the existing network of the businesses. The consequences would be high costs, vertical and horizontal integration, inflexibility and diminishing speed. (Goldman, 1998) (Child & Faulkner, 1998). On the other hand, the businesses could choose to open up and join networks the consequences of which would be opportunities to diminish costs, increase performance, and develop at high speed.

Therefore, in a high speed global market economy it was vital for SMEs to manage competences related to networks. SMEs had to decided which competences were important to hold inside the Business and which competences it was more preferable to "buy" outside or "network". As previously mentioned, it was the thesis of the research project that networks and the ability to do networking – also globally – was one of the enablers of high speed product development.

It was in this context that NB HS NPD were examined.

1.3 Overall Definitions in the Book

The high speed product development focus in literature was until 2003 centred on the lower part of the stage gate product development model (Cooper, 1993) – the process development phase as seen in Figure 1.2.

Focus and tools to speed product development had mainly been on product modularisation, process optimisation, cost improvement of prototyping and prototyping models, improvement of product performance and creation of variance (Hansen & Thyssen, 2000) (Sanchez, 1996) (Bohn & Lindgren, 2002).

The focus had been more fragmented on the upper part of the product development model and process – the innovation and the first part of the product development phase. Also the phase before the "formal" innovation phase had not in particular been in focus of high speed. Due to more creativity,

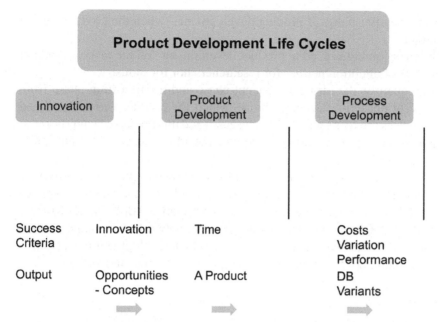

Figure 1.2 High speed in product development until 2003.

more unstructured-ness, and more complexity, the need for flexibility in these stages and gates had not until now been subject to high speed (MacCormack & Verganti 2002) (Verworn, 2002). Many businesses had "closed their eyes" looking into this area of high speed product development possibility because no models and no frameworks had until then been available. Additionally, until 2003 the management trend had been that this stage could not be a subject for high speed.

> *"Product development at the idea and concept stage always takes its time and it always will."* (Engineer employed at LSI Denmark 2002)

However, many business cases (Case No. 1 Zara, Case No. 11 Rossflex) had shown the importance of focusing on time and high speed in the initial phase of the product development stage. It was important in the initial product development phase – the idea and concept phase – that the businesses organise for high speed in PD (MacCormack & Verganti, 2002) in order to be able to speed product development both in a short term and a long term perspective to stay competitive in the market. However, the innovation phase was strongly related to the after phase of product development – when the product was on the market.

The focus on high speed in product development "on the market" had not been centre of attention neither for researchers nor for industry.

Consequently, the idea and concept phase and the transition phase from idea/concept phase to the phase during which the product was realized had been chosen as the focus point of this research and this book (see later) because time and speed was especially critical here (MacCormack & Verganti, 2002) (Sanchez, 2000).

The above-described problem required the use and implementation of new product development processes and models, which could at one time manage high speed, lowest possible cost, best performance (Bolwijn & Kumpe, 1998) (Boer, 2001) and product development in network. A strategic outlook characterised by special consideration for product development management in when product development was under pressure of time and speed.

1.3.1 The Product

There can be a tremendous difference between the nature of a product development process and the product which is meant to be created through the process.

On the basis of the above discussion, we have in this book chosen to define the product as follows:

> *"a business to business product that can be offered to a market for attention, acquisition, use or consumption, that might satisfy a want or need both tangible and intangible."*

Accordingly, it has been chosen to pay main attention to business-to-business products; both the material and the immaterial parts of the product. In this way the book and the research encompasses the entire business-to-business product.

1.3.2 Product Development Model and Process

In a previously published article (Bohn & Lindgren, 2000) as well as in Chapter 3 of this book various product development models are described. It was also decided to describe the development since the 1960s until 2003 in order to understand and describe the research in product development processes and models. On the basis of this work, the following analytical framework for NB HS PD emerged.

At an early point in the process it became apparent that the framework had to contain two basic elements:

1. The functions involved in the NB HS PD – i.e. both the internal and external departments/functional areas involved in product development
2. The core of the NB HS PD – i.e. the mission, the objectives, the strategies, and the resources controlling the product development project.

Thus, the initial generic processes are the focus of this book and its research. It is also among these processes that according to Cooper we found the greatest need not only for improvement but for continuous improvement and learning (Cooper, 1993) (Sanchez, 2000) (Bessant 1999) (Ulrich & Eppinger, 2000).

Relevant literature also described the entire product development course as an overall process (Booz, Allen & Hamilton, 1982) (Cooper, 1993) containing certain generic activity stages. As previously mentioned the project also focused on the first activity stages in the entire product development process, i.e. mainly the idea and concept phase.

1.3.3 Network Based Product Development Model

The framework of this research project and book determines that the point of focus should be on network-based product development. The network perspective is interesting in that study of literature (Ulrich & Eppinger, 2000) (Wheelwright & Clark, 1992) (Goldman & Price, 1998) (Caffyn, 1998) (Baker & Hart, 1999) as well as most of the cases (Part 7) indicated that product development subject to time pressure were employed and toke place in networks. Thus, the use of network could be the explanation for higher speed in product development but networks could also be the explanation for a lowering of speed in product development. Networks can be defined as both physical, internal networks within the Business, external networks established together with customers and suppliers – or the whole supply chain. Networks can also be defined as digital and virtual networks and the mixture of these kinds of networks.

1.3.4 High Speed Enablers

Through the PhD project the explorative analyses of secondary case businesses claiming that they did network based high speed product development found that the businesses used different enablers to speed product development.

These enablers were used in businesses to speed product development at different levels and at different intensity in the product development as shown in Figure 1.3.

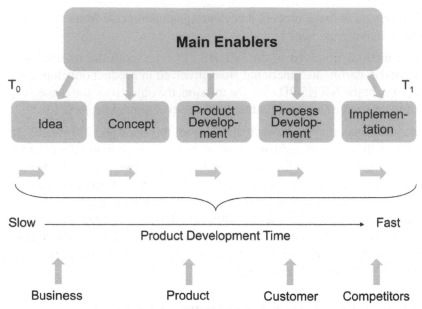

Figure 1.3   Differences in product development time.

The research and book investigate the existence of and the difference in importance and use of the high speed enablers through the product development phases.

1.3.5 The Success Criteria

It was decisive for the businesses that they were aware of how to find new product ideas and concepts first, how to get the concepts down "the product development funnel" (Wheelwright & Clark, 1992) and finally how to introduce new products to the market at high speed (Sanchez, 1996).

However, it was also important to be able to "produce" and "introduce" the new products with less costs, and with the best performance (Rosenau, 1993). Thus far, it had been evident (Roseneau, 1993) that product development of the Business must match at least these three success constraints. The match of these success constraints was vital – otherwise the customers would find other suppliers to patronise (Goldman & Price, 1998) (Sanchez, 2001).

Furthermore, in a long term perspective it was the hypotheses that businesses focused on other success criteria. These success criteria could be continuous improvement (CIM), continuous innovation (CI) and learning

(Gieske, 2001) (Smeds, 2001) (Corso, Martini, Pellegrini & Paolucci, 2001) (Boer, 2001) (Bohn & Lindgren, 2002). The different product development components of the businesses had to be improved continuously to match the competition on the market. CI must be brought into focus because the businesses needed continuously to innovate products to the markets (MacCormack, Verganti & Iansiti, 2001) because Businesses were expected in the future to not be able to survive with ad hoc innovation activities. As a consequence of the above, learning in product development had to be brought into focus as a major success criteria because future CIM in product development and CI demands development of knowledge and competences of product development (Child & Faulkner, 1998) (Nonaka & Takeuchi, 1995) (Gieske, 2001). However, the learning success criteria was expected in the future to be gained from network partners and within networks. The research project and book focus on all success criteria but with particular emphasis on the time and speed criteria in product development.

1.3.6 Industrial Conditions and NB HS PD

The conditions for product development have always been influenced by the industrial conditions. However, during recent years up to 2003 the conditions had turned out to be more dynamic and unstable than ever seen before.

Changing Industrial Conditions

Industrial conditions had changed radically during the last 15–20 years. In this time technology, market conditions, and customer demands changed at a speed and in a direction barely seen before e.g., dynamic market fragmentation, shrinking time in market, increasing product variety and production to customer specifications, reduced product lifetimes, and globalisation of production (Riis & Johansen, 2001).

Diminishing PLC

The change and pressure on speed in the development process were, i.a. due to the fact that the product life cycle had been considerably diminished and delimited (Fine, 1998).

Pressure on Existing PD Models and Processes

This challenge put pressure on existing product development models and processes to find faster and new product development models and processes that could match the demands for the future global market. High speed product development became increasingly central in the global competition. Businesses that were able to continually develop and focus on value adding elements or improvements to their products with high speed would be able to attract and hold customers and thereby gain a competitive advantage.

Businesses like Microsoft, Ryan Air, and the Zara Inditex group were international showcases to this argument in 2003.

Focus on Continuous Innovation and Continuous Improvement

By continuous incremental and radical innovation (CI) and by continuous improvement (CIM) of their product development models and processes, the above-mentioned businesses had fulfilled their shareholders' demand for profit, survival and growth.

Pressure on Competences of Businesses

The pressure on the businesses ability to develop new product had at the same time created a pressure on businesses competences. The change of markets and technologies were so dynamic and fast that one single businesses competences seldom were enough.

From Hierarchy Structure to Network Structure

Successful businesses changed their way of organising product development from a rather hierarchical structure to a network based structure.

The businesses tried to solve the product development challenge by increasing their network activities in a global perspective. Outsourcing, joint development and improvement on the supply chain had been elements in the network activities of the businesses.

The barriers to international trade and investments had changed and minimised at the same time thanks to WTO. Thus, WTO had increased and supported the trend and ability to develop products in global networks.

Product Development Turned into Global Networks

Product development had therefore been characterised by an increasing amount of openness and by many network operators possessing diversified qualifications and kinds of competence (Hamel and Prahalad, 1994)

(Goldman & Price, 1998) (Child & Faulkner, 1998) (Baker & Hart, 1999). Businesses realised that some kinds of competence had to be procured from outside the Business by means of networks. Therefore, the present years witnessed the establishment of many new and often dynamic network oriented high speed innovation environments, innovation models and high speed innovation processes. It was therefore undoubtedly important that SME businesses became active participants in such global networks.

Focus on Network Based Knowledge Transfer in Product Development

In order to conduct faster product development, businesses would realise that access to external competence was essential. The Businesses would be met with a demand for the surrender and transfer of competence and knowledge to other product development environments (Child & Faulkner, 1998) (Corso, 2001). Such aspects emphasised the importance of analysing and describing how businesses in a strategic comprehensive view were able to take part in such network based product development processes and in a long term perspective to learn or establish knowledge transfer from these processes (Boer, 2001) (Corso, 2001). The combination of network based high speed product development and learning were therefore a major challenge for SMEs in 2003 to gain continuous improvement, continuous innovation and continuous learning to secure high speed product development.

However, the above-mentioned tasks and challenge initially demanded a new understanding of network based high speed product development models and processes because of the new global context of product development.

Summarising on the above-mentioned context for product development showed the following Table 1.1.

Context of Product Development 2003 and in Future

As can be seen in Table 1.1, the global industrial context was changing and was preparing for more speed and more network in product development. Both researchers and industrial businesses were convinced of this challenge, and focus on speed and network in product development was therefore increasing.

However, not many had defined, observed, and analysed the relationship between high speed and network based product development or the impact of high speed on network based product development. Not many in 2003 had

Table 1.1   Context for product development

Context for Product Development	Until 2003	Trends for the Future
Market	National Stable Common	Global Fragmented Dynamic Individualised
Technology	Single technology Expensive Data power low Unstable	Mix of technology or multi-technology Cheap Data power over capacity Stable
Network	Closed networks Stable networks	Open networks Dynamic networks Virtual networks Global networks
Competences of the businesses	Stable competences Competences developed inside the Business or in narrow networks	Dynamic competences Competences continuously under development and pressure Competences developed with network partners and in open and many networks
The product	Mostly physical products and to some extent immaterial products The product is stable The product is used in the same way	A mixture of physical, immaterial digital and virtual products The product is continuously developing and changing form The product is used in many new ways
The product development model	Stable models Stage and gate models	Many product development models Stage and gate models Flexible models Dynamic models Process models
Success criteria	Speed, cost and performance	Speed, time, cost, performance, efficiency, Quality, CI, CIM, Learning

investigated the effects of high speed product development or answered the question why high speed was required in network based product development.

> *"Speed is the new competitive weapon"* although *"I will never recommend cutting corners of executing in a sloppy fashion in order to save time – it just does not pay off" "Speed is important, but it is only one component of our overarching goal of profitable new products."* (Cooper, 1993)

Furthermore, not many had investigated what enablers businesses could use to speed product development and finally what success criteria businesses could use to measure network based product development.

Cooper stressed very well the importance of managing a business's product development activities with more than one success criteria in mind – speed and time.

Management of Product Development was Changing

Product development management had until 2003 focused on short term management thereby success criteria such as time, cost, and performance. The major part of available literature on product development showed a major focus on the management of the product development process within the process at a relatively tactical and operational level (Wind, 1975) (Wheelwright & Clark, 1992) (Cooper, 1993) (Ulrich & Eppinger, 2000) (Baker & Hart, 1999). This focus was merely due to the practical and theoretical challenges to manage the product development process through the product development process from idea to market introduction. The management focus had been on the process within the development process and very seldom on other product development projects inside or outside the Business the centre of attention. Management had not paid particular attention to the management of product development in network because the possibilities, conditions, and the "field of product development" had not applied to this management constellation until 2003. The research project focus on the management area indicated above because the global market for the future as can be seen would apply for this network management in future product development projects.

1.3.7 Theoretical and Practical View to Network Based Product Development

Looking through literature and practitioners' statements on network based product development we found in 2003 some explanations of the appearance of network based high speed product development. The main context and the "product development game" in 2003 could be characterised as shown in Table 1.2.

A dynamic and "interactive picture" of the four main components on "the field of product development" – markets, technology, networks, and competences of the businesses playing in "the field of product development" was therefore central.

Table 1.2   The shape of the main components in the product development game

The Main Context of Network Based High Speed Product Development	Characteristics	Example of Markets 2002
Market – (Sanchez, 1996)		
Stable markets	Stable market preferences	Food industry, Furniture industry
Evolving markets	Evolving market preferences	Agriculture industry, environment industry
Dynamic markets	Dynamic market preferences	Software industry, Bio and gene industry
Technology (Sanchez, 1996)		
Stable technology	Stable and known technologies	Audio and video technology
Evolving technologies	Evolving technologies	Biotechnologies
Dynamic technologies	Dynamic and mixed technologies	Nanotechnology
Network (Child & Faulkner, 1998)		
Stable networks	Networks mainly based on physical and stable networks often internal and dominated network	Industrial groups, branch groups
Evolving networks	Networks based on a mix and evolving system of networks: Physical networks, ICT networks, virtual networks	PUIN network group, EU community
Dynamic network	Networks based on a mix of dynamic networks with high degree of dynamic where network partners constantly comes in and goes out. Often there is no formal network leader.	Virtual network groups, Ambia's
Business competence context (Hamel & Prahalad, 1994)	Support competences Complementary competences Core competences	

The interaction of market and technology in product development had been known for several years (Wind, 1975) (Hein, 1985) (Ulrich & Eppinger, 2000). What was new was the interaction that market and technology had with different types of networks and their relation and influence to the competences of the businesses as shown in Figure 1.4. This interaction created "the future field of product development". It was therefore central to overview and understand the main components in "the field of product development". Each main component could be of different shapes both prior to the product development project and during the product development project.

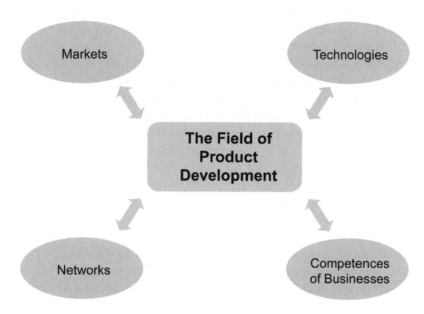

Figure 1.4 The context and main components of "The field of product development".

Until now it seemed as if many businesses had developed a "blind" high speed product development strategy seeking speed, uniqueness, and innovativeness without reading and analysing the characteristics and the development on characteristics of the component in "the product development field". To a large extent, the management of product development had concentrated their efforts on the establishment of high speed product development and on being an innovative business. This was partly due to the tendency to regard such a strategy as the survival in future competition (Bolwijn & Kumpe, 1998). Unfortunately, the combination of high speed product development, uniqueness, and being the network based innovative business had proved to be far more difficult to implement than originally expected (Bessant, 1999) (MacCormack, Verganti & Iansiti, 2001). Many businesses had realised several failures and problems with a strong focus only on high speed in NB PD.

1.4 NB HS NPD in the Future Global Market

Network based high speed product development was a necessity to compete in the future global market **Network** because the competition in the 21st century were expected not to be business against business but network against network (Goldman & Price, 1998), **high speed** because customer demands for new products would continue to increase, and **product development** because product development was an important solution for businesses to solve such heavy demand and competition on high speed on the product development area for the future (Grunert & Harmsen, 1997).

Although there was strong evidence that market wants and needs for new products, increased levels of product variety, and accelerating rates of product changes were different (Fine, 1998) (Sanchez, 1996) (MacCormack, Verganti, Iansiti, 2001). The differences in the needs for speed of market introduction of new product were due to differences in product/market contexts, technology contexts, network contexts, and in the competences of the businesses.

I claimed therefore in 2003 that the choice of speed was critical to the success of the product development project of the business.

As can be seen, all focus both in research and practical terms was mainly on speed or high speed in product development up to 2003. I claim that high speed in network based product development was not the issue. When products entered the market too early, the alternative cost of e.g., waiting for the market, showing the competitors the product, or "repairing" the unstableness of the technology inside the product causing failure was too high on the other hand. When products entered the market too late, the alternative costs of e.g., not being able to enter the market and not being able to harvest the market were fatal to the Business.

I therefore claimed that businesses must find an optimal speed. I call this "right speed and right time in network based product development". If Businesses could find the optimal time to do the market introduction not once but continuously, a major competitive advantage had been gained.

Some businesses had managed to find this critical point of time and speed in product development. The result on market share and turnover but more interesting net profit had been significant. A case for this example is the clothing industry where the Spanish textile Business Inditex by its retail chain Zara had set new standards to right speed and right time of product development.

The hypothesis in 2003 was that businesses who continuously find the right time and right speed of product development would develop a core competence which would be difficult to compete.

Speed and time in product development is therefore more complex than both researchers and Businesses realised in 2003. I claimed that in industry in 2003, speed and time were very much related to cost.

"The winners are concentrating on cash-generation, cost controls and subscriber margins." (Tim Burt, Financial Times, Monday 25 November 2002)

I claimed that it should be focused both on value and cost instead of only on cost as seen in below.

I also claimed that value and cost had to be defined by the individual business related to the look of "the field of product development". The characteristics of "the field of product development" must define the value of speed and time of network based product development.

When businesses speed product development, the consequences were fairly well known in the traditional product development world. However, when high speed was added to network based product development, the effects were mainly unknown. Therefore, high speed and time also had to be defined in relation to network based product development. Furthermore, the impact and the enablers to high speed on network based product development should be verified and investigated.

However, there were different views to speed and time in product development in 2003. The market (customers and competitors), the technology, the network and the Business had different views on this issue. I claimed that the new marketing view should be the optimal view on speed and time in product development. In the new marketing view perspective speed and time in product development are related to bringing the customer into the Business via a strong network relationship with the customer. Thereby the customer becomes a part of the Business's strategic product development foundation. The customer hereby develops the future new products together with the Business and its networks partners (supplier, other network partners and organisations, and even competitors) by focusing on the value of the customer related to wants and needs.

The Business hereby formulates the speed of product development together with the businesses they collaborate with.

I therefore claimed that high speed in product development was not the issue and not always advantageous. It can even be advantageous to "hurry slowly" when characteristics in market, technology, network, and the competences of the businesses are in a certain position.

The question of speed was therefore in 2003 more complicated than outlined by former researchers and practicians. The claim was that during the product development process the speed sometimes has to be increased and sometimes has to be decreased. The main components in the field of product development can turn out to influence and make radical changes to the optimum choice of speed in product development. Therefore, the hypothesis was that businesses have to alter speed during the product development process.

Until 2003 there were only fragmented knowledge of and research about the types of speed and speed tools that were available and appropriated in different situations of network based product development. My claim is that there were more enablers, tools, and views to speed in the product development process than verified in 2003. Learning had to be established in all areas of high speed product development to find enablers and tools to speed in NB NPD.

The critical issue before talking about speed in product development was however, the ability of the management to analyse "the game of product development" and learn from one product development project to another how to define the speed advantageous to this specific product development situation. Even more critical was the ability of the product development managers to learn throughout the product development process. The last learning area concerns the development process from idea to market introduction as well as the span of time before, during, and after market introduction as shown in Figure 1.5.

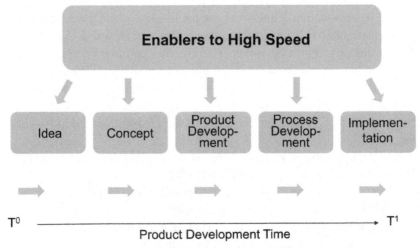

Figure 1.5   Speed in network based product development.

My claim was therefore that learning should be the optimum success criteria in network based high speed product development seen in a long term perspective. Furthermore, businesses should move their focus from short term success criteria to long term success criteria. My hypothesis was that a focus on long term success criteria in network based product development will develop right performance, right cost and right time and speed to the businesses product development activities.

How to establish learning of speed in network based product development across networks in the product development process was therefore an essential and important question. However, it will not be a focus area in this book. Future research was planned on this topic in 2003 and can be seen in my publication up to 2016.

1.4.1 Technical Field of Study and Initial Research Purpose

A strategic match of product development in time, with optimal costs and with the right performance within a dynamic competitive network environment was of interest to my research in 2003. More specifically, the time perspective and the high speed perspective of network based product development were of interest of my research. More specifically what was the right time and right speed related to right performance and right cost as shown in Figure 1.6.

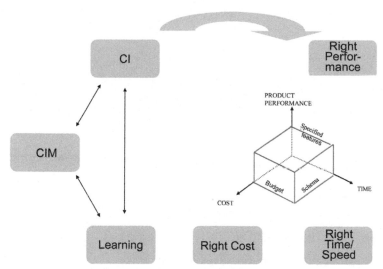

Figure 1.6 Relationship between long term success criteria in network based product development.

Source: Lindgren & Bohn, 2002.

In other words, it was the hypothesis of my research that the businesses must possess the competence and the abilities of incorporating high speed into their product development processes. It was therefore important to analyse if and how the strategic, tactical and logistic elements of high speed product development are and could be incorporated into the idea and concept phase.

It was the hypothesis of the PhD project that normative check lists for guiding a close interplay in networks during the product development process or during the product preparation phase together with the forming and draft of models for network based high speed product development were an essential factor in the success of future businesses. Among other things, the above-mentioned aspects justify the problems of the thesis and why the study of the above-described focus point: – network based high speed product development processes and models – is the prime objective of this research project and book. Furthermore, the final discussion for the research purpose of the research project can now be carried out in accordance with the framework outlined in Figure 1.7.

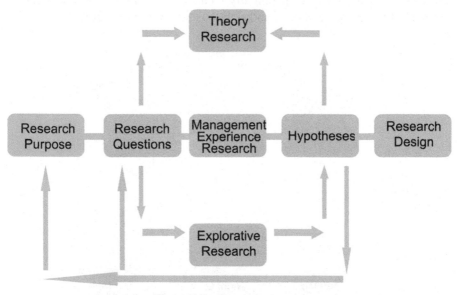

Figure 1.7   Research purpose.

1.4.2 Perception of Problem in Detail

This paragraph will deal with the motivation for the problem and its formulation. This paragraph will define the research purpose and the research question and research subquestions as shown in Table 1.3.

Table 1.3 The question for the research on NB HS NPD

Key Questions of the Book on NB HS NPD
What is the product?
What is network based high speed product development?
What is a high speed product development model and which NB HS product development model can be identified?
What is a high speed product development process and which NB HS PD can be identified?
What is a high speed product development enabler and which enablers can be identified?
What are the success criteria for network based high speed product development?

Initially, the research project therefore focus on defining the "key words" of the book "Network Based High Speed Product Development Models and Processes".

What Is NB HS PD?

The first of the Research Questions was very often asked in 2003. The hypothesis was that high speed related to product development could be seen from different points of view – e.g., the macro environment view, the market and customer view, the technology view, the network view, the Business view, the product view or the competitive view – as sketch out in Figure 1.8.

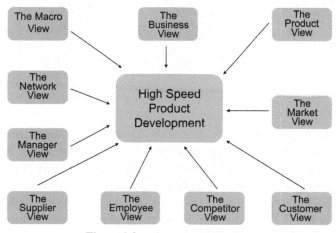

Figure 1.8 Views on HS NB PD.

This book will look at high speed product development formulating thesis from a market view – combined with a maximization of net profit – the new marketing view.

The hypothesis is that high speed is not only a question of developing new products fast to the market and the customers but a question of developing new products at and within the right speed and right time – this means when the market, the technology, the network and the Business are ready for product introduction.

The hypothesis is that high speed to NB PD can be seen both in a vertical, a horizontal, and a network perspective. **The vertical perspective** covers high speed PD from idea to implementation of a product on the market at the highest possible speed – with less amount of time. **The horizontal perspective** covers high speed PD for as many or as complex PD projects as possible from idea to implementation on the market. **The network perspective** covers high speed PD in network both vertically and horizontally as shown in the model in Figure 1.9.

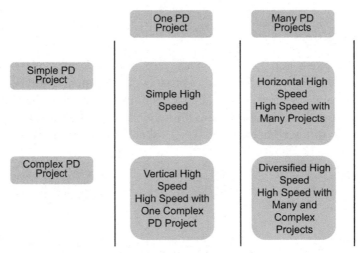

Figure 1.9   High speed product development matrix.

However, the book will mainly deal with PD projects in the incremental, vertical, and network based high speed area.

As can be seen until 2003 product development has primarily focused on the product development process inside the product development model – from idea to market introduction. However, speed in product development is also very much related to the activities taking place before and after the product development process.

It was my hypothesis that speed in product development is influenced by the handling and activity of the Business before and after the product development process. Still, this is not the major focus in the present book.

Identifiable High Speed Enablers

Case studies, Business interviews, and literature search showed numerous ways to gain high speed in product development as sketch out in Figure 1.9. These enablers to high speed can be seen as catalysts to speed product development as sketch out in Figure 1.10.

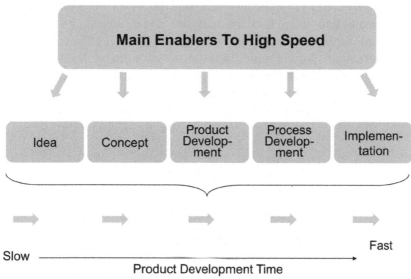

Figure 1.10   Enablers to high speed.

Some SMEs use modularization (Hansen & Thyssen, 2000), other electronic development tools – E-development (Lindgren, 2001), and other new product development models and networks (Cooper, 1985) (Goldman & Price, 1998) (MacGormack & Verganti, 2002).

The hypothesis was that there are more enablers to speed the product development. The research and the therefore aims at finding and verifying enablers to high speed product development.

Additionally, the hypothesis was that the enablers will play a different role depending on which product development situation the particular Business faces. The role of the enablers are not a main focus of the book but when it is possible to register such roles, they will be described.

When the definitions above have be clarified and the enablers of high speed PD have been found, the book will proceed to find a generic framework for models and processes in network based high speed product development. The idea and concept stage will be in prime focus in the book.

Measurable Models and Processes

The research on product development was intense and manifold (see Chapter 3 for further elaboration) and showed some generic models for product development and product development processes.

The book tires to bridge the domains of former NPD and network research/theory by taking pre-existing conceptual models of NPD and network theory, and rebuilding them into an analysis model which can be used to describe and explain the application of HS within NB PD models and processes. There are two main stages in this process of model development.

First, the existing PD model is re-interpreted theoretically in terms of a NB and HS context, with reference to literature and case investigations as can be seen in Figure 1.11 with only a single Business including internal networks.

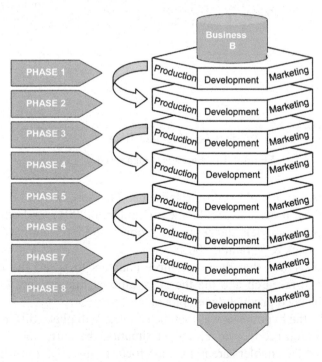

Figure 1.11   Analysis model for product development in networks.

Then data generated using a triangle of research methods are analysed to provide a picture of the application of HS models and processes within NB PD practice.

The current context will be shown as a research for NB HS PD models and processes, focusing on the need for finding and verifying NB HS PD models. The hypothesis is that businesses follow a different PD model when developing at high speed.

The problem perception process which has been largely iterative constitutes an initial process for the perception of a hypothesis model for network based high speed product development model and processes as shown in Figures 1.11 and 1.12. The first part of the model was already shown as

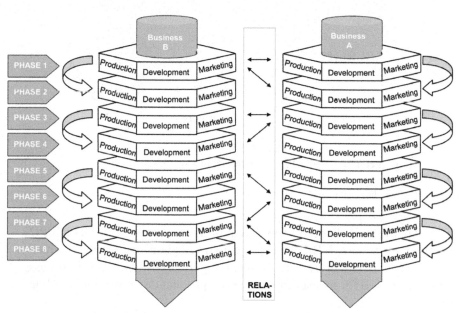

Figure 1.12   Product development in networks.

Source: Bohn & Lindgren, 2000.

in Figure 1.13 and also described in articles (Bohn & Lindgren, 2000).

Subsequently, additional development and research of this model on behalf of further studies were the objectives of the research project.

Firstly, with the above-mentioned in mind, the hypothesis was that there was one generic model for radical NB HS PD and one for incremental

NB HS PD. Radical PD takes place when the PD project faces new markets, new customers, new technologies etc. Incremental PD takes place when the Business faces related markets, related customers, related technologies as shown in Figure 1.13.

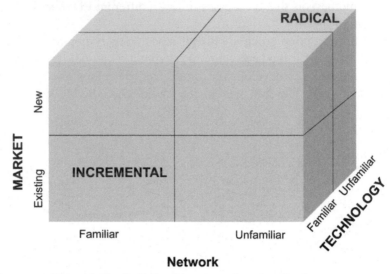

Figure 1.13   Radical and incremental product development.

Source: Inspired by Balachandra, 2000.

The hypothesis is that the formulation of success criteria of NB HS PD is related to the generic NB HS PD project – radical or incremental – which the SMEs face. It is therefore relevant to ask the question: "What success criteria can be used for measuring high speed product development based on networks?"

Possible Success Criteria for Measuring NB HS PD

The hypothesis was that the time success criterion – to achieve high speed or to shorten the time by which the product development process is done – was essential. However, the hypothesis was also that performance and cost are still essential success criteria, also in NB HS PD, and that a pressure on time will have impact on the other two success criteria either by improving earnings, reducing costs, and improving performance (Goldman & Price, 1998) or the opposite. The research wanted to clarify when high speed will have positive effects on profitability and performance.

The hypothesis was also that there are more success criteria for NB HS PD than time, performance and cost. On a short term and a long term basis there must be different success criteria. The final aim of the research is to find the success criteria dependent on NB HS PD project and find their importance both on a short and a long term basis.

The hypothesis was that network based high speed product development should be evaluated in a short term perspective with the same success criteria as product development had been evaluated until 2003. However, in a long term perspective NB HS PD should be combined with other success criteria such as:

1. continuous improvement (CIM)
2. continuous innovation (CI)
3. learning (L)

Otherwise, a Business using network based high speed product development will not be able to reach and meet the success criteria – time, cost and performance neither in a short term or a long term perspective. This is shown in Table 1.4.

Table 1.4   Short and long term NB HS PD criteria

NB HS NPD Success Criteria Short Term Perspective	NB HS NPD Success Criteria Long Term Perspective
Time	Right Time
Cost	Right Cost
Performance	Right Performance
	Continuous Improvement and Continuous Innovation
	Learning
	Right Speed

The research wanted to verify these success criteria in the businesses involved in the research.

The above-mentioned key overall research questions finalise the first part of the phase of formulating the research questions. The second part where research questions in details are elaborated, will now be presented.

1.5 Research Questions – When?

The final question to ask is: when is it preferable to use NB HS PD? The answer to this question must be fully fathomed in order to understand the essence of NB HS PD.

The research project thesis is that the use of NB HS PD must depend on:

- Market characteristics
- Technology characteristics
- Product characteristics
- Network characteristics
- Business competence characteristics

The research will try to map out and give an answer to the question: When was NB HS PD used and under which circumstances should NB HS PD be used? The research questions were formulated by theory research, management experience and explorative research. This will be commented on at a later point in this chapter.

Summing up on the hypotheses to be tested, the following plan in Figure 1.14 shows the overall research questions and related theses formulated and shown in Table 1.5.

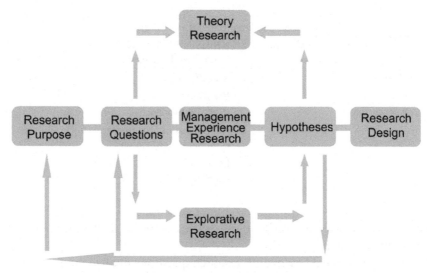

Figure 1.14 Research hypothesis.

Table 1.5 Overall research questions

Overall Research Questions	Hypotheses to be Tested
1. What is network based high speed NPD?	1. HS NPD can be seen from different view (Macro environment, Business, product, market, customer, technology, competitive and network view) 2. HS NPD is a matter of right speed and not high speed.
2. What enablers to NB HS PD can be identified?	1. Businesses use different HS enablers. 2. HS enablers are identical to the 10 enablers – 1–10 3. There can be more than these 10 enablers to HS PD. 4. The enablers will play a different role according to the PD situation and project (Secondary focus) 5. The customer enabler, the network enabler and the PD model enabler plays an important role in the upper phase of the HS PD phase.
3. What framework models and processes in the idea and concept stage/gate of high speed product development based on networks can be measured?	1. The HS PD projects can be divided into to radical and incremental PD projects. 2. The radical and the incremental PD projects follow different generic HS PD models and processes and can thereby be described by different generic frameworks. 3. A HS PD model follows another PD model than the normal PD model of the Business.
4. What success criteria can be used for measuring high speed product development based on networks?	1. The success criteria for HS PD are dependent on the specific PD project – radical or incremental. 2. HS PD success criteria can be formulated as short term and long term success criteria 3. Time, cost and performance are central success criteria in a short term perspective 4. Continuous improvement (CIM), continuous innovation (CI), and learning are central success criteria in a long term perspective to reach right time, right cost and right performance in NB HS PD.

1.6 Methodology

This paragraph discusses the structuring of the problem, gives a description of the scientific methods regarding the problems, and explains how the methodology is applied.

The book is divided into six parts. The first part formulate the Problem Statement. The second part will describe the theoretical foundation of the research project and will present an empirical explorative secondary Business case part, which will help to construct a framework for the research on network based high speed product development models and processes in SMEs in accordance with the preliminary theoretical foundation and analytic approach.

The third part – Analysis Model – will elaborate the research methodology, empirical methods and analysis tools. The fourth part – Empirical Results – will describe and analyse the research findings, and the fifth part – Comparing Theoretical Framework and Empirical Results – will discuss the application of NB HS PD together with criticism and implications. Finally, the sixth part – Conclusion – will summarise and present an agenda for future research. Please see the structure outlined in paragraph 1.9.

1.6.1 Problem Identification and Theory

In the second part of the book a thorough study of relevant literature on the subject – **Network Based High Speed Product Development Processes and Models** up to 2003 – is carried out together with an explorative secondary case analysis of the product development system of generic industrial businesses including related models and processes. Hereby the book endeavours to describe, make an identification of the problem, and produce a theoretical model apparatus for network based high speed product development processes and models. The preliminary process for the study of relevant literature and the explorative analysis of specific case businesses were characterised as a problem conception which followed an iterative perception of problem, processes and models.

In the third part of the book a further theoretical study endeavours to reach an understanding of the problem area. This understanding formed the necessary basis for a final identification of the problem and for a final delimitation as regards the theoretical framework of the empirical part.

The second and third part form the basis for the fourth part – the empirical research.

1.6.2 Empirical Part

The theoretical model outlined above was subsequently empirically tested in different ways as recommended by Wind (1975) and Aaker & Day (1980).

The sources of the hypotheses were both theoretical research, management experience (senior researchers and industrial managers dealing with product development) and explorative research as shown in the model in Figure 1.15.

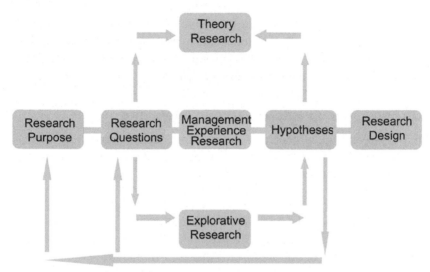

Figure 1.15   Sources of hypothesis.

1.6.3 Theory Research and Experience

Specific Case Businesses with Explorative Focus

The empirical part comprises an in-depth case study of Danish and international case businesses which claimed to have established network based product development processes and which employ such high speed product development types. These cases are presented as secondary cases. An overview of the cases can be seen in the collection of cases.

Literature Research

The literature research comprises an in-depth literature research of Danish and international publications on network based product development carried out both in Denmark supervised by associated professor Kim R. Bohn, Poul Kyvsgaard Hansen, Professor Harry Boer and associated Professor Frank Gertsen, associated professor Poul Dreisler together with at my stay at the Polytechnico di Milano, Italy supervised by Professor Roberto Verganti and Professor Mariano Corso.

Specific Case Businesses with Explorative Focus

The empirical part comprises an in-depth case study of Danish and international case businesses which claim to have established network based high speed product development model and processes and which employ such product development types. These businesses were Lyngsø Industires, Lindholst, TLC, GSI Lumonics and AKV Langholt.

In-Depth Focus Group and Newsgroup Interviews

This analysis was supplemented with in-depth focus group and newsgroup interviews with 13 different case businesses. The representatives and businesses in these focus group interviews were:

- Managing Director Svend Lindholst, Linco Trading A/S
- Managing Director Henrik Olesen, Wolfking Scanio
- Development Manager Kristen Laurbach, Wolfking Scanio
- Managing Director Frank Sørensen, Ansager Møbler A/S
- Managing Director John Chr. Aasted, AKV Langholt AmbA
- Sales Manager Kjeld Ole Nielsen, Lyngsø Industries
- Solution Manager Nils Bundgaard, Tele Danmark Internet
- Personnel Manager Lars Thomsen Tele Danmark Internet
- PT-Project Manager Wolfgang Schröder, Grundfos A/S
- Planning Manager Erik Lou, Grundfos A/S
- Technology Manager Peter Karlsen, Grundfos A/S
- Product Manager Simon Whitley, GSI Lumonics, Rugby England
- Consultant Boris Wortmann, Dansk Teknologisk Institut
- Project Manager Steffen Sørensen, NEG Micon A/S
- Product Development Manager John Sahlertz, LEGO SYSTEM A/S
- Development Manager Bjarne Gedsted, Bang & Olufsen A/S
- Strategic Manager Carsten Christensen, Danfosss Controls

This research was both a management research and an explorative research.

Research PUIN and DiSPU

The research of PUIN is documented in Chapter 11 together with the research of DiSPU in Chapter 13. A survey research called PUIN together with a research called DiSPU carried out at the Danish Technological Institute and Centre for Industrial Production are documented here.

The Book – Network Base Product Development

The result of the work in PUIN and DISPU mentioned above became a book "Netværksbaseret højhastighedsproduktudvikling" (Bohn and Lindgren 2013). In this work some of the hypotheses could be verified for the PhD project.

The TIC Network

The above-mentioned empirical study objective has contributed towards an explanation to processes and models for product development carried out at high speed and deeply rooted in networks. These studies have among others been documented and tested in a number of SMEs in Viborg Amt together with TIC Viborg described in details in Chapter 12.

Joint Research with Polytecnico di Milano

A joint research together with Polytecnico di Milano in the Italian and Danish SME industry has been carried out in relation to "The TOM Project" and the "SALSA" project.

Observation and Action Research with TIP Project (The Aarhus School of Business)

Research on a high speed joint product development activity carried out by students from The Aarhus School of business, Arkitektskolen i Århus, Ingeniørhøjskolen i Århus and their professors together with major industrial businesses in Denmark (www.tipprojektet.dk). 18 different product development projects were established and documented.

Observation and Action Research with Industrial Businesses in Denmark, Germany and Czech Republic

Through the research I had the opportunity to observe and carry out action research within different industrial businesses developing new products in the areas of E-business projects (www.bestcom.dk, www.langcen.dk, www.deluca.dk, www.design.dk, www.nap.dk, www.damb2b.dk). The experience and observation are documented in this section.

1.6.4 Indication of "State of the Art"

The book presented in 2003 a new analysis framework of network based high speed product development processes and models.

The project presented a new descriptive framework model for network based high speed product development in a process perspective and developed an analysis foundation which could form the basis of understanding network based high speed product development processes and models at the upper part of the PD model – the idea phase and the concept phase.

Additionally, the research project and the book presents main enablers to network based high speed product development and analyses such enablers important to create state of the art network based high speed product development specifically in the idea and concept stage in the product development process in 2003.

The research project and book also presents generic short and long term success criteria which can be used for measuring NB HS PD. This study toke as its reference point both the research programs for the Centre for Industrial Production and the holistic production concept programme (Strategy for Centre for Industrial Production, 1999).

Analysis and Thesis Forming Part

On the basis of the above-described studies, the book will finally feature an analysis and thesis forming part.

Thus, the above constitutes the general preliminary methodological framework for the actual research project, which was defended as a PHD project in 2003. The model in Figure 1.16 illustrates the phases and time schedule of the aforementioned research project.

All experiences and results obtained from the first year of the research were described in books, articles, conference papers, and working papers, which are included in this description as enclosures. An overview of these can be seen in the following list.

1. Hypoteser til forståelse af netværksbaseret produktudvikling under høj hastighed Center for Industriel Produktion Aalborg Universitet, Kim Bohn & Peter Lindgren (2002).
2. Begreber i netværksbaseret produktudviklingen under høj hastighed Kim Bohn & Peter Lindgren (2000).

3. Product Innovation in Networks – Analysis Framework for Case Studies Center for Industrial Production, Aalborg University Kim Bohn & Peter Lindgren (2000).
4. E-development – Center for Industriel Produktion Aalborg Universitet Peter Lindgren & Kim Bohn (2001).
5. Right Speed not High Speed (Lindgren & Bohn, 2001).
6. Continuous Improvement and Learning in Network Based High Speed Product Development (Bohn & Lindgren, 2001).
7. Right Speed in Network Based Product Development and the Relationship to Learning, CIM, and CI (Bohn & Lindgren, 2002).
8. Knowledge Management and Product Development (Lindgren, 2002).
9. DiSPU research (Bohn and Lindgren, 2002).
10. Produktudvikling i Netværk (Bohn & Lindgren, 2002).
11. Inter-organisational Project Management in SMEs (Hørlück, Kræmmergaard, Nielsen & Lindgren, 2002).

1.7 Time Schedule for the Research Project

The accomplishment of the project was related to a preferable time schedule which is schematically shown in Figure 1.16.

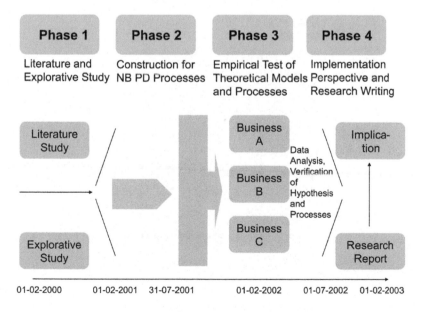

Figure 1.16 Time and phases schdedule for PhD project.

The time schedule for the research project was suggested to last for three years to commence on 1st February 2000 and to be completed on 1st February 2003. The model in Figure 1.16 outlines the phases and time schedule of the project.

As can be seen from the Figure 1.16 the research project was fixed to pass through four phases. Each phase contained specific tasks and demands which had to be satisfied to reach the overall objective and time schedule of the project.

1.8 Presentation of Book

The book is organized by means of the overall table of contents shown in Figure 1.17.

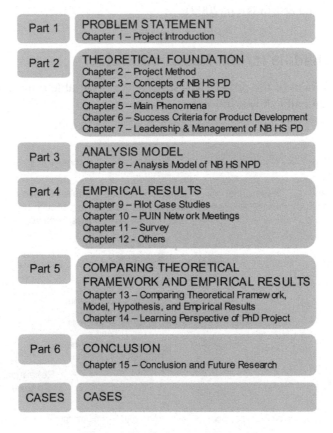

Figure 1.17   Structure of the book.

The book can be supplemented with books and articles written during and after my PhD project, particularly "Produktudvikling i netværk – Refleksioner omkring produktudvikling i høj hastighed" (Bohn & Lindgren, 2002), "Organizing for Networked Information Technologies – Cases in Process Integration and Transformation" (Hørlück, Kræmmergaard, Rask, Rose & Lindgren, 2001), Bohn & Lindgren (2002) and Hørlück, Kræmmergaard, Steendahl Nielsen & Lindgren (2002). For further information, see list of references in Chapter 15, my CV and Linkedin profile.

PART II

Theoretical Foundation

This part presents the theoretical foundation of the research and book. The present part introduces a number of basic concepts for the understanding of product development. The objectives of the part are to clarify and put into perspective the contents and distinctive features of the basic concepts of Network Based Product Development. Each chapter strives to apply such foundation to the present book and also to work as a framework for the research project. Finally, each chapter seeks to reflect on the concepts of network based product development in 2003 and for the time beyond. The objectives of the part is to put forward a theoretical concept of network based product development to elaborate the foundation of a descriptive model for network based high speed product development in a global perspective defined in Chapter 7.

2

Project Method

2.1 Introduction

One of the main demands to a research project is a verification of the scientific foundation on which the research is elaborated – often called state of the art analysis. To answer this question it is necessary to verify the use of scientific view and research methods. A proper interpretation of the above require that this book must contain an account of the reflection of scientific point of view.

Both the scientific point of view and the research method constitute a function of the initial problems and questions. As a consequence, this chapter will be constructed on the basis of the five focus areas of the research:

- Concepts of product development
- Concepts of Network
- Concept of speed and time related to product development
- Concepts of product development models and processes related to high speed product development
- Concepts of enablers to high speed product development

Sections 2.1 and 2.2 focus will discuss thoroughly firstly the scientific view of this book, the scientific ambition and dilemma together with the research methods used.

2.2 Scientific Concept

If one should discuss the question *"What is the scientific view of this research project?"* one has firstly to clarify and verify the concept of science.

Science should be considered a procedure used for answering questions, solving problems, and developing increasingly efficient ways in which to answer questions and solve problems. Thus, it is the process which is regarded as science and not the actual amount of knowledge accumulated by the process.

On the basis of this definition or view on science it can be stated that science can be established via a process of putting questions, answering questions, solving problems, and developing ever more effective methods and procedures to keep on answering questions and solving problems.

On the strength of the above, the scientific view of this research project will be based.

2.2.1 Scientific View

The scientific view of this research project can be traced back to the French philosopher and mathematician Réné Descartes (1596–1650). Without breaking with the Catholic church Descartes claimed "the right of Reason to correct the doctrines handed down by the Christian faith".

Descartes said, that man has a right to trust in the Reason given to us by God even when such reason leads to results which are at variance with the beliefs of the authorities. On the basis of ...... (the concept of Reason in Latin is rational), this philosophical movement was later known as Rationalism.

The methodological treatment of economics is based on rationalism and can be traced back to The Austrian school of thought with among others Carl Mengen and Ludwig von Mises.

The Austrian school of thought departed from the positivism thesis of the method-monoistic point of view which claimed that there is no principal difference between the natural sciences and the social sciences – there is only the scientific method corresponding to the one found in the science of physics. In their argumentation the Austrian school of thought stressed the fact that the objectives of social science are to understand the human acts which originate from certain non-observable concepts.

The rational scientific view is based on two fundamental assumptions.

Firstly, it is the intellect – the researcher's common sense – which decides the final decision of theory. On the basis of a critical discussion with his colleges the researcher has to develop a list of arguments which support the specific research. This list of arguments must not only be numerous but also of a high quality.

Secondly, the researcher's choice must be guided on the basis of a function of criteria, which demand the weighting of the above-mentioned arguments.

On the basis of the above, the research presented in this book is fundamentally based on Rationalism.

2.2.2 Scientific Approach

As regards the question of basic scientific research as opposed to application oriented research, such a distinction is founded on two alternative approaches.

The objectives of basic scientific research are to produce new knowledge "for the sake of knowledge itself".

Consequently, basic scientific researchers disregard the practical efficiency of their research, whereas application oriented researchers are able to describe their science as a subject which implement basic scientific experience in the solution of practical problems.

In Table 2.1 the characteristics of the basic scientific research and the application oriented research are shown.

Table 2.1 Basic scientific research as opposed to practice oriented research

	Basic Scientific Research	Practice Oriented Research
Aim/Motivation	To understand and explain the phenomenon better	To improve out control of the phenomenon
Types of Problem	Scientific problems (guided towards maximum scientific contribution)	Technological problems (guided towards the reaching of a function with instructions for a process)
Paradigm Dependency	Yes, the paradigm of the subject	To a lesser extent. Is multi-paradigmatic in a narrow sense (cross scientific and paradigmatic)
Role	Basic realization of the present time	Communicator across scientific and practical areas
Problem Generator	The researcher society (from the inside)	Practice (without the researcher society)
Criteria for Result Evaluation	Increased insight and scientific durability	Increased efficiency, achievement of function and increase of problem solution preparedness

Source: Bohn, Kim (adapted from Frank Gertsen, 1989).

The objectives of the research presented in this book are to "increase the awareness and knowledge of network based high speed product development and processes".

2.3 Research Method

At the beginning of the research project several methodical models for the solution of the problem areas were discussed. In this connection, I came to realise the necessity of addressing the four main areas methodically:

- An explorative methodical design
- A new way of doing research in cooperation with the industry

- Reflection on what – why – when – how
- An ambition to reach a triangulation in the methodological design of the research

2.3.1 An Explorative Methodical Design

Inspiration from Literature on Product Innovation

The novel nature of the project encouraged the researcher to use explorative and Delhi-like methods in the initial research phase on the concept of network based product development to clarify the problem area and the analysis framework.

Existing literature on product development refers to this method especially in the case of product innovation assignments in unstructured areas or in areas where no or very little knowledge exist (Wind, 1975) as shown in Table 2.2.

Furthermore, this method is recommended in cases where we desire the development of new models or at least new ways to approach the area in order to produce something innovative. Thus, Wind (1975) suggests the following Table 2.2 as a solution to the problem area outlined above:

Table 2.2   Structured and unstructured PD tasks

Consumer-Based		Expert-Based	
Unstructured	Structured	Unstructured	Structured
Individual in-depth interviews a) non directive b) semi-structured/ focus individual analysis	Ned/benefit segmentation	Brain storming	Problem/opportunity analysis
Motivation research	Problem detection studies	Synetics	Morphological analysis
Focus group interviews	Market structure analysis/gap analysis	Suggestion box	Growth opportunity analysis
Consumption system analysis	Product deficiency analysis	Independent inventors and licensors	Environmental trends analysis
Consumer complaints			Analysis of competitive products
Projective analysis			Search of patents and other sources of new ideas
Observations			

Source: Adapted from Wind, Yoram J. (1975).

The researcher chose to regard his research assignment as a product development assignment equivalent to those set in industrial life where they fall into the category "rather radical product development". This means that the market was known (Balachandra, 2000), the technology was rather familiar (Balachandra, 2000), whereas the innovation degree was rather unfamiliar (Balachandra, 2000). In this particular research area the market – the business-to-business industry was known, the technology – the product development models and processes – were fairly well known but the innovation degree – the pressure and impact on NB PD as a consequence of a pressure on higher speed in product development was unfamiliar and not known before. Furthermore, the enablers to high speed were unknown and not verified.

However, the choice of an explorative design of research methods gave rise to major considerations about generalisation, validity and reliability which will be discussed later in this chapter. Furthermore, the explorative design and the chosen research method differed from the general research ambition and policy on the CIP centre. The analysis of the research object and challenge forced me into another design of the research method compared to the general CIP research tradition from 2000–2003.

Consequently, an explorative design was set out as a profound preliminary study of existing literature followed by a thorough series of interviews with experts and chosen players in the research area in question. Such experts and players included:

- Representatives of the product development management of the industry
- Representatives of national and international research environments focusing on product development

In this way, the methodical demand for a wide-ranging board of experts was met.

Inspiration from International Research Partners at Centre for Industrial Production

The network formed at the Centre for Industrial Production brought a major work of research carried out at the University of Brighton, England to the attention of the research group. This work of research applies a research method as outlined in Figure 2.1.

The basic idea of this method is to encourage researchers to take their starting point in problems specific to the industry and thus to focus on problems known to exist in industrial life.

Subsequently, our research intends to clarify and analyse the known problem in cooperation with representatives of the industry. Finally, the joint

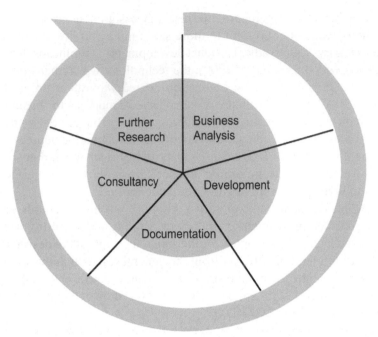

Figure 2.1   Benneth's research model.

Source: Based on model presented by John Bessant, (2000) University of Brighton.

work is substantiated in a combined writing process taking place between the industry and the research institution. In this way, we ensure that the research combines practical relevance and obedience to our chosen research focus with theoretic validity and reliability.

According to the models, the results of such research will be subsequently handed over to other interested parties, including e.g. consultancies, to be further exploited and used by other lines of businesses. Thus, the contents and contribution of the research can be spread to a much wider industrial audience and to a much wider environment than would have been possible if such an assignment had been undertaken solely by the university.

This method of doing research has proven particularly popular at several European universities, whereas American universities have chosen to carry on the classic research tradition and method. The type of research described above has been successfully used during recent years in particular by a series of British research environments in cooperation with small and medium-sized businesses in South and Central England.

The cooperation between Centre for Industrial Production and universities abroad is one of the first results of and contributions to mutual inspiration and use of new research methods and models for cooperation between industry and research environment.

2.3.2 New Way of Doing Research in Cooperation with Industry

On the basis of the above-mentioned overall discussions and considerations of the research design I was highly inspired to try a new way to do research in cooperation with the industry.

The actual methodical framework of the research project in hand was therefore worked out as indicated in Figure 2.2.

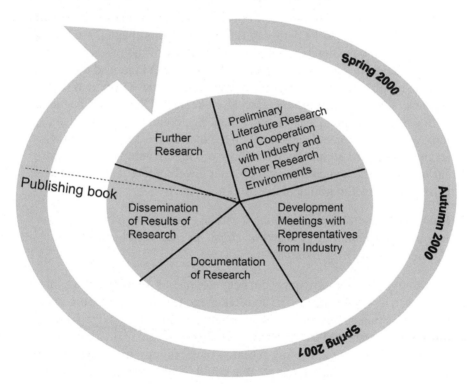

Figure 2.2 Actual research model.

In the following paragraph I will comment on the different phases of the research.

Methodical Framework Phases of Research

The present research project has been divided into six central phases:

- Determination of the methodological ambition of the research
- Preliminary search for literature and cooperation with the industry and other research environments
- Developing meetings with representatives from the industry
- Substantiation of the research work
- Dissemination of the Results of the Research
- Further research

Determination of the Methodological Ambition of the Research

The final determination of the methodological ambition of the research will be commented on in a separate paragraph at the end of this chapter. I will now explain the general considerations of the methodology of the research.

Initially, when I began to do research on NB HS NPD I realised that I was dealing with an area which had not been researched upon before and which seemed to me to be a radical product development task. Therefore, I began to search for a methodological way or framework that could help me to plan the research. As a consequence, I came up with the idea that as I was dealing with product development I should focus on how product development literature and real life deal which such a radical product development task. I therefore put the first initial methodological question to my self

> *How would a product development manager or a product development team handle a product development task which could be characterised as rather radical?*

Looking through the literature of product development I found that the explorative approach is what Wind (1975) and Aaker and Day (1983) propose when researchers and businesses deal with product development and marketing research tasks that are radical and new to the market area. Bearing this in mind I decided first to organise my research methodology in an explorative way but still to develop an architecture that could secure a structured research process that did not let the research and research object slip out of my hand.

I therefore once again investigated the possible structure models for my research. Once again I turned to the product development literature to find a model that would secure this task. The choice fell on Cooper's stage-gate model as a framework model recommended both by researchers and industry as a best practice example of a product development model.

As in a product development project running according to the stage-gate model, each main phase of this research project was therefore divided into various development phases and processes. Such phases and processes have contained the development pertaining to the phases and to the subsequent screening phase in which the development of the preceding phase has been evaluated. During this phase, the objectives of the subsequent phase have been determined.

As in a modern product development process it has been possible in this research project to develop certain areas in parallel and simultaneously. This has particularly been the case during the last part of the research project, but also during the initial phase it has been possible to carry out parallel and simultaneous work.

However, it has also been necessary to attend a rather flexible design of the research with a strong focus on a research architecture that offered the possibility to maximise flexibility. The research architecture focused on the central research questions whereas the research activities opened up for different research method observations, case research, focus group interviews, surveys and some kind of action research. This rather flexible design enabled me to reach an ambition of trying to achieve a triangulation in the final research result. This subject will be commented on later in this chapter.

Each separate phase is described below, and a drawing of the research process is shown in the overall research model below. This model has both a vertical and a horizontal process which is shown in Figure 2.3 and Figure 2.4.

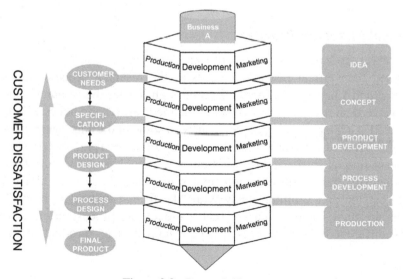

Figure 2.3 Research idea stage.

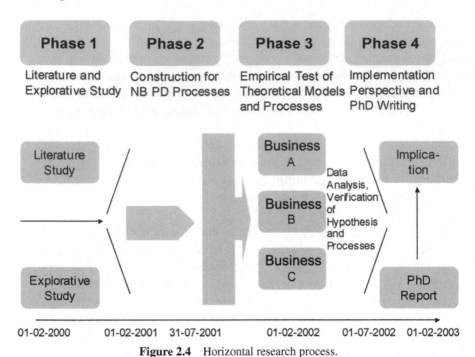

Figure 2.4   Horizontal research process.

The process and the thoughts about what has been going on in the research process will now be described.

The Vertical Research Process

The vertical research process consists of stages and gates:

I defined the vertical research process as the research process stages and gates which I had to consider and go through during the research time period. The vertical research process can be planned beforehand but as will be demonstrated at a later point I could not adhere to some of the stages and gates as the research turned out to have the same characteristics as a practical product development project in the industry. Often in this book I will demonstrate that the industry has to develop informal stages and gates and informal processes outside and simultaneously with "the formal stage-gate model". Therefore, the final – and actual – research model and process often turns out to be different from what was initially planned. The "product of the research process" turns out to be influenced by the research environment and continuously developed and improved through the research process until the "encapsulation of the PhD project" is decided upon. Hereafter, the research process will continue as I will comment on later in this chapter.

The vertical research process initially looked as illustrated in Figure 2.3.

Research Idea Stage

The research idea was initiated by the Center for Industrial Production and the thesis was formulated as

Network based high speed product development models and processes

From this general, overall idea I recognized and generated an explorative approach to the research object and task. This process was divided into three generations; one on the research view, one on the research approach and one on the research methods.

The research view considers various points of view such as the business economic view, the technical view, the customer view etc. The research approach considers the approach which a researcher can take on network based high speed product development – the action research approach, the case research approach etc. Finally, the research method considers different methods e.g. the explorative method, the semi-structured method and the structured research method.

These considerations were all possible ways "to handle" the research task. This process generated the first ideas to the research design. In this process I discussed and interfered with several internal and external network partners and for the first time I had to depart from my research development model because I joined and participated in informal product development processes outside my own research process to find ideas to solving my research task.

Research Idea Gate

The different "research ideas" and the different possibilities were considered and evaluated in the research idea gate. The ideas turned out to be decided upon and planned as an explorative research approach.

At first I decided on a multidimensional view on the research objective with an overall focus on a business economical optimal view as the optimal point of entry. This view was chosen because of my business economical background, because the research was carried out for and to the industry and because the research was done together with industrial businesses who focus on the business economical optimal point of entry.

Secondly, I chose a multitude of research methods, however, mainly with a focus on case methods and focus group and survey research methods. This choice was made on the basis of and very much inspired by the product development literature (Wind 1975) which claims that when a business faces a radical product development task, a multitude of development methods are appropriate for the generation of the product idea and concept. Furthermore, an ambition for trying to generate and reach a generalisation on the research object via a triangulation approach was also a main decision criteria. I will later comment on the ambition of triangulation for which I was very much inspired and supported by Professor Mariano Corso and my stay at Polytechnico di Milano. During my stay at Polytechnico di Milano I really understood what I was doing in my research and suddenly I could put words to the research approach.

Concept Stage

After having determined which research idea I was follow and focus on, my research turned into a kind of conceptualisation and break down of the original research idea. This work was fairly easy because the architecture and the focus of the research formulated at the idea stage was very clear and narrowly defined. The conceptualisation phase resulted in a research concept plan.

In this concept generation I was very much helped by my research colleagues and industrial businesses via discussion and inspirations to concepts to the research plan.

Concept Gate

The concept of the research was evaluated and only minor corrections to the final research plan were added. I hereafter began to develop the research.

Research Development Stage

The research development stage turned out to involve simultaneous research activities and it was therefore possible to do carry out these research activities at one time. This of course put pressure on the research architecture and on the management.

Research Development Gate

The research development gate was a very brief phase and much of the research development was carried out as a kind of "on the market" development or as a direct process development. The reason to this was simply that there was a high time pressure on the research. I will later in Chapter 15 comment on the sources to faults in the research when the research is under such a time pressure. Initially, however, I can already state that I tried to diminish faults in this area by sticking very strictly to the previously elaborated research architecture and also by continuously evaluating and continuously improving the research development process.

Research Development Process Stage

The research development process stage is very much identical to what I will describe in the horizontal research process in the paragraph below. However, the process was actually similar to the "production process" of all the empirical data which formed the basis of this research project.

Horizontal Research Process

The horizontal research process of the research project is as illustrated in Figure 2.4. Each separate phase will be described in the following.

Phase A – Preliminary Search for Literature and Cooperation with Industry and Other Research Environments

Taking his starting point in the above methodological framework the researcher began a major search for literature by means of books, databases of academic papers, newsgroups etc. in order to uncover the framework of network based high speed product development (Please see reference list in Chapter 1).

Such a search for literature also served to provide the researcher with a definition of concepts, a historical view of existing literature on product development as well as a preliminary model for the academic framework of the project.

Historically, the research project takes its starting point in a collection of models developed during the 1960s. Subsequently, the researcher has analyzed and mapped out the theoretical and practical work of models from the 1960s and until 2003.

The study of literature has been directed towards the prevailing literature on product development during this time, ranging from the stage-gate models, the functional models, the department models and to process oriented product development models and product development models based on simultaneousness and flexibility. Further the study of literature has been directed towards the prevailing literature on network during this time, ranging from literature on physical internal and external network to digital computer network to virtual network.

In terms of method, it can be argued that the initial research into product development focused on finding a theoretical framework model to be tested in the industry. The framework model was based on a hypothetic model which had not been tested in the industry.

The main focus of the product development model was to map out the separate phases of a product development sequence – the stage and gate models – in order to determine which phases the course of product development ought to comprise (Wind, 1973) (Cooper, 1986) (Myrup & Hein, 1985) (Eppinger, 1996) (Hart, 2000). As researchers in the United States focused on theoretical studies and on developing theoretical product development during this time, European researchers concentrated their efforts on studying and interpreting the implementation of the results and models of product development.

At a later point in time, a wider selection of researchers such as e.g. Cooper and Eppinger tried to test theoretical models in real, industrial product development life. Consequently, a need arose to look into which functions ought to be taken care of in each separate phase of product development. The result of this was a shift in research in terms of method; researchers and industries approached each other and the functional and departmental models came into existence (Hart, 2000).

Later, concurrently with the methodical approach of research towards the industry especially at the beginning of the 1980s, the parallel and simultaneous models gained a footing (Myrup and Hein, 1985) (Eppinger, 1993) and others. Among other things, this was due to a growing awareness of the apparent inability of previous theoretical as well as practical product development models to explain the product development processes sufficiently. Furthermore, it became apparent that the classic stage-gate models (Wind, 1975) (Cooper, 1986) were too slow compared to new market and technology demands.

At this point in time, however, in terms of method, research was still characterized by having the industry as its object and researchers as mere observers.

Up to 2003 there had been a shift in the methodical character of research. Thus, the application of the action research method particularly among European researchers had become popular (MacLoughland, 1998). Among other things, this brought about a close cooperation between a growing number of researchers and the industry.

This cooperation meant a growing recognition of the fact that the old product development models were not capable of explaining all product development processes in the industry. Action researchers fully realized that this research method influences the processes and the models according to which the industry attacked the complex of product development problems.

The action research method was chosen as the overall research method and framework of research method at the Centre for Industrial Production. The practice of the action research method can be compared to other practices as is shown in Table 2.3.

The present research project did not offer me the opportunity to carry out research as dictated by the pure action research method. In practice. The present research was carried out as a mix between the committed method and the action research method; however, with the greatest importance attached to the committed practice.

Consequently, the research project cannot be considered an action research project but rather an initial explorative study of literature and business cases based on a combination of theoretical literature cases, real explorative confrontation with select businesses, and a subsequent survey containing descriptive and analytical comprehension of the problem area.

At a later point in time, it was expected that it would be possible to carry through actual action research in the industry on the basis of this project.

In terms of models, the project had attempted to advance a hypothesis model for this research project. Such a model were elaborated to represent a combination of the stage/gate models and the department/functional models as an overall framework. The argument for this choice can be seen in Chapter 8.

The research project focused its efforts on the initial processes of the preliminary stage and gate models and on the models and processes used in NB HS NPD. Such an approach produces the overall analytical framework shown in Figure 2.5.

Table 2.3   Research methodology

	The Relevant Method	The Committed Method	The Action Researcher
Choice of Problem Method	Prior to data collection Positivistic	In relation to field of research Interpreting	In relation to field of research Interpreting
Starting Point for Research	Theoretic categories and hypotheses	Met-theoretic guidelines	Meta-theoretic guidelines
Research Strategy and Analysis Unit	In terms of quantity	In terms of quality	In terms of quality
Empiric Collection Methods	The field is examined independently of context	Phenomena are studied in relation to context	Phenomena are studied in relation to context
Interaction with Actors in the Field	Disassociated relationship	Interact without intending to bring about changes	Active participation in the field intending to bring about changes
Pitfalls	Ignorant, *a priory*, unknown conditions	Risk of getting lost in details	Risk of getting lost in details
Completion of Study	Sufficiently representative	When inner understanding suffices for an outer understanding	When inner understanding has been reached and the problem solved
Method of Presentation	Normative and explanatory	Descriptive and understanding	Descriptive and understanding

Source: Translated from Pernille Kræmergård – Centre for Industrial Production, 2001.

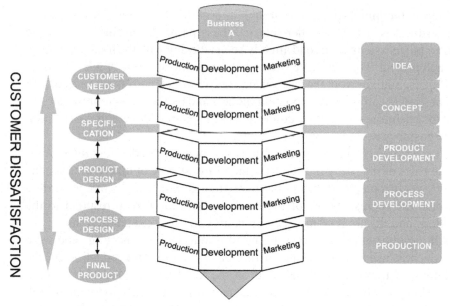

Figure 2.5 Analysis model for network product development.
Source: Bohn & Lindgren, 2000.

During my preliminary study of literature I examined a series of business cases described in literature (Please see the collection of cases). The cases were all selected on the basis of the literature search where it was found that some businesses and researchers claim that some businesses were developing at high speed and in networks.

The results of my case study were a collection of 71 cases which all represent one or more high speed enablers to network based high speed product development. Methodically, each case has been registered and at a later point analysed in relation to i.a. product development models, product development type, and identification of the main phenomena which govern high speed product development.

The analysis of these cases was performed as follows. By way of introduction the high speed enabler of each case was exploratively registered. Subsequently, crosswise of the cases, the frequency of the occurrence of the enablers was registered. This has been elaborated on in Chapter 5 of this thesis.

Next, the analysis was applied in the project as a help in putting forward the hypotheses of the research focus of the project. A description of all cases can be found in the Collection of Cases.

During the initial round of thorough interviews the researcher visited five case businesses. The researcher selected these particular businesses because the case businesses in focus had already been working with the focus area of his research project. The objective was to get a explorative view of NB HS NPD.

Representatives from the case businesses described and characterized their own businesses product development models and methods. As representatives from the businesses, managers connected to the product development function of the business were chosen (see Chapter 9).

The interviews can be characterized as semi-structured, direct interviews. On the basis of my study of literature and of my survey of business cases, the interviewer had prepared a semi-structured questioning framework, according to which the representative of the business was allowed to respond to the questions and the areas of my questionnaire with the least restrictions.

All my interviews were committed to minutes of the meetings and were subsequently used to contribute to further discussion and reflection on the modelling of hypothesis models and on the identification and confirmation of the HS enablers in my efforts to understand product development at high speed. Appendix 1 outlines my preparation of interview questions and NB HS PD hypothesis models.

Representatives from national and international research environments and projects have been actively involved in this research project. Thus, several different methods have been employed as shown in Table 2.4:

Table 2.4   Research activities and empirical activities

Research Activities	Activity	Referred to in Chapters	Publication	Appendix
Scenarios	Discussion with Professor John Bessant, Professor Marianno Corso, Professor Roberto Verganti, Professor Spina, Professor Harry Boer, Professor John Johansen, Professor Jens Riis, Professor Richard Leifers, Professor Mogens Myrup, Associated Professor Poul Dreisler	Several Chapters		(Appendix 2)
Conferences	Eiasm Conference Anti Polis CiNET – Aalborg CiNET – Helsinki	More Chapters	Right speed in NB HS NPD	(Appendix 3)

Table 2.4 Continued

Focus Group Discussions	PUIN Group DISPU Group	Chapter 10	Product development in Network	(Appendix 4)
Survey – internet based	PUIN DISPU	Chapter 11		(Appendix 5)
Meetings with experts on specific areas	Development director Preben Meyer TDC Internet,	Chapter 12		(Appendix 6)
Participation in various lectures and seminars relevant to the project	PhD Summer school in COMO	Chapter 12		(Appendix 7)
Action research laboratory project – TIP Project	Aarhus School of Business Economics; School of Danish Architecture in Aarhus; and School of Engineer Århus	Chapter 12	Network based High speed product development ISBN	(Appendix 8)
Participation in other research groups	(SMER, Loknit, Pitnit, Dispu, PUIN, RESME, SALSA, The TOM Project)	Chapter 12	SMER Loknit, Pitnit Case Book	(Appendix 9)
Participation in national and international projects	www.bestcom.dk www.viborg.dk www.loknit.dk www.smer.dk	Chapter 12	Process report 1, 2 and 3 Bestcom toolbox, case book and theory collection	(Appendix 10)

In connection with the first explorative examination of the problem area, a cooperation with Syddansk Universitet and DTU was initiated with a view to discussing the problem sphere and possible solutions to the assignment.

Thus, the researcher employed both telephone and direct focus group meeting techniques during his cooperation with Syddansk Universitet on specific sources of inspiration from a parallel research project on small and medium sized businesses in the southern part of Denmark.

Furthermore, the researcher took the opportunity of presenting and discussing with a major research group the relevance of and the approach to the problem sphere at the annual Fuglsø seminar in which DTU and Aalborg University both participate. In this context the product development groups of DTU headed by professor Myrup had been of interest and central importance in my efforts to define the problem sphere and as a guide in my study of literature.

In addition, certain parts of the project had been discussed by the research group PITNIT, who had been particularly interested in discussing the complexity of network and the HS enabler of e-development (www.loknit.dk) and book from the PITNIT project.

In connection with the PITNIT project it had been possible for the researcher during the initial phases to discuss the complex of problems with other small and medium-sized businesses when meeting for the LOKNIT arrangements. During this research project LOKNIT had provided the PITNIT researchers with the opportunity of discussing the research results with the industry.

Finally, internal seminars and meetings at the CIP and at the institute for production enabled me to discuss the problem sphere with other researchers working at the CIP. Such meetings had involved me in particular discussions on the learning organization and on main enablers for high speed product development to which discussions Professor Jens Riis and John Bessant had been contributing. Questions on speed, time, and logistics had been discussed with Professor John Johanssen. Continuous improvement and learning had been discussed with Professor Harry Boer and with Associate Professor Frank Gertsen. On the subject of product modulation and product modelling, Associate Professor Poul Kyvsgård Hansen had been consulted. In addition, the actual composition of models as well as the marketing related aspects of product development had been discussed with Associate Professor Kim Bohn.

Associate Professor Poul Dreisler from the Århus School of Business had discussed the processes of network based product development under high speed in a student project called the TIP project (Århus School of Business, Arkitektskolen and Ingeniørskolen in Århus and Herning (www.tipptojektet.dk).

Phase B – Development Meetings with Representatives from Industry

Subsequent to the initial search for literature, to the initial analysis, to the preliminary visits to selected industrial businesses, and to the development cooperation with other research environments, I decided to institute Phase B – Development Meetings with Representatives from the Industry. This phase entailed a series of more thorough focus group meetings in which a wider circle of industrial businesses took part. These businesses included:

- B&O
- NEG Micon
- Linco Trading
- Lyngsø Industri

- SCANIO
- AKV Langholt
- Ansager Møbelfabrik
- Tele Danmark Internet
- Grundfos
- Danfoss
- Dansk Teknologisk Institut
- GSI Lumonics (invited)

Our meetings took place from Autumn 2000 and until Autumn 2002 approximately once a month. Consequently, ten meetings were held. The meetings lasted between four and five hours and aimed to present and discuss details of the product development models of the selected businesses. At the same time, we aimed to approach the processes which the businesses passé through in the course of their product NB HS PD models. (See Appendix 4 + minutes of the meetings).

In addition, observations and their consequential hypotheses could be tested and discussed with a view to trimming the immediate validity of the hypothesis models.

All participating industrial businesses were offered the opportunity to present their formal and informal product development models as well as the processes attached to their model. Each meeting was assigned a main and unifying theme. Previous to each meeting, the participants had received the papers prepared by the researchers together with selected articles which could support and put into perspective the unifying principle of the meeting (Appendix 4 and Chapter 10).

The selection of industrial businesses was drawn from a wide range of businesses in terms of size and line of business as can be seen from the list of participants. This was a conscious and disciplined decision on the part of the researchers and should serve the purpose of having as many different product development situations and aspects represented as possible.

Phase C – Substantiation

Already before the development meetings were completed, Phase C – Documentation could be initiated. Methodically speaking, such an act was an explorative attempt at getting researchers and representatives of the industry to write together. These efforts resulted in the book "Product development in network, 2001" and the articles Network based product development a question of right speed not high speed.

Methodically, it was of the utmost importance to the researcher to attempt to analyse, describe, and develop a research area in a new way. Accordingly, it was significant to me to attempt a closer working relationship with representatives from the industry, among other things by writing directly in cooperation with participants from the industry.

As a result, the I forced to adapt myself to a new way of writing and under altered conditions. Additionally, the high speed concept had suddenly turned more relevant and had drawn nearer to the research environment as the representatives of the industry have had specific expectations to the performance of the "product" – or rather to the results of the book – to the degree of topicality, and to taking their own complex of problems as the starting point. At the same time, our chosen method had the result of involving the industry to a large extent.

On the other hand, the representatives of the industry had been forced to confront the researchers' theorizing of i.a. the concepts behind the theme of the project. Such a scenario gave rise to a continuous immersion into and consideration of the application of actual models and processes. As an example, the following statement giave cause to further research and preoccupation:

> *"The product development models you present are not used in our business at all! – We just do it. We just develop a prototype"* (SCANIO)

> *"If we ask our employees to slow down during a short span of time, everything stops. Therefore, a product development course must be run at high speed throughout the entire course"* (Grundfos)

> *"Generally speaking, we only perform incremental product development – in 95% of the cases. Advanced models are only used in connection with large, radical product development assignments"* (Danfoss)

These comments were central to the interpretation of the problem and have resulted in among other things the book "Product Development in Networks" (Bohn and Lindgren 2013).

The researchers have written Chapters 1 and 2 of the above-mentioned book by themselves, however, with representatives from the industry as critical reviewers. Chapters 3 to 6 have been written by researchers and industrial representatives in cooperation. A researcher has been attached to the writing of Chapters 3 and 4, and another researcher to Chapters 5 and 6. Each of these four chapters have had two or more industrial representatives as co-authors.

Such a cooperation was most fruitful to both parties and in terms of process it has been a most exciting method of working.

Phase D – Survey

On the basis of the results of the primary case research and the focus group interview a survey was elaborated. It was not possible to have all results ready from the case and the focus group interviews because the case and the focus group interviews were running simultaneously with the survey. However, most results could be used to elaborate the final survey as the survey was "produced" very late in the research project. In Chapter 11 a more detailed description of the survey and the survey results can be seen.

Phase E – Other Research Activities

Parallel to the formal, planned research activities other research activities which I joined came up (see Chapter 12). I chose to participate in these research activities because I wanted to observe other ways of doing NB HS NPD in other environments.

Phase F – Dissemination of Results of Research

The dissemination of the results of the project has been of a different character. However, the major dissemination toke place at the publication of the PhD project. During the PhD project some publications had been developed – a book, conference papers, articles e.g. Furthermore, several presentations and discussions toke place with internal colleagues, research network partners, managers of industrial businesses and others.

Phase G – Further Research

A series of new articles on the subject in question was planned after 2003. Such articles have attempted to discuss in detail a number of related subtopics to this research project. Please take a look into my profile at Aathus University (http://pure.au.dk/portal/en/persons/id(244bfceb-2a4c-4ef0-8c5a-34ad8238b5eb).html)

Furthermore, the research project provided the research environment with the possibility of involving several international researchers.

Finally, the group behind the research project decided to carry on their network cooperation in a number of small or large groups with the object of discussing and exchanging experience and activities related to the subject.

Our research project also initiated a series of students' projects at the study of MSc Engineering at the Institute for Production at Aalborg University.

2.3.3 Reflection on What – Why – When – How

One of the most important scientific discussions today is the paradigm discussion. The paradigm discussion concerns the analytical object, the theoretical foundation and the theories along with the scientific method. The paradigm discussion was firstly introduced by Thomas Kuhn (1962) who claimed that the paradigm provided models from which particular coherent traditions of scientific research are developed.

Kuhn explains how paradigms are generated and how paradigms vanish. Kuhn claims that in a time when a particular line of science seems to evolve with stableness with verifications of findings, analyses of findings, and development of normative guidelines, the scientific work and activity take place.

The stable time is characterised by Kuhn as normal scientific time phases and in this time phase some anomalies – some unexpected and unexplainable differences may turn up. As the anomalies become more and more expressed and increase, the researchers begins to loose faith in the paradigm, and the scientific area enters a time of crisis. In the time of crisis more competing paradigms can evolve and the paradigm which best explains the new phenomenon turns out to be the dominant paradigm. This time is called the revolutionary time. After this time, science moves into a new stable era of normal scientific time, where focus shifts from development of core scientific definitions to detail development of models and methods in the new, dominant paradigm.

The picture of the above-mentioned process is shown in Figure 2.6.

The paradigm is an overall definition in the development of science. Abnor and Bjerke (1977) have elaborated further on Kuhn's work. They talk about a working paradigm which breaks down the dominant overall paradigm and focus on an actual specific research area. The overall paradigm is relatively stable over time whereas the working paradigm is continuously changing in relation to the characteristics of the specific focus areas, to the evolving methods etc.

Abnor and Bjarke claim that the philosophic and scientific statements can be gathered in a methodological view where the working paradigm builds a

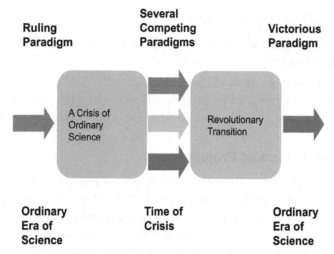

Figure 2.6 Kuhn's theory of paradigm.

bridge or a connection between the methodological view, the methodological learning and the focus area. A model of the above-mentioned is shown in Figure 2.7.

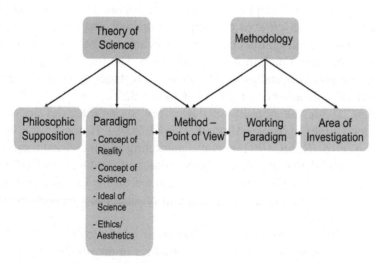

Figure 2.7 Working paradigm of scientific area.

It is the responsibility of the individual researchers to formulate the working paradigm to establish consensus between the methodological view, the methodological learning and the research area as shown in Figure 2.7.

2.3.4 Relating Theory and Method

To clarify the working paradigm of this research project, the following three areas should be described:

1. The normal scientific view of the project
2. Network based high speed product development as a new research area
3. The scientific view and scientific ideal of the focus areas

Normal Scientific View of Project

The normal scientific view of the research project is related to the business economic scientific area, in particular the product development research area. The focus of the area is development and implementation of new products to the market in an business economically efficient way. As an independent research area it is a relatively new research area which came into being in the 1960s with the American effort to develop a spacecraft that could reach the moon before the Russians. There was major focus on product development and how to do product development in an effective way.

In the mid 1970s a few teaching and learning books were published (Wind 1973) and in the early 1980s a stable academic discipline was introduced.

The normal scientific discipline was very much divided between a purely technical and production view to product development and a sales view to product development. The conflict between the technical/production view and the sales view was very much in focus at this time. The focus was primarily based on an internal view and primarily focused on stages and gates. Please see later for a more detailed description in Chapter 3.

In the late 1980s the normal scientific view of product development was developed further in detail (Abnor & Bjerke, 2009). However, the view experienced a crisis (Kuhn, 1962) because marketing and increased focus on customer needs and demands entered the discussion of product development. To a large extent, the discussion was concentrated on the conflict between market needs and wants and the technical and production competences and capacity of the businesses.

In the middle of the 1990s new research came up related to the process view of product development and the need of continuous improvement in product development. Increased focus on time and speed in product development came up, however, mainly focused on internal improvement and internal diminishing of time to market. In the late 1990s and early 2000 the continuous innovation and the network view came up related to product development along with the theory of learning in product development. These new trends came up with the increasing acknowledge of the need to develop a new theory on the product development area concerning:

1. The need to see the business in a new way – e.g. a network business
2. The need to manage the product development of the businesses in a new way
3. The need to do product development in network.
4. The need to do product development faster and more customized
5. The need to evaluate the product development process

The development on the product development area has given rise to new scientific areas:

- continuous improvement
- innovation theory
- learning theory
- network theory related to product development

The focus in this research project is on network based product development with special emphasis on speed.

Network Based High Speed Product Development as a New Research Area

Network based high speed product development must be said to be a new research area from 2000–2003. Many researchers had carried out important research in network and product development. However, it was new to science to combine network and product development. Furthermore, it was also new to combine speed and high speed related to network based product development.

The Center of industrial production at the University of Aalborg decided in 2000 to focus on this scientific area as one important research area and strategy of the centre. This was due partly to a huge interest in gaining more knowledge on this specific area, partly to the desire of businesses related to the centre to gain more knowledge about NB HS NPD.

Scientific View and Scientific Ideal of Focus Area

The Scientific work follows two essential but different goals:

- The first goals is to find new knowledge to increase the amount of available knowledge on the scientific area – basic science
- The second goal is to give clearness and systematic to the scientific area – applied science

This project can be classified in the area of applied research which goal was to develop new theory that explains and supports specific practical goals. Therefore, the project is strongly related to businesses and their product development models and processes.

2.3.5 Reliability, Validity and Generalisation

A discussion of Validity, reliability and generalisation is always important and relevant to all academic research. They are even inevitable when analyses, verifications, conclusions and recommendations are based on empirical data which include many personal interviews, registrations of group discussions, survey results, observations from group interaction etc.

The validity assessment concerns the planned phase of the research method and is focused on the interview, focus group discussion, survey and observation phase that created the data. The reliability assessment focuses on the data collection and the data analysis that precedes the final analysis, development of conclusions and recommendations.

Validity

Validity in any research or evaluation means that

> *"the research instrument measures what the individual using the instrument wishes to measure"* (Philips, 1996)

Even though this definition is simple, the concept of validity may include the four different aspects shown in Figure 2.8.

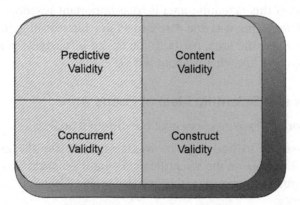

Figure 2.8   Four aspects of validity.

A validity discussion in relation to NB HS PD is not considered relevant to all four validity aspects.

Predictive validity which evaluates the ability of a research instrument or method to predict the future, is not considered relevant, since the purpose of the research on NB HS NPD was to grasp the present status of businesses involved and to paint a picture of NB HS NPD in up to 2003.

Concurrent validity refers to the extent to which the results of an analysis correspond to results of similar analyses made at approximately the same time. For NB HS NPD, no standard of reference exists. Some researchers have elaborated important results on product development and to some extent on product development in network. However, until now nobody has focused on NB HS NPD.

Another argument for the irrelevance of concurrent validity is that NB HS NPD is dynamic, and results from later research – observations, interviews, focus group discussions etc. – would necessarily be carried out differently in e.g. another business focus group or another TIP student group. However, I claimed that the generic results and research findings will be the same. On the basis of the research architecture which seeks to meet the demands of a triangulation research method I claim therefore that the generic results would be the same if the same analysis was carried out at the same time.

Content validity is important for the management of the research on NB HS NPD, since it refers to the extent to which the analysis is representative. An analysis which is not considered representative has a low content validity whereas a high content validity means a good balance between analysis and reality. With regard to the research on NB HS NPD, the content validity is only to some extent obtained.

The case interviews were carried out in five businesses with a general and a specific case analysis in each business. In most case businesses the case research was carried out with an interview with two or more persons responsible for or involved in the business product development activities. Of course it would have been better to have all actors involved in the product development activity presented in the interview but due to time limits, this was not possible. Furthermore, it would also have been better to follow a case from beginning to end but there is a danger that the memory of the respondents can be under the influence of time, policy or other components. However, I had no such impressions that this was the case when speaking to the respondents.

Obviously the case interviews can only be representative within the single business but combined with other cases a stronger representation can be obtained. Of course it would have been better to carry out more than five case researches but this was not possibly within the time given.

The focus group interviews were carried out with people responsible for product development in ten different businesses. The selection of businesses was carried out carefully with a view to business-to-business businesses and a view to different lines of business. The latter criterion was selected because it was essential to the research in an explorative perspective to get as many aspects of NB HS NPD as possible. Major differences in lines of business turned out to result in many aspects of NB HS NPD which surely could not have been obtained if all businesses had been in the same line of business.

The survey was sent out to a representative amount of SMEs on the business-to-business market. However, as will be verified later in this project, the amount of answering businesses was not large but it still allowed me to form an explorative picture of how NB HS NPD was carried out in SMEs in 2003. However, it was not possible to make general conclusions on specific lines of business.

Other research activities have been carried out on the subject NB HS NPD. None of these activities can be said to match the criteria of representativeness. However, neither was this the intention. Instead it was the intention to find add-on information to NB HS NPD and to support findings in some of the other research activities. Furthermore, especially the TIP project offered the possibility of coming closer to NB HS NPD and to make close observations on the process of NB HS NPD.

The visit to Italy added new dimensions to NB HS NPD although this could not be characterised as representative add-on to the research.

Summarising on the content of validity it could be verified that each individual empirical research activity cannot be said to be representative. Nevertheless, some of the activities had more representativeness than others. However, this was not the intention as the research was carried out in an explorative perspective.

Altogether the research activities try to match a triangulation on the research focus NB HS NPD. In this perspective the general generic findings must be said to be representative.

The construct validity refers to the extent to which the analysis maps the construct that it is intended to map. The construct validity is a complex matter and forms a general idea of something formed in the mind by combining a number of pieces of information. This can be defended by expert opinion.

The expert opinion related to the construct validity of my PhD project came primarily from my supervisors of the research project, from colleagues at the CIP centre and from discussions with professors and seniors at the Polytecnico di Milano. Secondarily, it came from the individuals evaluating the research project.

The explorative and semi-structured interview approach is exactly what Wind (1973) and Aaker and Day (1983) propose when researchers and businesses deal with product development and marketing research tasks that are radical and new to the market area.

Reliability

Reliability in any research or evaluation means that

> *"the research instrument is consistent and the outcome of subsequent measurements produce approximately the same results".* (Philips 1986)

If there is a significant difference each time a research instrument is used, then the instrument is considered unreliable. Differences or errors can be explained by fluctuation in the mental alertness of participants, variations in conditions under which the instrument is administered, differences in interpreting the results from the research instrument, and random effects caused by the personal motivation of the participants (Philips 1986).

With regard to reliability and instruments used in the research on NB HS NPD, the question is whether or not another research group would have developed a significantly different action plan for researching on NB HS NPD. The argument defending the reliability of the NB HS NPD models and processes of the PhD project are found in:

1. The data collection which included not only a few businesses but nearly 180 individual businesses
2. The strong involvement of businesses, participants, students in case research, focus group interviews, survey interview, student NB HS NPD projects etc.
3. The strong involvement of management of product development in businesses involved in the research project
4. The strong involvement of other researchers in the research project

The data collection included all phases of the product development process though with a strong focus on the upper part of the product development process. A high degree of verification was attached to e.g. the enablers to high speed product development due to the number of participants involved in the research process.

The strong involvement of businesses is explained by the degree of the actuality of the project and intense discussions in business networks, internal businesses and in research – how to perform high speed within network based product development.

The strong involvement of product development management was important in the discussion of relevance and priority of the suggested hypothetical models and processes to NB HS PD. Since cost, benefit and priority of each suggested model and process to NB HS NPD were discussed or considered in several meetings with the persons responsible of product development and the management, only the significant models and processes survived and were verified.

Therefore, it was not considered likely that a similar research would have given significantly more or different findings and suggestions than the findings concluded in Chapter 15.

All in all, it can be concluded that the research undertaken in the research project NB HS NPD fulfils the aspects regarding validity and reliability, which support the credibility of the research on NB HS NPD and the results obtained.

2.3.6 Ambition of Triangulation on Research on NB HS NPD

Previously I commented on the ambition to do a triangulation on the research on NB HS NPD. As it was argued before I was inspired during my research project to try to fulfil this ambition. The research work was therefore focused on trying to come as high as possible on the generalisation according to Jick's (McGrath et al., 1982) model of the different aims of a research project, see Figure 2.9.

> *"Ideally, a scientific study should reach all three aims – at least traditional Natural Science studies. However, within the world of management, the three objectives will constrain each other. In McGrath's (1982) interpretation, the model becomes a "three-horned dilemma" in the sense that a researcher can only reach two or three objectives in a study – leaving him or her vulnerable to critique for not reaching the third objective – the third objective (whatever that may be). Thus realism, precision and generalisability are impossible to reach at the same time for researchers."* (Drejer A. et al., 1998)

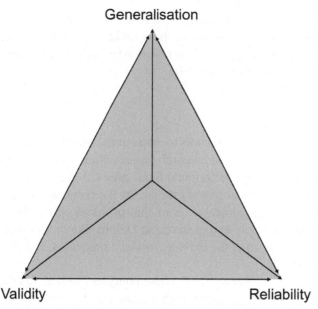

Generalisation

Validity   Reliability

Figure 2.9 Jick's model for trade-offs in doing scientific research.
Source: Based on McGrath et al., 1982+.

With the above-mentioned in mind I was left with a difficult challenge. I chose to go for a maximum of generalisation and reliability. Via the ambition to do a triangulation in the research (case research, focus group interview, survey etc.) my research project had the possibility to reach a high level on both areas whereas the validity area could suffer from this chosen research focus and ambition.

The project differs from other research projects at the CIP Center. Normally, the research projects focused on one or two cases which normally give the researcher a possibility to reach a high level of validity. In this research project I chose to do a multitude of research activities and was very much inspired by the UK tradition of research (Drejer A. et al., 1998). This resulted in a strong empirical content and a combination of more cases, focus group interviews, surveys and other empirical results.

This meant that initially my research project was confronted with the challenge to meet a high level on the arrow of generalisation and realism. On the other hand, this choice resulted in problems in the validity area because it was not possible to perform very thorough research in all case businesses and other research activities. Therefore, I had in 2003 to accept a lower validity to fulfil my ambition on generalisation and realism. I tried to cover this area as already explained above in the paragraph on validity. In my final conclusion I will comment on how the results of the research turned out seen in the context of validity, reliability and generalisation.

2.4 Summary

The purpose of the present chapter was to map out the scientific and theoretical framework and foundation of the research project. Similarly, the purpose was to determine and explain the methodical focus areas of the research project.

Consequently, it was established that due to the newness and practically non-existing degree of documentation of this problem sphere in 2003, the research project had to apply explorative and Delphi-like methods in order to gain insight into the nature of the existing models, processes, and phenomena of this problem sphere.

As a result, the researcher used a 7-phase analysis frame as his theoretical basis. During the first phase, a) a thorough literature and case study was carried out in combination with b) semi-structured business visits and interviews. The results of such preliminary research were continuously discussed and documented via c) various working papers with national and international research environments and via d) individual meetings, e) focus group meetings and seminars together with f) a survey and g) other research activities.

Special attention must be drawn to the focus group meetings consisting of representatives from the industry and from the research environment at Centre for Industrial Production, Aalborg University. Meetings between the two parties had the purpose of discussing each separate area of the project in detail and of activating the subsequent documentation and writing process between the industry and the research environment. Among other things, the results of our cooperation has been a book on network based product development.

Methodically, my research project has been based on a multitude of research methods with both a business case method and on a combination of engaged and action-like research methods. This choice was mainly due to the need to carry out an explorative research and to the research ambition to try to reach a level of generalisation via triangulation. If this has been achieved and the sources of faults will be commented on and analysed upon in the final conclusion.

References

[1] Abnor & Bjerke (2009) Methodology for Creating Business Knowledge SAGE Publications Ltd.

[2] Blantern, C (1994) the Learning organisation (organization of learning) and the emerging contribution of computer networks.

[3] Bower/Hout – Learning Loops Fast – cycle capapility for competitive power.

[4] Booz, Allen & Hamilton, 1982, 'New Products Management for the 1980s', New York.

[5] Cooper R. G. (1993), Winning at New Products – Accelerating the Process from Idea to Launch, Perseus Books.

[6] Drejer, A. & Riis, J. O., 2000, 'Competence Strategy', Børsens Forlag, ISBN 87-7553-740-0.

[7] Pedler Mike, John Burgoyne and Tom Boydell (1997) Den lærende virksomhed.

[8] Wind, Y. (1973), "A New Procedure for Concept Evaluation," Journal of Marketing, 37 (October), 2–11.

[9] Wind, Y. (1981), "Marketing and Other Business Functions," in Research in Marketing, Volume 5, Sheth (ed.), Greenwich, Conn.: JAI Press, pp. 237–264.

[10] Wheelwright and Clark (1992), Revolutionizing Product Development. Quantum Leaps in Speed, Efficiency, and Quality, Free Press, New York, NY.

[11] Hart, Suaan

[12] Kræmmergård, P. (2001), Table Translated from Pernille Kræmergårds PhD – Centre for Industrial Production.

[13] Bessant, J. (2001), Presentation at Center for Industrial Production and University of Brighton.

3

Concepts of Product Development

3.1 Introduction

There is a certain coherence between Chapter 3 and Chapters 4, 5, and 6 which define the other concepts of network based high speed product development models and processes. My study of literature and the secondary cases form the basis of Chapters 3, 4, 5, and 6. Additionally, the said chapters should be seen as a continuation of the case studies carried out under the previously defined analysis framework described in the articles "Network Based Product Development – Analysis Framework for Case Studies" (Bohn and Lindgren, 2000) and "Network Product Development – Main Phenomena in Network Based High Speed Product Development" (Bohn & Lindgren, 2000) written for the sub-project on network based product development at the Centre of Industrial Production, Aalborg University.

The preliminary research objective on the concept "network based high speed product development" demanded that the researchers should carry out in-depth research on the concept of *Network Based High Speed Product Development*. During the work it became apparent that a series of concepts were related to network based high speed product development and needed to be accurately defined.

Consequently, the objective of such definitions is to clarify basic concepts of the research work and to put into perspective such concepts in relation to the role that they play in today's and future network based product development. Such concepts will help to form the final framework of the research project.

Specifically, the framework for the definition and understanding of the research project includes the following definitions and answers to questions:

1. The product perspective – What is the product?
2. The product development perspective – related as well as new product development perspective – What is product development?
3. The product development model – What is a product development model?

81

4. The functions of the NB HS PD – Which functions are involved in the product development process?
5. The product development process – What is a product development process?

3.2 Development of Concepts

3.2.1 Introduction

For a long time up to 2003, concepts of product development had been very stable. Both researchers and industry used the same concepts about e.g. products and product development models and processes. However, as the pressure and focus on speed in product development intensified, some incremental changes came about concepts of product development in the last 5 to 10 years. Consequently, such gradual incremental changes resulted in radical new understandings and definitions of the concept and definition of product development.

Researchers realised that the existing concept of product development did not sufficiently explain the march of events in product development. Additionally, the industry realised that existing concepts of product development cannot match the new demands for product development on the global market.

Concepts of product development therefore changed, increased and were "mixed" in a way hardly ever seen before up to 2003. Concepts of product development were used in new and other ways and the boundaries of the concept of product development change, were developed, and used in ways we had never seen before.

In the following paragraph I will show how such changes in concepts developed both theoretically and in practice.

3.2.2 The Product Perspective

The Theoretical Approach

Central to the course of a product development process is the creation of a product. If it is reasonable to claim that a product development process has an end or an output, it is the final creation of the product. Thus, we must address the issue of the concept of product and ask the following questions:

- What is a product?

and

- What does the concept of a product comprise?

Several authors have addressed this issue.

> *"You may have noticed by now that the new products process essentially turns an opportunity into a profit flow. It begins with something that is not a product (the opportunity) and ends up with another thing that is not a product (the profit). The product comes form a situation and turns into and end."* (Crawford, 1992)

Crawford's point of view takes its point of departure in the marketing and customer perspective. In such a perspective the marketing point of view is the motivation of the business – "to create a profit" – whereas the customer point of view is to create maximum satisfaction of the customers' needs.

However, Crawford's dimension lacks three important points of view, viz. the design point of view and the technical point of view of the product. The technical and the design point of view are often closely related and will be dealt with collectively later in this chapter. Furthermore, Crawford's point of view should be seen in the light of his anchorage in the business economic theory conception, where the technical and design point of view are quite different.

To decide the nature of a product, we must therefore take our point of departure in at least five points of view:

1. the marketing point of view (Philip Kotler, 2001) (Jiao and Tzeng, 1999)
2. the customer point of view (Philip Kotler, 2001)
3. the technical point of view (Eppinger, 2000)
4. the design point of view (Verganti, 2001)
5. the network point of view (Haakonson & Johanson, 1982) (Child and Faulkner, 1998)

To all appearances there is considerable difference between these five points of view and between their view of the nature of a product. We will come back to these different views later in this chapter. However, we will concentrate on the way in which a product is "born", "dies" or leaves the market.

In the area between the these five points of view, the new product commences, is developed, and its future is decided upon. Many authors have dealt with the "birth" of a product.

> *"In the course of a product development the product definition will often be prepared as a cooperation between marketing, customer and designers."* (Jiao and Tseng 1999)

In their model outlined in Figure 3.1, Jiao and Tseng show the birth of a product as well as the continuous refining.

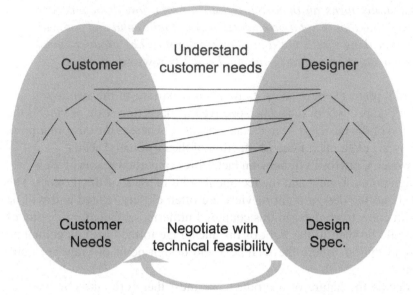

Figure 3.1   The Elaboration/Refinement process and product definition.

Source: Copied from Jianxin Jiao and Mitchell M. Tseng "A requirement management database system for product definition"; Integrated Manufacturing Systems 10/3 [1999] 146–153.

The authors indicate that the product is continuously being formed and defined during the product development process in a close teamwork between customer, marketing, and designers. It is worth noticing that Jiao and Tseng do not operate with more than three partners; the technicians have been left out as single players but have been included in the design function.

Crawford says that a product has a beginning and an end. However, several examinations show that researchers and businesses do not exactly know when the product started – "the big bang" – or when the product is quite finished. Many cases have verified that it is also difficult in the course of a product development process to make the customer and often the technicians agree on a "final start of the product" and on a "final end and completion of a product". This issue is often subject to major discussions and "fights" between customer and supplier. However, it can be stated that this might be a hopeless discussion. Several businesses and several researchers in 2003 considered therefore the concept of a product to have a propensity to be "floating", and they believe that a product is a process without a beginning or an end or with many beginnings and many ends (Corso, 2001).

The Practical Approach

The case businesses used for the present research project mentioned that the decisive characteristic of the course of a product development process was the ability to "freeze" the product as late in the product development process as possible. Businesses like Glunz and Jensen (Case No. 30) and ScotIT (Case No. 55) confirmed this perception. The product is a process with many beginnings and many ends. The business can choose to encapsulate – "freeze" and start its beginning on the market. The product is by "nature" dynamic and offers the possibility of "developing" over a span of time pari passu with changes in market demands. Furthermore, the initiation of a product often offers the possibility of several product ends (Corso, 2001) and "encapsulation".

Thus, the above scenario contradicts Crawford's point of view with the product having a beginning and an end unless we consider the "frozen" product the end of "the product". However, the "frozen product" only forms part of the total possibilities of the idea or concept of a product which the business chooses to "encapsulate" and market. By encapsulating or finishing the product, the business has given the product its final definition and have thereby applied a series of consequences and limitations which may at a later point in time turn out to have positive effects as e.g.:

1. Being first in the market
2. Forcing the competitors to improve their products
3. Changing the competitive conditions
4. Possibility of quickly making many variations

However, such effects may also turn out to be negative:

1. First mover disadvantage – the product had hidden, serious flaws which made the customers mistrust the product (GSI Lumonics)
2. Major consequential costs – repairs, upgrading etc. (MV)
3. The customers did not use the product in the expected way
4. The product architecture does not allow changes or the quick making of variations

By "freezing" the product, the developer or the product developing organisation has opted out of some possibilities once the decision has been made. Subsequently, the developer or the developing organisation will have to live with the characteristics of the product (Case No. 54 Grundfos) until a new product has been developed. However, the "encapsulation" also offers new possibilities. Therefore, it is vital to make the product architecture right at the idea and concept phase so that a maximum of flexibility can be obtained at a later point in the product development process and on the market (Verganti, 2001).

The above picture emphasizes the importance of addressing the two main areas of product definition:

- What is a product?
- When do we consider the product to begin and to be final or finished?

In the present research project, the latter question could easily be answered by defining the finished or the final product as:

> *a product is considered finished at the exact time when it is introduced on the market*

A product could also be defined as shown in Figure 3.2 presenting the transition from the product development phase to the market introduction phase.

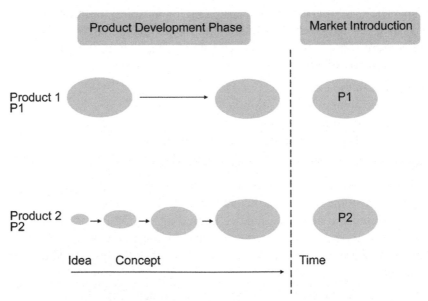

Figure 3.2   Birth and maturity process of product.

Source: Peter Lindgren, 2001.

In such situations the product development group will generally hand over the developed product to the sales and production department for further market adaption and production (Hein and Myrup, 1985).

Thus, from idea to market implementation, the product undergoes a "birth", a "maturity process", and a "market introduction" – with many concepts and product prototypes before final product is decided upon as shown in Figure 3.3. In the terminology of the present product development theory the

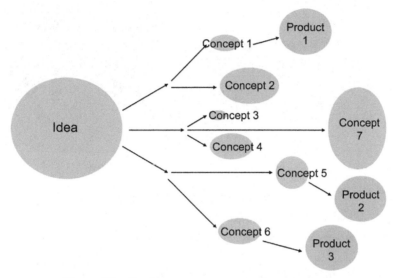

Figure 3.3 One idea – many concepts, fewer products.

marketing phase designates the final finishing of the product or the completion of a specific product development process. However, we recognize the fact that specific product development processes pave the way for new product development processes and also pave the way for an adaption of the final product during its life cycle in the market – the adaption of parameter mix as shown in Figure 3.4.

The present research project will not focus on the market introduction phase as the project deals with the actual development of the product prior to market introduction, i.e. from idea to market introduction. In other words this project will not in particular address the problems or the product development adaptations related to products which have already been introduced to the market.

However, we therefore found ourselves between two possible definitions of a product:

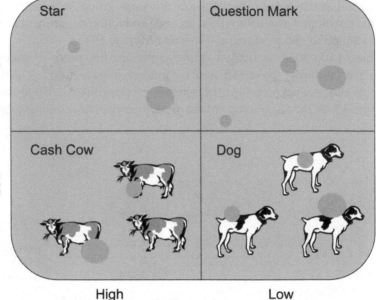

Relative Market Share

Figure 3.4   The Boston consulting group's market share.

Source: Heldey. B., "Strategy and the Business Portfolio", Long Range Planning.

- a product is only a product when it is introduced to the market
- a product is a product already when it is an idea or a concept offered to some customers but which has not yet been "produced" to the market

A lot of products in 2003 were sold to customers as the latter definition. The definition of a product therefore depended on:

- the viewer's approach to the product – an "encapsulated" product or an idea or concept
- the viewer's focus on before or after the market introduction

In the PhD project the main focus is on the idea and concept phase where the product has not yet been "encapsulated".

The Practical Approach

The case analysis (Case No. 11 Rossflex, Case No. 30 Glunz & Jensen, Case No. 1 Zara) showed very clearly that the businesses defined the product as an "encapsulated" product introduced to the market. The businesses did not regard the product as a product before its introduction on the market. However, the customers regarded the product as existing once the idea and concept were "borne" and they did not see an end to the product because it was always there when needed. The technicians regarded the product as finished when the product is technically "encapsulated" and the network regarded the product as existent when the idea and concept were defined.

To finish the answer to the approach to the viewer on – What is a product? – we must therefore discuss the different approaches a viewer could have to a product up to 2003. I saw the following views on the product – the marketing, the customer, the technician, the design, and the network view.

To answer the question of defining a focus before or after the market introduction a discussion on the model and process of product development was central. This will be done in Section 3.4.

3.2.3 Marketing and Customer Perspective

The Theoretical Approach

The marketing and customer perspective are addressed collectively as these two issues are closely related in that they both take their point of departure in need and want satisfaction or which value the product can give the customer. Yet, the results of the two perspectives differ significant.

In relation to the marketing and customer perspective McCarthy and Perreault define product as follows:

> *"the need satisfying offering of a firm or anything than can be offered to someone to satisfy a need or want."* (McCarthy and Perreault, 1990)

McCarthy and Perreault take the need or want of the customer as the starting point; *need* being what the customer actually needs, whereas *want* is a manifestation of the customer's desires and is thus a fickle and flighty need characterised by uncertainty and dynamics. As can be seen from the above, *need* can be accurately defined unlike *wants* which leave the product definition with a high degree of uncertainty. *Need* is often connected to existing products or to "incremental change" of existing products, whereas the satisfaction of *wants* can often be categorised as "radical change" where new demands are made on the product, new questions are asked of the shape of the product.

The Practical Approach

The above explanation illustrates how dfficult it was to define the concept of a product accurately. This was partly due to the experience that the desires of the customers during the initial phase of product development or even after the completion of the product development process often could not be fulfilled neither by the customer nor by the supplier. Needs and wants of the customer often changed during the product development process which means that the needs and wants of the customer often were significantly different at the beginning of the process compared to his needs and wants at the time of market introduction (Case No. 30 Glunz Jensen) (Case No. 56 Amanda project). Consequently, during a product development process customers often saw the product as a dynamic concept of the product, and the definition of a product could therefore often be characterised as a process towards an "encapsulation" of the final product which was subsequently introduced to the market or to the customer. However, this "encapsulated" product was the start of a process of new wants and needs. The product development process was therefore a driver of continuous innovation.

The above scenario may be the result of technological changes in the market or of changes in customer needs which arise while the product development is in process. The product may even be out of date or not answering the needs and wants already while the development is in process (Eppinger, 2000) (Hart, 2000). From a traditional business economic view this situation is not easy because it was practically impossible for a business in 2003 to earn money if the product never finishes or if the product is never "encapsulated".

The case research showed that businesses selling goods would provide against such a situation by means of a system requirements specification or of a written agreement of what is the final product or the "encapsulated" product. As opposed to this, the customer would try to safeguard his interest by stipulating that the product was future-orientated or by including open standards in the agreement to make subsequent adaptations to the product (Case No. 57 TDC, Case No. 38 Lyngsø).

The above discussion left the buyer and the seller up to 2003 locked in a situation which required an answer to the question "Does the product have to have a beginning and an end?" If the product has an end, such an end must be the "freezing of small incremental bits the product" and the "market introduction" of "encapsulated bits" of the product.

The question is, however,

- Is it interesting to finish the product?

From a business relationship perspective it is better to have a continuous process "going on" with the customer. It is more interesting and better to regard the product as a process where there are several "encapsulations" going on.

If the product is finished, there is no need for the supplier and customer relationship any longer. In most secondary cases we studied and due to the radical and dynamic environment mentioned earlier, it showed that there is a continuous need for incremental and even radical improvements to what has been developed as "the final product" at a certain time. So the reality is that the product is a process or an ongoing process. Many researchers and many businesses were still "stucked" in 2003 in an old belief that the product was to be finished instead of finding out how they could do business on a product which was a process.

This stressed the necessity of viewing the product and the development of the product as a process in the future which may never reach its end. Such an argument was supported by several cases (Case No. 50 Microsoft, Case No. 30 Glunz Jensen) (Corso, 2001) (Verganti, 2002). In these cases the businesses had passed directly to prototyping perhaps realizing that the final product definition could not be made until the product had been introduced on the market and maybe not even until the customers had tested the product or had found new use for the product.

The present discussion also explained why some businesses in 2003 (Case No. 50 Microsoft) continuously market beta-versions or very early beta-versions of their products (Case No. 50 Microsoft) The business had realised that it will not be possible and it was not business economically advantageous to "finish" the product at a certain time. The business realised that product should not be finished because customers always wanted to change the product and always had a want for new product features (Case No. 55 Scooter IT, Case No. 50 Microsoft, Case No. 30 Glunz & Jensen).

The discussion also explicated the growing demand for and movement towards service, adventures and other not well defined products in 2003 – virtual products which were characterized exactly by having been created as an on-going process or ongoing product development in cooperation with the customer. Such products were generally produced in the "here and now" and "on the market" which meant that the final results were difficult to predict.

Additionally, it was often a matter of "trial and error" before the final product definition could be determined.

This of course accentuates the dynamics and the possibilities as well as the element of uncertainty of the product. It therefore seemed to be necessary to change the classical and traditional view of the product in 2003 and change our view of the product to a process view which defines the product as:

> *"the product is regarded as a process in which the business creates a product – or a platform – with many solutions, many possibilities of extension and many starts and ends."*

When this definition is the case then a discussion of tangible and intangible products is relevant.

3.2.4 Tangible and Intangible Products

Theoretical Approach

In spite of the above mentioned discussion several authors put another dimension to classify the final product. Dibb defines it as follows:

> *"everything, both favourable and unfavourable, that one receives in an exchange"* (Dibb et al., 1991:208).

Obviously, Dibb gives a very open and vague definition of the product. At the same time, he designates the consideration against which the seller will produce the product. Such consideration indicates one of the differences between the marketing and the customer perspective. It is obvious that there is a potential conflict of different views on this whether it is the marketing or the customer view that is to be used. The optimum of product development is of course to create a product which satisfies both views. Then it is possible to make the exchange and conduct the business.

Stanton makes a more specific definition of the concept of a product by indicating that the product consists of tangible as well as intangible attributes:

> *"a product is defined as a set of tangible and intangible attributes, including packaging, colour, price, quality, brand, and the services and reputation of the seller. A product may be a tangible good, service, place, person or idea."* (Stanton et al., 1991:168–9)

Thus, according to Stanton the concept of a product includes tangible and intangible attributes which are the features on which the global market is focusing today. There is a strong trend towards less tangible products and more intangible products.

Practical Approach

Stanton's definition of a product corresponds to the developments which had taken place in industry up to 2003. Practically all products contain tangible and intangible attributes, and intangible attributes make up an increasing part of the products today. (Case No. 59 Nokia, Case No. 71 Nokia (perceived value).

In Figure 3.5 the product dimensions are described and the above-mentioned development on the basis of the increasing influence of products changing from "encapsulated" final physical products to "encapsulate mixtures of physical, digital, and virtual processes.

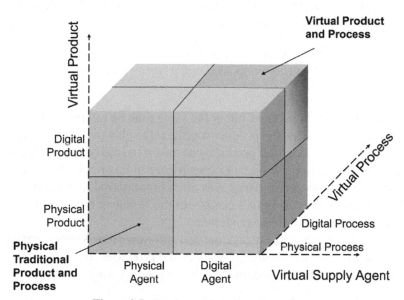

Figure 3.5 Product and process dimensions.

Source: Inspired by Turban 2001.

As can be seen from the Figure 3.5 my hypothesis was that there was a move in industry in 2003 from the physical product to a mix of physical, digital, and virtual products. Moreover, the model indicated how the creation of a product moved from being a physical process to becoming a digital process and end up being a virtual process. Additionally, the product moves from physical supply agents to a mixture of physical, digital, and virtual supply agents. All combinations of the above will therefore be the possibilities and the potentials for the future global market. As can be seen there was a huge unused potential of new products and processes.

As a consequence, this research project had chosen to regard the concept of a product as a total mixture of total products or total processes consisting of both tangible and intangible attributes and processes (Kotler, 2000) (Verganti, 2000).

3.2.5 Needs and Wants of a Product

Theoretical Approach

Our definition of a mix of a total product and process was to some extent similar to the marketing definition of a product. Philip Kotler (2002) describes the concept of a product as follows:

> *"a product is anything that can be offered to a market for attention, acquisition, use or consumption, that might satisfy a want or need."*
> (Kotler, 1994: 432)

Kotler emphasized *want* and *need* but at the same time he referred to the use of the product or the role that the product was meant to play for the customers. In other words, the function of a product was meant to satisfy specific wants and needs with the customers. As can be seen later in this book, the success rate of a product or of its performance was often proportional to the ability of the product to satisfy the customer's needs – or the "voice of the customer" – or the value a product gave to the customer.

However, my definition of the product was different from Kotler's definition because it stressed the mixture of product and processes, in particular the process element. Therefore, my definition of the product was:

> *"a product is a mixture of anything physical, digital, or virtual that can be offered to a market for attention, acquisition, use or consumption, that might satisfy a want or need"*

This research project focused on the need, demand, and value which a product gave to the customer because it was in this area that businesses had the possibility to make business. The wants of the customers were one of the important drivers to new products and processes. However, the business had to carefully analyse such wants to fulfil their purpose of making a profit.

Bradley emphasized Kotler's focus on the customer perspective in which the customer attempts to satisfy his needs with the product:

> *"a product is therefore, anything that satisfies the customer and increasingly it is something which has embodied in it a high level of service. The consumption of products and services is the way in which users attempt to satisfy needs."* (Bradley, 1995)

The needs and demands of a customer could be related to the value perceived by the customer. Such a value can be quite different from what the business, the technicians, or the network think are the needs and wants of the customer. The Nokia (Case No. 71) and Ford (Case No. 72) cases showed this very clearly.

The customer perspective was instead defined as the point of view of the customer where the customer's wants are fulfilled regardless of the customer's real needs and demands.

The new marketing view was defined as the point of the customer's perceived wants and needs fulfilled regardless of the customer's real needs and demands. This meant that the product can easily be "constructed" less complicated than the technicians, the network, and the customer think it should be as long as the product and the process fulfil the customer's perceived value. An example of this was verified in Nokia (Case No. 71), Ryan Air (Case No. 62), and Zara (Case No. 1).

The new marketing view and the customer view do not necessarily contradict each other. In the future, seen from a business economic outlook, the challenge in 2003 would be to maximize profit and/or reduce costs of production or product development with regard for the maximization of the customer's perceived value.

Consequently, the seller also has needs which he wishes to satisfy by participating in and entering into a product development process. Such needs are not necessarily merely of a short-term business economical character but can also be of a long-term strategic character.

Most often the customer perspective will take the side of the customer or will regard the product as a means to obtaining the highest degree of satisfaction for the customer without regard for the business economic interests of the seller. The marketing perspective on the other hand takes its starting point in the needs, wants and perceived value of the customer and with a regard to business economics and profit maximization as shown in Figure 3.6.

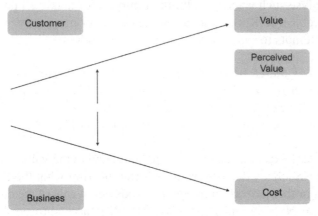

Figure 3.6   Cost value.

The field of tension on perceived value to the customer is decided between seller and buyer in the framework for product design from idea to market introduction.

Thus, the product related point of view of this research project toke the new marketing outlook as its starting point. This outlook also formed the basis of the analysis framework of the product concept as illustrated in Figure 3.7.

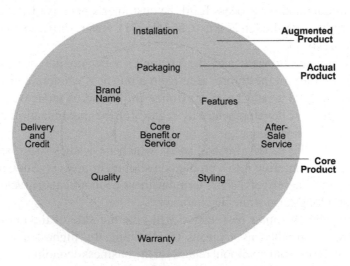

Figure 3.7   Three levels of a product.

Source: Kotler, P., 2001.

The model takes its point of departure in the consumer perspective but the basic elements of this perspective were of central importance in the further analysis because B2B products contain the same basic elements, however, seen from a business to business point of view.

As the research project deals with business-to-business products, the products can be classified according to the model shown in Figure 3.8.

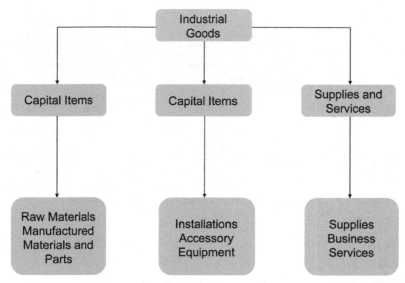

Figure 3.8 Classification of products.

Source: Kotler, P., p. 436–438 Marketing Management 8th edition 1994.

To throw light on the problems of this research project, I choose to consider all classes of possible B2B products as relevant. As will appear later in this book, most of the cases and most of the empirical data were concentrate on products in the business-to-business market.

3.2.6 The Design Perspective

The design perspective had already been thoroughly described by Eppinger in his model reproduced in Figure 3.9.

The design perspective differs from the technician's perspective by not only focusing on functions but also and in many cases more on meaning (Verganti, 2000). Verganti went as far as to talk about design driven innovation. Thus, he put forward the idea that a product consists of three main components:

	Generic (Market Pull)	Technology Push	Platform Products	Process Intensive	Customized
Description	The firms begins with a market opportunity, then finds appropriate technologies to meet customer needs.	The firm begins with a new technology, then finds an appropriate market.	The firm assumes that the new product will be built around an established technological sub-system.	Characteristics of the product are highly constrained by the production process.	New products are slight variations of existing configurations.
Distinctions with respect to generic process		Planning phase involves matching technology and market. Concept development assumes a given technology.	Concept development assumes a technology platform.	Both process and product must be developed together from the very beginning, or an existing production process must be specified from the beginning.	Similarity of projects allows for a highly structured development process.
Examples	Most sporting goods, furniture, tools.	Gore-Tex rainwear, Tyvek envelopes.	Consumer electronics, computers, printers.	Snack foods, cereal, chemicals, semiconductors.	Switches, motors, batteries, containers.

Figure 3.9   Summary of variants of generic development process.

Source: Eppinger, 2000.

- Language
- Message
- Meaning

The design perspective was very much related to the perceived value of the customer.

The design perspective was also closely related to the performance criteria or the design because the "soul" of the product is often related to and created by the designer. In terms of performance an excellent design can mean the difference between the life and death of a product. Additionally, an excellent design will make it exceedingly difficult – and in some cases impossible – for the competition to plagiarize or copy (Verganti, 2001).

Verganti makes a well-defined distinction between the design perspective and the technical perspective. Thus, Verganti maintains that the technician generally worries about functions which the designer sometimes does not necessarily do.

Products with meanings but without function (Philip Starke)

Practical Approach

The B&O case (Case No. 2) reveals how design and protection of a good design could be extremely valuable to a business in 2003. B&O focus on design and perceived value at the customers'. This positions the business in an extremely attractive competitive advantage situation in 2003 where it was possible to gain a high price for their products.

Nokia (Case No. 71) was another example of how focus on design left the business in a competitive situation where Ericsson and other mobile telephone businesses found it very hard to compete up to 2003. Nokia focused on perceived value, and cost leadership production left the business with a significant net profit margin with which none of the competitors could compete at this point in time – 2003.

3.2.7 The Technical Perspective

The technical perspective is not in focus in this PhD project. However, I stress the importance of this perspective and its role as one of the main reasons for high speed and time pressure in "the field of product development". Furthermore, the technical perspective made it possible to move businesses or products into new areas of digital and virtual products and processes.

Technically, everything the customers want seems possible either in a short-term or a long-term perspective. Most technical equipment became less and less expensive up to 2003. This meant that costs were no longer a limit to implementation of new technical features and products. However, there were still technical limits to new products as Richard Leifers (2002) described very well in his book about radical product development. Customers often did not adopt new technical products and processes in the way which the technicians had intended.

Therefore a close coordination between market, network, business competences, and the technical perspective of the product and the process were vital and important. The Amanda case (Case No. 56) and the Lumonics case (Case No. 37) show clearly the possible outcome when the technical perspective were forgotten and not well integrated in the product development project.

3.2.8 The Network Perspective

This perspective will be dealt with in Chapter 4.

3.2.9 The Total Product and Process Concept

On the basis of the above discussion, this project had chosen to define the product as follows:

> *"a mixture of business to business product and process that can be offered to a market for attention, acquisition, use or consumption, that might satisfy a want or need both tangible and intangible"*

Accordingly, I choose to pay attention to business-to-business products and processes from idea/concept, to product development, to drivers to market introduction, and back to a new idea and concept development in a continuous product development process. This is illustrated in Figure 3.10.

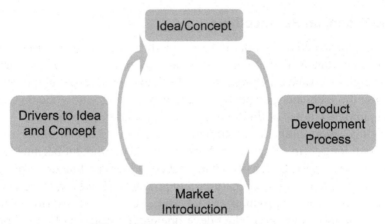

Figure 3.10   The total product concept.

3.2.10 Other Product Perspectives

A Strategic Product Perspective

Some authors (Abell, 1980) saw product development in a much wider perspective than the ones outlined above. Thus, market development and market penetration were considered product development just as integrative growth was considered an extensive form of product development. Such an outlook is illustrated in Figure 3.11.

The above-mentioned PD strategy is relevant because the boundaries of product development strategy are integrated into and blended with the other three strategies. At the same time the PD strategy was under high

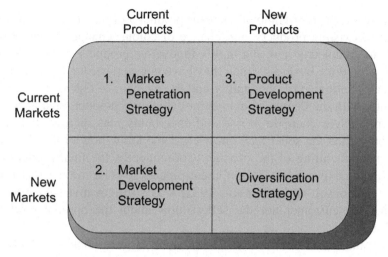

Figure 3.11 Three intensive growth strategies.

Source: Ansoff; Product/Market Expansion Grid.

pressure for change. In this research project I choose to deal primarily with product development processes which were taking place in the shaded area in Figure 3.11. As a result of this, I decided not to include pure market development or integrative growth. Market penetration of existing products has been left out where such market penetration includes only promotion, price, or distribution adaption.

3.3 The Strategic Product Development Perspective

3.3.1 Theoretical Approach

There can be a tremendous difference between the nature of a product development process and the product which is meant to be created through the process. Ansoff described the differences of the innovative elements or the degree of innovation in product development in his model reproduced in Figure 3.11.

Ansoff considered the product development assignment less complicated when the business was developing for existing customers or for existing markets because the customers and their characteristics were known to the selling business. Thus, Kotler 2002 maintains that the element of risk and the degree of uncertainty of the product development process and of the selling business had diminished when operating in this area.

Likewise, the degree of innovation in the product oscillates from variations on known or existing products, e.g. change of colour, change of size, new functions, to completely new and highly innovative products to the market, the customer and the business which they have not seen before. Other things being equal, the degree of uncertainty and the element of risk will increase concurrently with the degree of innovation which the product must achieve or perform. Such an increase in risk and uncertainty may increase because both the buyer and the seller lack knowledge and know-how of the product or of the manufacturing of the product. Consequently, the final product of the most innovative product development process – "The trouble shooting situation" (Håkonson and Johanson, 1982) – is so innovative and radical that neither the customer nor the seller know about the optimal product (Leifer, 2002).

Leiferd stated that there are four main types of uncertainties on radical innovation as shown in Figure 3.12:

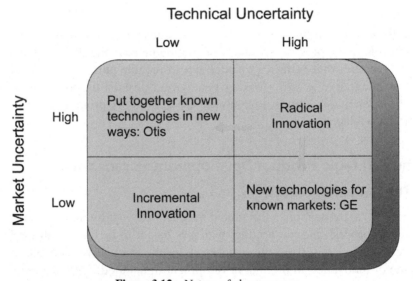

Figure 3.12   Nature of phenomenon.

Source: Ricard Leiferd.

- Market Uncertainties
- Technical Uncertainties
- Organisation Uncertainties
- Resource Uncertainties

Furthermore, he claimed that the uncertainty is high in all four areas in cases of radical innovation.

However, Lefierd (2002) claimed that businesses had to intensify their involvement in future radical product development to gain competitive advantage and continuous survival. Under such circumstances the parties will often agree on an experimental design of the product development process to reach a common goal or to achieve features that both parties want from the product (Verganti, 2001). The final product becomes flexible and dynamic right until the market introduction. The above of course influences the product development model.

The innovative element of product development can also include that the market and the customer are variables and unknown to the selling business. This area is named "the diversification area" and covers both product development as well as market development. Different degrees of diversification can be stressed

1. Concentric diversification
2. Horizontal diversification
3. Conglomeratic diversification.

It is apparent that the element of risk and of uncertainty is increased according to the degree of diversification contained in the product development process. This had been proved in several surveys (Wind, 1973) (Abell, 1980) (Leifers, 2002).

Abell (1980) underlines and extends the assessment of the strategic risk by showing the consequences it will have when a business involve themselves in diversification and especially in conglomeratic diversification. When businesses force themselves into the area of diversification, they move outside their existing SBU with a change to customer needs, customer groups, and customer technology. This is shown in Abell's Figure 3.13.

Balachandra takes it even further when repeating the above but at the same time seeing the product development task specifically in relation to the innovative element of a product development process.

Thus, Balachandra suggested that the various product development processes can be described on the basis of three contextual variables as shown in Figure 3.14.

1. The nature of the innovation – incremental, radical
2. The nature of the market, existing, new
3. The nature of the technology – familiar, unfamiliar

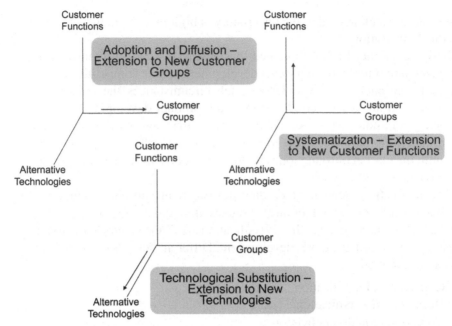

Figure 3.13   Defining the business – market evolution in three dimensions.

Source: D. F. Abell Defining the business The starting point of strategic planning.

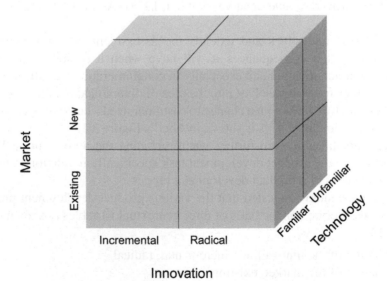

Figure 3.14   Elaboration/Refinement process and product definition.

Source: Balachandra, R. "An expert system for new product development projects" Industrial Management & Data Systems 100/7 [2000] 317–324.

Richard Leiferd (Leiferd 2002) supports Balachandra's definition by his definition on incremental and radical product development. Leiferds talks about radical product development as developing new to the market products.

Consequently, there were at least four strategic areas on which a product development process and a new product development project could be judged. These four areas are:

- The Customer Need Dimension – definition
- The Market Dimension
- The Technological Dimension
- The Innovation Dimension

Yet, the Market Dimension ought to be divided into two separate dimensions, the customer group dimension – the micro level (Abell, 1980) and the market dimension – the macro level (Albaum, 1994). Furthermore, the Technological Dimension should also be divided into two separate dimensions; the customer technological dimension (Abell, 1980) and the pure technological dimension. The former can be defined as the production technological method by which the manufacturer chooses to satisfy the wants and needs of his customers. The latter can be defined as the collection of technologies needed to meet the product development challenge.

Comparing the theories of Abell and Balachandra more closely, we realized that both of them characteristically emphasize the strategic dimension and the customer/market dimension. Furthermore, both authors underline the technological perspective, whereas the innovative perspective is considered differently by the two theorists. Balachandra considered the innovative perspective in the light of the product development perspective, whereas Abell considered it in the light of the strategic business perspective. However, both authors agreed that the element of uncertainty, risk and dynamics are radical when developing for the radical, new, and unfamiliar area. The handling of this area and the insight into the unfamiliar, the new, and the radical area became increasingly important during these years when the market life of a product was continuously diminished (Hamel and Prahalad, 1994) (Sanchez, 1996) (Leiferd, 2002). Such a move offered the possibility of radical competitive advantage. However, this research project and book will not put particular focus on the radical product development area.

In order to establish whether the theories outlined above would influence the speed of product development after 2003, I attempted to place and characterize each separate business case and each business joining the survey of this research project according to Abell's three dimensions, according to

Balachandra's innovative dimension, and according to Sanchez' classification of stable, evolving and dynamic product Market context. The characteristics are outlined in Table 3.1 below.

Table 3.1   Characteristics of product development assignment

Product Development Characteristics	Variable 1	Variable 2
Customer Needs (Abell, 1980) (Kotler, 2002)	Old	New
Customer Group (Ansoff, 1980) (Balachandra 2000) (Abell, 1980) (Kotler, 2002)	Existing	New
Customer Technology (Abell, 1980) (Balachandra) (Sanchez, 2000)	Existing	New
Production Technology (Abell, 1980) (Sanchez 2000)	Familiar	Unfamiliar
Market (Ansoff, 1980) (Balanca, 2000) (Abell, 1980) (Kotler, 2002) (Leifers 2002)	Existing	New
Technology – One or Few (Balachandra, 2000) (Sanchez, 2000)	Few	Many
Innovation – Product Development Assignment (Ansoff, 1980) (Abell, 1980)	Incremental	Radical
Innovation – Strategic Area (Abell, 1980) (Ansoff, 1980) (Balachandra, 2000)	Existing	New
Process (Boer, 2001) please see later	Old	New
Competition (Porter, 1980)	Low	High
Network (Coldmann & Price, 1995) (Child and Faulkner, 2001) please see later	Old	New

Table 3.1 primarily summarizes the characteristics of the product development assignment and presents the structural framework characteristics to the task of the product development project. This will be commented on later in Chapters 7 and 8. Such a summary, however, does not deal with the choice of product development model or of the product development process. It is therefore necessary to discuss the product development model in detail.

3.4 The Product Development Model

In a previously published article (Bohn & Lindgren, 2000) as well as in Chapter 1 of this research project various product development models were mentioned. We described the development in product development models since the 1960s in order to understand and describe the research in product development processes and models. On the basis of this work, the following analytical framework for NB HS PD emerged.

At an early point in the process it became apparent that the framework had to contain four basic elements:

1. The types of product development models.
2. The functions involved in the NB HS PD – i.e. the departments/functional areas involved in product development.
3. The core of the NB HS PD – i.e. the mission, the objectives, the strategies, and the resources controlling the product development project.
4. The phases in the product development process.

In the following I will discuss items 1 and 2 together.

3.4.1 PD Models and Functions in NB HS PD

Apart from the contributors mentioned in the article, initial inspiration to the above model had been Sarens (1984) and Hart et al. (1999). The latter two classify the product development models into five categories.

1. Department Stage Models
2. Activity Stage Models
3. Decision Stage Models
4. Conversion Process Models
5. Response Models

Department Stage Models

> *"These describe the NPD process by focusing on the departments or functions that hold responsibility for various tasks carried out"* ...
> *"The ideas are often assumed to arise in the R&D department; the engineering function will then "make" the prototype, after which production will become involved to plan and carry out the launch."*
> (Hart, 1999)

Hart presented the following criticism:

> *"These representations are rather outmoded as it is now accepted that the "pass the parcel" and "relay" approach to NPD from one department to the next is too time-consuming and does foster ownership of and strategic responsibility for the new product and there is no market feedback since marketing is presented with the product to market."* ... *"These models have been abandoned by the literature which examines NPD and by major businesses as more NPD cases show that everything happens mostly simultaneously."*
> (Hart, 1999)

Activity Stage Models

"This type of model improves on the department model-stage models through its focus on actual activities carried out, which include various iterations of product development and market testing." (Sarens, 1984) (Hart, 1999)

An example of this model is shown in Figure 3.15.

Figure 3.15   Activity stage model.

Source: Cooper, 1993.

Hart presented the following criticism:

"These models have been criticized for still promoting a pass the parcel approach to NPD since the activities are still seen to be the responsibility of separate departments or functions." (Takeuchi and Nonaka, 1986)

Decision Stage Models

"The model describes processes consisting of stages of activity, followed by review points, or gates, where the decision to continued (or not) with the development is made. These NPD models are known under a number of different names, depending on their origin: (Hart, 2000)."

Such models include i.a.:

• Phase Review Process model
• Stage-Gate Process model
• Pace model
• And others model

An example of these types of models are shown in Figures 3.16 and 3.17.

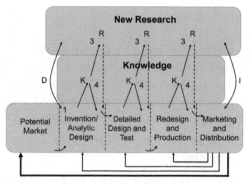

Figure 3.16 Chain model.

Source: Christensen, Jens Frøslev; Produktinnovation – proces og strategi.

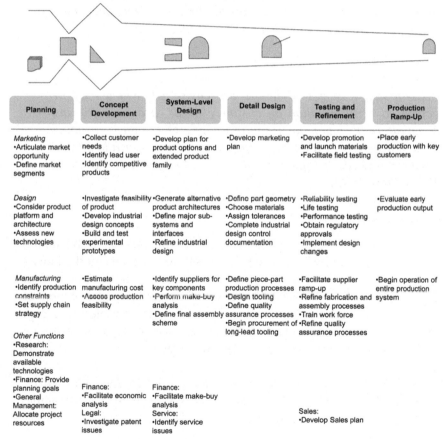

Planning	Concept Development	System-Level Design	Detail Design	Testing and Refinement	Production Ramp-Up
Marketing •Articulate market opportunity •Define market segments	•Collect customer needs •Identify lead user •Identify competitive products	•Develop plan for product options and extended product family	•Develop marketing plan	•Develop promotion and launch materials •Facilitate field testing	•Place early production with key customers
Design •Consider product platform and architecture •Assess new technologies	•Investigate feasibility of product •Develop industrial design concepts •Build and test experimental prototypes	•Generate alternative product architectures •Define major sub-systems and interfaces •Refine industrial design	•Define part geometry •Assign tolerances •Complete industrial design control documentation	•Reliability testing •Life testing •Performance testing •Obtain regulatory approvals •Implement design changes	•Evaluate early production output
Manufacturing •Identify production constraints •Set supply chain strategy	•Estimate manufacturing cost •Assess production feasibility	•Identify suppliers for key components •Perform make-buy analysis •Define final assembly scheme	•Define piece-part production processes •Design tooling •Define quality assurance processes •Begin procurement of long-lead tooling	•Facilitate supplier ramp-up •Refine fabrication and assembly processes •Train work force •Refine quality assurance processes	•Begin operation of entire production system
Other Functions •Research: Demonstrate available technologies •Finance: Provide planning goals •General Management: Allocate project resources	Finance: •Facilitate economic analysis Legal: •Investigate patent issues	Finance: •Facilitate make-buy analysis Service: •Identify service issues		Sales: •Develop Sales plan	

Figure 3.17 Generic product development process.

Source: Eppinger, Steven D.; p. 16.

Hart gave the following criticism:

> *"This approach clarifies the reality and importance of feedback loops, which although not impossible within the framework of the simpler activity-stage models, are usually not highlighted either. With the decision-stage models, each stage is viewed in terms of its potential output."* (Sarens, 1984) (Hart et al., 1999)

Conversion Process Model

> *"These NPD models provide little insight into the NPD process, since they view it as a "black box" in an attempt to get away from the imposed rationality of departmental, activity, and decision based models. The alternative conversion process is a collection of the unspecified tasks which may or may not be carried out, depending on the nature of the innovation."* (Hart, 1999)

Hart (2000) gave the following criticism:

> *"Essentially, a series of inputs are envisaged, which may be composed of information on customer needs, a design drawing or an alternative manufacturing procedure. Over time, depending on a multiplicity of factors, including human, organizational and resource related, this input is converted into an output."* (Sarens, 1984) (Hart, 1999)

Response Models

These models take their starting point in changes taking place at the beginning of the NPD (Becker and Whistler, 1967). These models focus on the individual's or on the organisation's response to changes such as new product ideas, or R&D project proposals in terms of acceptance or rejection of the idea or project. A number of factors influencing the decision to accept or reject the proposal are helpful to the extent that they provided a new angle on what might otherwise be called the screening stage of the NPD process.

The two models that had been most widely used in and validated by research until 2003 were the decision- and activity-stage models. An example of the activity-stage model is shown in Figure 3.18.

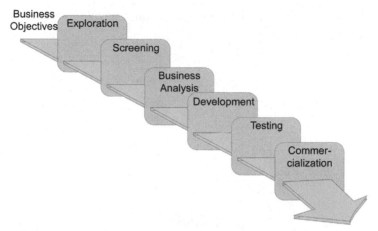

Figure 3.18 The Booz, Allen and Hamilton activity- and stage model of new product development.

Hart gave the following criticism to the model:

The NPD process is idiosyncratic to each individual firm and to the new product project in question. Further if a new product concept fails the concept test, then there is no guidance as to wat might happen next. Another crusial issue related to the activity- and decision-stage models is that the models do not adequately communicate the horizontal dimensions of the NPD process. For example if, at the product development stage, production people have a problem which pushes production cost up, this could affect market potential. The marketing and technical assumptions need to be reworked in the light of this new information.

These shortcoming of the activity-stage models resulted in the advancement of the idea of parallel processing, which acknowledges the iterations between and within stages, categorizing them along functional configurations (Baker and Hart 1999).

Flexible Models

Roberto Verganti stressed the importance of developing the existing product development models e.g. the stage gate models because of their lack of flexibility in relation to strong, dynamic, and rapid changes on the global markets. Stage gate models seem to be too inflexible when businesses entered a market of rapid change and major dynamics. Stage-gate models turn out to be too expensive for businesses focusing on the dynamic and unstable "field of product development".

Verganti's suggestion of a more flexible product development model is shown in Figure 3.19 below.

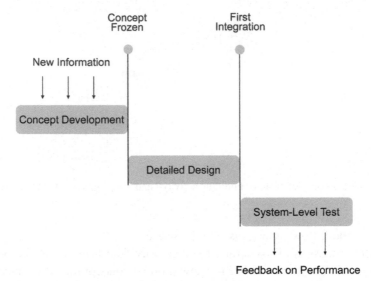

Figure 3.19   Flexible model of product development.

Source: Verganti, R., Developing Product on "Internet Time", 2001.

Verganti argued that flexible models would have some cost advantages when moving into uncertain and dynamic markets and technical areas as shown in Figure 3.20 below.

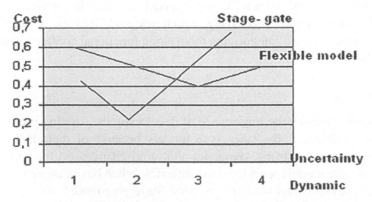

Figure 3.20   Costs of flexible and stage-gate product development models related to uncertainty and dynamics. Inspired by Roberto Verganti.

At the same time Verganti stressed the important issue of creating a product architecture at the very beginning of the product development process. Otherwise, he argues, the product development project would end up in chaos which is not the intention of the flexible models. A case to show the use of flexible models is the Microsoft case (Case No. 50).

The criticism of the flexible product development models maintains that if the models are not chosen and treated in the right way, they will create chaos in the product development project and the businesses will suffer a major loss of money. Flexible models were argued to be best used in flexible and dynamic product development environments.

"On the Market" – Product Development Models

A large Italian research carried out by Mariano Corso and others at Polytecnico de Milano on the Italian SME industry showed new perspectives of product development models in 2002. Case research (Case No. 55 SCOIT) showed that the major part of product development was carried out as an "on the market" product development process between customer and suppliers. The product development was mainly focused on incremental "on the market" product development activities. The best result of these product development activities and ideas were carried back to the businesses afterwards to be built into new products or more radical product development projects. An example of such a product development model is seen in Figure 3.21.

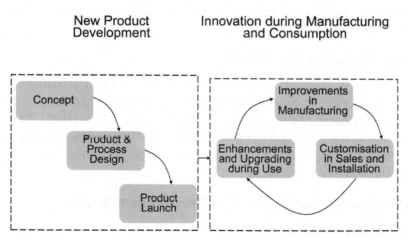

Figure 3.21   Process of continuous porduct innovation at single product level.

Source: Inspired by Corso, M., 2001.

The "on the market" product development model was one of the first attempts to integrate network partners in the product development model. I will elaborate on this type of model in Chapter 8.

The research project tried to verify the product development models of the case businesses. In this way I hoped to be able to analyze the businesses product development model and product development process and to verify which NPD model was actually being applied.

The criticism of the "on the market" product development model was that there had to be a very tight leadership on which product development can be attended "on the market". Otherwise, – as I have discussed earlier – the product development activity can turn out to be too highly based on the customer view. This may result in businesses not earning money.

Practical Approach

The secondary cases showed that it was the decision and activity stage models that were the most frequently used models and the best documented of the known NPD models. Hart (1999) and Biemans (1992) verify this result and my hypothesis was that this will also be the result of the empirical data on businesses joining the primary research. Hart (1999) and Biemans (1992) also maintain that the decision stage model was really an extension of the activity stage model and can be adapted to incorporate input from third parties. This makes it potentially useful as a means of integrating other players in the NPD process such as network partners, suppliers, customers, and others. Such an integration is the next important focus of this research project in regard to the network perspective.

3.4.2 The Core of the NB HS PD

The decision on a strategic core to the product development activities are critical activities prior to the beginning of the product development process (Wind, 1973) (Cooper, 1993) (Clark & Wheelwright, 1993) (Hart, 2000) (Verganti 2002). The product development project and the activity can "stick" to the core and know about the mission, goals, strategy, organisation, and boundaries to network partners.

3.4.3 Informal and Formal Product Development Models

Theoretical Approach

Most literature on product development maintained that businesses had a formal product development model in accordance with which they carried out their product development (Bohn & Lindgren, 2002).

An increasing number of cases in product development literature including my own explorative case research verified and stressed the existence of another model – the "informal product development model or the informal product development process" which runs parallel with the official product development model and process. This hypothesis was indicated both in theory and in practical product development especially when radical product development projects were carried out (Leifers, 2002). The research project wanted to verify this and its impact on high speed product development.

Practical Approach

The secondary cases showed that the case businesses would often display their "official" product development model (Case No 73 Lego; Cases No. 38 Lyngsø) but when looking into the case it was strongly indicated that an informal product development model and process was taking place in the business.

My hypothesis was that the classic stage-gate models and official models could not explain the entire course of the product development. It was maintained that when it comes to "gaining time", increasing the speed of the product development process, and developing such products as the customers want, such informal product development models were often more decisive than the formal product development models.

At the same time my hypothesis was that the "informal" product development model and process influenced the very early phases of the product development process to a large extent and maybe even decisively. See Figure 3.22.

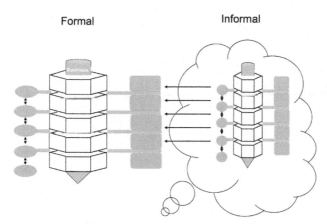

Formal   Informal

Figure 3.22   Informal product development model.

Source: Lindgren, P., 2001.

It was therefore important to this research project to analyse thoroughly whether informal models and processes such as the above existed or and if they had impact to high speed product development.

At this point in time in 2003 we did not know how such informal product development models looked or how they influenced and interact with the formal product development model. However, I put forward the hypothesis that such informal models *do* exist, and one of the objectives of this project was to confirm or disprove their existence.

The hypothesis of this research project was that such informal product development models and processes were important to the achievement of high speed product development. Part of the explanation of high speed might be found in the fact that informal product development models as well as informal processes exist in the businesses. However, my hypothesis was that this was not officially accepted by product development managers in the businesses as such a state of affairs would collide with the ISO9000 guidelines.

3.5 Product Development Drivers and Funnel

In most product development back in 2003 businesses saw the aim of any product or process development project to take an idea from concept to reality – or through the product development funnel Figure 3.23 – at as high speed as possible.

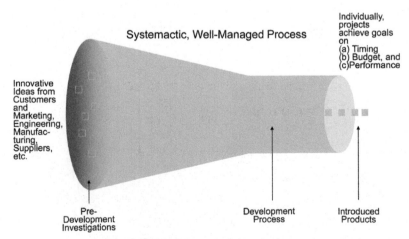

Figure 3.23   The NPD funnel.

Source: Wheelwright & Clark, 1993.

The converging NPD funnel is graphically illustrated in Figure 3.23 in its simplest form. The illustration is inspired by (Wheelwright & Clark 1993) but – as previously mentioned – it is also initially verified in the secondary case businesees.

The funnel is determined by the following dimensions of choice:

1. Sources of ideas

    a) Entry Points
    b) Direction
    c) Breath

2. Selection Process

    a) Purpose
    b) Criteria
    c) Structure
    d) People

There are many sources from which ideas may spring. In their model Clark and Wheelwright suggest some sources from which the ideas may come.

Table 3.2 shows clearly that ideas can be attributed to many sources and that as early as in the 1990s Clark & Wheelwright drew attention to the network of a business as a significant source from which to generate ideas.

Table 3.2   Sources of ideas

Innovative Ideas
Customers
Marketing
Engineering
Manufacturing
Suppliers

Source: Clark & Wheelwright, 1992.

The research project wanted to verify where the ideas came in at that time and verify the impact on speed in product development.

Clark & Wheelwright suggested another model according to which the management "forces" an idea through the development funnel. See Figure 3.24.

As can be indicated, the management is also a source of ideas but they are also decisive of the speed at which the product development progresses (Clark & Wheelwright, 1993).

It has also become apparent that ideas arise via various so-called drivers. Such drivers are described and examined in Section 3.6.

The More Common Reality

Haphazard, Loosely Mamaged Process

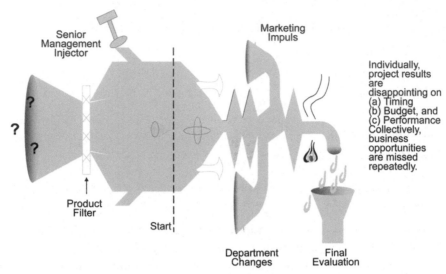

Figure 3.24   Actual development funnel.

Source: Clark & Wheelwright.

Practical Approach

The secondary cases showed different sources to ideas for product development as indicated in Table 3.3. Mainly such sources were registered as follows:

Table 3.3   Sources to product development ideas in general

Sources to Product Development Ideas in General	In Percent
Customers	
Suppliers	
Marketing	
Sales	
Leadership/Management	
Production	
Product Development	
Competition	
Others	

This research project wanted to verify the sources of product development.

3.6 Generic Product Development Process

The generic product development process or the generic product development processes and their background had been repeatedly described by several different researchers and authors, i.a. Abel, 1986; Eppinger, 1995; Hart, 1999, Conti & Spina, 1999.

When discussing the generic product development process, much of the discussion had been about the factor that initiates the product development process, the factor that "drives" and maintains the process, and the factor that completes the process.

3.6.1 Drivers of NB HS PD

Theoretical Approach

The driver or the feature which starts off the product development process has many aspects. My hypothesis was that the process is often characterised by a market-pull, a technical pull or a network pull situation. This means that "the field of product development" initiates the product development process when an opportunity arises in the market – when a strategic window opens (Abell, D.F., 1980). Subsequently, the business will employ all possible markets. The technological and network options required to satisfy the needs of the market (Eppinger, 1995). Thus, the market "pulls" the decision of development along (See Figure 3.10).

In addition to such a pull process on the market, there are another four variants or drivers according to Eppinger (1995):

- The Market Pull Process
- Technology – Push Products
- Platform Products
- Process Intensive Products
- Customized Products

These five generic product development processes are outlined in Figure 3.25.

However, the research showed a lack of one important dimension in Eppinger's model as it seems that the design perspective (Verganti, 2001) as an initiator of new products had been left out. This has been elaborated on in Figure 3.25.

	Generic (Market Pull)	Technology Push	Platform Products	Process Intensive	Customized	Design Driven
Description	The business begins with a market opportunity, then finds appropriate technologies to meet customer needs.	The business begins with a new technology, then finds an appropriate market.	The business assumes that the new product will be built around an established technological sub-system.	Characteristics of the product are highly constrained by the production process.	New products are slight variations of existing configurations	
Distinctions with respect to generic process		Planning phase involves matching technology and market. Concept development assumes a given technology.	Concept development assumes a technology platform.	Both process and product must be developed together from the very beginning, or an existing production process must be specified from the beginning.	Similarity of projects allows for a highly structured development process.	Language, Message, Meaning
Examples	Most sporting goods, furniture, tools.	Gore-Tex rainwear, Tyvek envelopes.	Consumer electronics, computers, printers.	Snack foods, cereal, chemicals, semiconductors.	Switches, motors, batteries, containers.	Philip Starch, Marie Bergmann

Figure 3.25 Generic product development processes – design driven (adapted from Eppinger and Verganti).

Source: Verganti, 2002.

Verganti refered to the design driver stating that this driver differs from the technical and market driven drivers in that designers are driven by a wish to design existing product better or design product with out functions but with meanings (Phillipe Starck, 1999).

This research project did not wish to exclude any of the six types of product development drivers from our field of research as they were expected all to be relevant to the research project. However, our primary focus was on identifying drivers but primarily the market pull and customized product drivers. All our cases were in accordance with Table 3.4.

Table 3.4 Drivers for the PD process

Drivers for PD Process	Idea Stage
Market Pull	
Technological Push	
Platform Products	
Process Intensive Products	
Customized Products	
Design Driven Pull/Push	

The drivers listed above describe the origin of the ideas at the entrance of the funnel. It was the hypo these that what happens in terms of process in the product development model at this stage had influence to the speed in the product development process.

Practical Approach

The secondary cases showed examples of different drivers to the product development process. However it seemed very clear that the market pull and the technological push drivers were the main drivers to product development.

3.6.2 Overall Processes in NB HS PD

When considering the generic processes in a PD project from a general point of view, such processes could initially be described as outlined in Cooper's (1993) model below in Table 3.5.

Table 3.5 Frequency of activities in new product process development project

Activity/Processes	Frequency (%)	Proficiency Quality Index Scores (0–10)	Need for Improvement Scores (0–10)
Initial Screening	92,3	5,27	5,48
Preliminary Market Assessment	76,8	5,47	5,37
Preliminary Technical Assessment	84,9	6,69	4,60
Detail Marketing Study/Marketing Research	25,4	5,74	5,83
Business & Financial Analysis	62,9	6,49	4,27
Product Development	89,1	6,55	4,47
In-House Product Testing	88,9	6,96	3,87
Customer Test of Product	66,3	6,69	3,99
Test Market/Trial Sell	22,5	6,86	4,59
Trial Production	48,9	6,79	3,66
Pre-Commercialization Business Analysis	34,5	6,26	3,95
Production Start-Up	56,0	6,31	4,37
Market Launch	68,1	6,36	4,44

Source: Cooper, 1993.

The process could also be described as Eppinger (2000) did in his model. The generic processes from 1–5 were the focus of this research project. It is also among these five processes that according to Cooper we found the greatest need not only for improvement but for continuous improvement and learning (Cooper, 1993) (Sanchez, 1996) (Bessant, 1999) (Eppinger, 2000).

Relevant literature also described the entire product development course as an overall process (Booz, Allen, Hamilton 1982) (Cooper, 1993) containing certain generic activity stages. As previously mentioned the literature also focused on the first activity stages in the entire product development process, i.e. mainly the idea and concept phase.

R. Cooper (1993) and others did not, however, deal with the processes pertaining to these stages in order to obtain their PD objectives. Until this point in time, this area had not been thoroughly dealt with.

Cooper also described the process as being visionary and claimed that ideas and concepts did not appear from the PD process. My hypothesis was that new ideas and concepts indeed appeared from product development.

Furthermore, continuous improvement and learning in PD posed an increasing problem to businesses and was consequently the subject of ever growing research efforts (Boer, 2000) (Corso, 2001). Such efforts were the result of an overall wish to improve each separate product development process. However, it will be much more interesting to improve the subsequent product development processes. This could only be achieved by edifying learning about and from the preceding and on-going product development projects and from other product development processes known within the network of the business (Gieske, 2001).

In other words, CI and learning should not only be sought and achieved by internal PD processes but also across PD processes and also across the network.

My hypothesis was that this would result in learning and double loop learning in NB NPD.

Until now research had not focused on NB HS PD and continuous improvement and learning. The case examinations showed that the businesses were primarily concerned with CI within the individual PD processes or PD projects. However, there were some underlying processes going on as Wheelwright & Clark verify in their process model.

The above mentioned Figures 3.22 and 3.25 signify a need for thoroughly examining the understanding of the processes and the partial processes of product development.

Hart et al. (1999) attempted such an examination by extending the discussion of generic processes by breaking the overall processes into small tasks. She also spoke of generic processes within the separate activity stages of the entire product development process; see Table 3.6 and also Figure 3.26.

Table 3.6 Processes within stages and gates

Stage of Development	Information Needed for Stage; Nature of Information	Sources of Information	Likely Output of Stage in Light of Information
Explicit statement of new product strategy, budget allocation	Preliminary market and technical analysis; business objectives	Generated as part of continuous MIS and corporate planning	Identification of *market* (NB. Not product) opportunities to be exploited by new products
Idea generation (for gathering)	Customer needs and technical developments in *previously* identified markets	Inside business: sales people, technical functions. Outside business: customers, competitors, inventors, etc.	Body of initially acceptable ideas
Screening ideas: finding those with most potential	Assessment of whether there is a *market* for this type of product, and whether the business can make it. Assessment of financial implications: market potential and costs. Knowledge of business goals and assessment of fit	Main internal functions: • R&D • Sales • Marketing • Finance • Production	Ideas which are acceptable for further development
Concept development: turning an idea into a recognizable product concept, with attributed and market position identified	*Explicit* assessment of customer needs to appraise market potential. *Explicit* assessment of technical requirements	Initial research with customer(s). Input from marketing and technical functions	Identification of key attributed that need to be incorporated in the product, major technical costs, target markets and potential

(*Continued*)

Table 3.6   Continued

Stage of Development	Information Needed for Stage; Nature of Information	Sources of Information	Likely Output of Stage in Light of Information
Business analysis: full analysis of the proposal in terms of its business potential	Fullest information thus far: • Detailed market analysis • Explicit technical feasibility and costs • Production implications • Corporate objectives	Main internal functions Customers	Major go-no go decision: business needs to be sure the venture is worthwhile as expenditure dramatically increases after this stage. Initial marketing plan. Development plan and budget specification
Product development: crystallizing the product into semi-finalized shape	Customer research with product. Production information to check "makeability"	Customers Production	Explicit marketing plans
Test marketing: small-scale tests with customers	Profile of new product performance in light of competition, promotion and marketing mix variables	Market research: production, sales, marketing, technical people	Final go-no go for launch
Commercialization	Test market results and report	As for test market	Incremental changes to test launch Full-scale launch

Source: Hart, 1999.

Hart stated that the process models had been the object of considerable criticism.

"The NPD process is idiosyncratic to each individual firm and to the new product project in question. Its shape and sequence depend on the type of new product being developed and its relationship with the firm's current activities." (Cooper, 1988, Johne and Snelson, 1988)

In addition to the need to adapt the process to individual instances, it should be stated that in real situations there is no clear beginning, middle and to the NPD process (Hart, 1999).

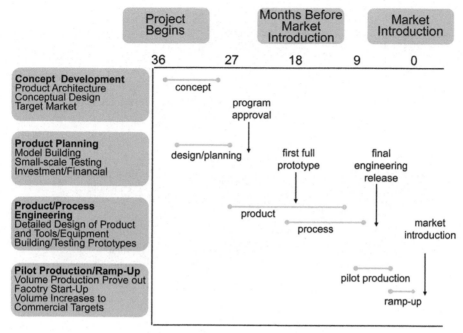

Figure 3.26 Underlying processes.

Source: Wheelwright & Clark, 1992.

At an earlier point in this research project, I asked when a product is final or when a product development course is final. As can be seen from the above, Hart (1999) confirms that it is extraordinarily difficult to determine the beginning and the end of a process as a product development process. Furthermore, each separate stage and gate in a product development process can have infinitely many beginnings, processes and sub-processes, and results as indicated in Figure 3.27.

Therefore it is imperative to put the question

• is it important to determine a beginning and an end of a product?

My hypothesis was that it is not important because businesses do not gain any value by defining the beginning and the end of a product.

A growing number of researchers and authors claimed at that time that with the development of a product development process, other challenges and opportunities in other functional areas or departments involved in product development come into existence (Hart, 1999). Consequently,

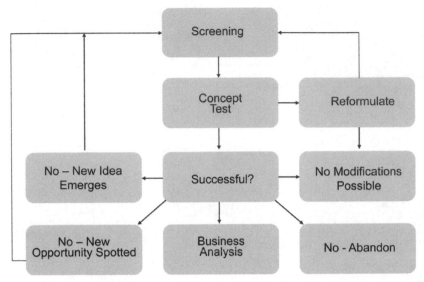

Figure 3.27   Related strands of development.

Source: Lindgren, P., 2002.

the argumentation that NPD must necessarily be based on interplays between the various departments and networks involved in product development arises. Additionally, it follows that product development is an iterative process, not only between stages but also within stages. The criticism directed at the activity and decision stage models claimed that such models:

> *"do not adequately communicate the horizontal dimensions of the NPD process"* (Hart, 1999).

As a consequence, the idea of "Parallel processing" had come into existence as indicate in Figure 3.28. This idea acknowledged the iterations between and within stages, categorizing them along functional configurations.

The idea of parallel processing was highly prescriptive: it advised that major functions should be involved from the early stages of the NPD process to its conclusion. This, it was claimed, allowed problems to be detected and solved much earlier and the entire process was much speedier (Hart, 1999). This phenomenon was one of the main phenomena of high speed NPD which was also the focus of this research project. This aspect is illustrated in Figure 3.29.

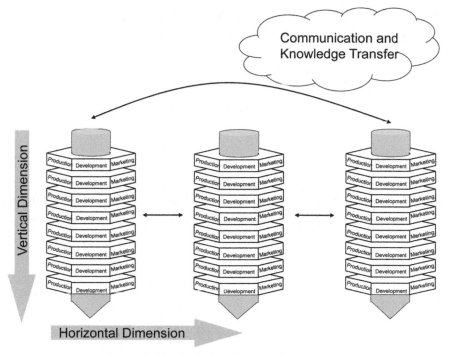

Figure 3.28   Vertical and horizontal processes.

Source: Lindgren, P., 2002.

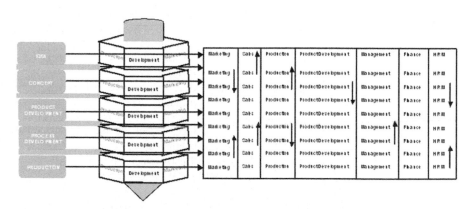

Figure 3.29   Stage gate functions which move up into the model.

Source: Lindgren, P., 2002.

As can be seen there are processes going on

1. Between functions
2. Between phases
3. Between functions and phases
4. Between networks

I will elaborate further on this framework model later in Chapter 8.

3.7 Summary

Concepts of product development had changed up to 2003 as we have seen above. Both researchers and the industry realised in the years up to 2003 how the concepts of the products and product development models and processes were changing even more as indicated in Table 3.7.

Table 3.7   Change in the concept of product and product development

Change in the Concept of Product and Product Development	Now	Future
Product characteristics	Physical products and immaterial products	From physical products to more immaterial products
	A product	A total product
	One final product delivered to the market	Multi products (Both physical, digital, immaterial and virtual products)
		Sequential encapsulation of a product – "never ending story"
		From product to process
		The product change to a process
	Stable products	Dynamic products
	Focus on function and value	Focus on perceived value and meaning
	The business finish the product	The customer change the product together with the business continuously
	The customer and the business does not accept trials and error	The customer and the business accepts trial and error
	Physical and digital process	Physical, digital and virtual process
	Physical service and to some extent physical processes	Physical, digital and virtual process

Table 3.7   Continued

Change in the Concept of Product and Product Development	Now	Future
The core of the product	Mostly stable and well defined core of the product Not interactive into all areas	Rarely stable or always under construction and dynamic Interactive in all areas
The core of the product development project	Mostly static	Rarely static – high degree of dynamic
The product development model	Mostly stage-gate model	All types of product development models
The Product development process	One process with one start and one beginning.	Continuous process with many starts and many ends.

The changes, the pressure, and the focus on product development was expected to intensify in future but it was also expected to turn out to be more complex and challenging when product development intensifies with more product development carried out in networks under even higher pressure on speed.

Product development in networks will be the issue of the next chapter.

4

Network Perspective

The framework of this research project determines that the point of focus should be on network-based product development. The network perspective is interesting in that our study of literature (Eppinger, 1993) (Clark & Wheelwright, 1993) (Coldmann & Price, 1995) (Caffyn, 2000) (Hart, 2001) as well as most of the secondary cases indicated that product development subject to time pressure employed and toke place in networks. Thus, the use of network could be the explanation for higher speed in product development. Moreover, networks could also be the explanation for a lowering of speed in product development. Networks can be defined as both physical, internal networks within the business, external networks established together with customers, suppliers and other network partners. Networks can be a mix of internal and external networks both physical, digital and virtual networks. Virtual networks come into existence when needed and when there is a task for the networks to accomplish.

The case examinations left us with the clear impression that there was a tendency for product development taking place in the businesses to move towards a wider and earlier use and interaction with external networks and network participants.

In other words, there was not only a trend towards several functions of the business being involved at an early point in the PD process but also a trend towards the networks of the businesses becoming involved at an early point in the PD process (Case No. 55 ScotIT, Case No. 13 Mayekawa, Case No. 11 Rossflex).

Some of the secondary case businesses also stated that participation of external networks and network partners were a prerequisite for the successful course of a product development process. It seemed that the businesses generally wished to involve in particular the customers and the suppliers in the network at a very early point in time of the product development process. The businesses also claim that network and the handling of networks were the two main reason for the success of the businesses of the Italian design industry (Verganti 2001). It seemed as if network based product development could add value to product development.

However, there can be many definitions of the concept of network, and in addition, the concept had been the subject of major research during the years up to 2003 (Håkonson, and Håkonson & Johansson, 1985) (Rind, 2000). (Nielsen, 2001).

Generally, I identified three central types of network which do not necessarily conflict but rather showed the different characters of a network. The different kinds of networks apparently allow for various possibilities. The businesses can achieve several synergy effects by combining such different kinds of network not least in relation to high speed product development. The network types included:

1. The physical network perspective – internal and external networks
2. The computer network perspective
3. The virtual network perspective

In the case research I identified 3 characteristics of networks as seen in the Table 4.1 which involved different mixtures of the above mentioned network types.

Table 4.1 Shape of the network components

The Main Components Context	Characteristics	Example of Networks
Stable networks	Networks mainly based on physical and stable networks often internal and dominated network	Industrial groups, branch groups
Evolving networks	Networks based on a mix and evolving system of networks – Physical networks, ICT – networks, virtual networks	PUIN network group, EU community,
Dynamic network	Networks based on a mix of dynamic networks with high degree of dynamic where network partners constantly comes in and goes out. Often there is no formal network leader.	Virtual network groups, Ambia

4.1 Physical Network Perspective

4.1.1 Theoretical Approach

In general terms, the network approach implies as shown in Table 4.1 that industrial activities (manufacturing of goods, product development, marketing, purchasing, etc.) are seen to take place within a network of exchange relationships between producing businesses, suppliers, customers, distributors, research organizations and other types of industrial actors. The members of the network performed complementary, competing or otherwise interrelated activities whereby certain resources are consumed and others transformed into new resources.

> *"The actors are linked to one another through various kinds of interaction processes in which resources are exchanged or used in joint activities. The network is thus the arena in which industrial actors compete and cooperate with one another in order to achieve their goals. For the individual actor the surrounding network constitutes an important part of the environment in which it operates. Its freedom to act is very much determined by the characteristics of the network, which give both opportunities and constraints."* (Laage-Hellmann 1989)

Actors, activities and resources were the three main groups of variables in the basic physical network model (Håkansson and Johanson, 1985; Håkansson, ed., 1987).

Actors are defined as those who perform activities and control resources. One can identify actors at several organizational levels, such as individuals, departments, divisions, firms, and groups of firms. Actors control resources of various kinds. Physical assets (raw materials, equipment and other types of goods), information, knowledge, competences, labour and money are some important types of resources used by industrial actors when they perform activities. Some of these activities are carried out under the control and management of one actor. New or improved products and resources were created by combining, refining or changing existing products, resources and competences to develop new products for the future. They were called transformation activities and could either be incremental or radical by character. Other activities involve two or more actors that were exchanging or transferring resources or performing other types of joint activities. They are called exchange activities.

Transformation activities can be manufacturing of goods and Research & development (R&D). The aim of a single R&D activity is to produce new scientific or technological knowledge or to develop new industrial products, processes or applications. These can either be radical or incremental, new products.

In industrial networks each actor's transformation activities are more or less dependent on a number of past or future transformation activities and tacit knowledge performed and developed by other actors. This dependency can be more or less specific. There is a specific dependency between two actors if at least one of them is dependent on the activities of the counterpart. However, dependency could also be more general in nature, when the dependency is not related to one specific counterpart, but rather to an involvement in the network.

The function of exchange activities is to handle the activity dependency between actors. Commercial exchange serves to link production activities performed by different businesses with the market. Technological exchange serves to link the product development activities of different interdependent firms technically and link R&D performing actors to each other to complement or develop their own resources with external competence.

The stronger the specific dependency between actors, the more they will be inclined to develop and exchange relationship with each other. A relationship can be said to exist when two actors are aware of each other and perceive each other as counterparts or partners in an exchange process. Relationships can be either occasional or short-lived e.g. some types of virtual networks. Such relationships do not lead to any ties or bonds between the interacting parties. Lasting relationships, characterized by a higher or lower degree of stability and closeness can also be the case. The intensity of interaction can vary over time. If the tie is strong enough, a relationship can even persist over long periods without any exchange taking place. The relationship can be said to be "sleeping"; that is the two actors know each other and, if needed, are prepared to recommence the exchange – virtual networks.

Exchange may lead to mutual adaptations and commitments being made by the parties. This leads to a strengthening of the relationship and to a further increase in the specific dependency.

In contrast to transformation activities, exchange activities are never totally controlled by only one actor. The fact is that at least two actors must be involved in order to produce an exchange situation. This does not imply, however, that the influence is evenly distributed. One of the actors may be more powerful than the other and therefore be in a position to exert a higher degree of influence on the exchange process.

All existing industrial actors are normally members of a global, worldwide network of relationships. Each individual actor is not directly linked to every other actor in the global world but are each actor will be directly related to a small subset of all potential counterparts. These relationships may be built on business or other types of exchange and imply a larger or stronger dependency among the interacting parties. The industrial network is thus characterized by certain of its members being linked through strong ties, while others are linked through weak ties. Still others are not directly linked at all, but only related to each other indirectly, i.e., via one or several intermediate actors, or through general dependencies.

This way of viewing the global market means that the interest of the researcher is focused on the structure of the network in network based high speed product development and the pattern of interaction and interplay among the actors when product development is carried out under pressure of time and speed. The network model shortly summarized above provides a set of concepts which can be used when approaching a problem or question from a network perspective.

The network approach has been chosen as a starting point for the empirical research. However, there are certain methodological problems associated with the usage of the network model.

One concerns the delimitation of the network. It is impossible, of course, to consider the entire global network of product development relationships. For all practical purposes, the network has to be delimited given the context and the specific issues or problems of the product development project to be addressed. The focal network in this research is defined as those actors which are tied together in relation to a certain product development project.

Attention is then focused on the interplay which takes place among these actors. What happens outside the network boundary is left aside or treated as part of the general environment.

Another problem has to do with the choice of actor level. The individuals who represent a particular product development organization do not necessarily act "as one man". Each of them have their own personal goals, ambitions and perceptions, and these do not necessarily coincide. Furthermore, different individuals may have contacts with different types of counterparts. For example, the marketing and R&D people may have parallel and more or less separate contacts with a customer firm. Such parallel person-to-person relationships may be strongly connected, but they may also live their own lives without influencing each other too much. The complete exchange relationships which develop between various businesses and organizations in the industrial network may thus contain several more or less independent personal relationships.

In this research such latter relationships are called "informal product development models and processes" based on informal product development networks.

In the empirical part of the research the aims was to verify and document these informal processes and clarify their influence on time and speed in product development. The focus was mainly on the internal physical informal networks as the external informal networks will demand a much wider research plan. However my hypo these was that these external informal networks have big influence to network based product development.

In this research the focus was on product development in networks. As stated in Chapter 1, there were strong reasons to believe that a very large share of the product development toke place within inter-organizational relationships. In Håkansson (ed., 1987, Ch. 1) three arguments were put forward as to why this should be the case. These arguments were related to knowledge development, resource mobilization, and activity coordination respectively (Corso, 2001).

It was well known that product development often emerges at the interface of different knowledge areas. Product development exchange in relationships could therefore contribute to knowledge development by bringing together different bodies of knowledge. Firstly, an "interactive effect" could be obtained when the needs and competences of one business were combined or confronted with the possible solutions known by another business. The second type of effect is related to the fact that new product or process development requires the combination of several different technologies and knowledge of fragmented markets, which is the case in to days product development as stated before in Chapter 1. When two or more actors with complementary competences join forces in order to develop a new product or process a "multi-competence effect" is produced.

The need to mobilize resources is another reason for actors to interact with one another during the innovation process. In order for an invention, e.g. a new product, to be used and turned into an innovation, the actors involved, in the capacities of producers, users etc., have to adapt in different ways. For example, the actors must learn how to use the invention and combine it with other products. The invention also has to be adapted, revised, and redesigned in order to be useful in different applications. The innovation process, thus, has elements of learning, adaptation, and socialization which require the investment of resources.

Here, the interaction among the actors involved plays a vital role in ensuring that enough resources are mobilized. The mobilization process often takes place in a context where the resources are scarce and where other activities compete for the same resources.

The high degree of specialization in industry and fragmentation on the global market required that product development activities in different parts of the network had to be coordinated. The business had to gain a coordinated overview over the possible network activities in all types of network.

The importance of these phenomena made it fruitful to use a network approach when dealing with questions related to high speed product development. By identifying and examining the relevant network, one could increase the understanding of how high speed product development occured in a certain product development field. The pattern of product development in a specific network could be characterized in such terms as:

1. What types of actors participate in product development? How are the roles divided? Which actors dominate and control the development?
2. The degree of concentration, i.e. are the Product development resources and activities controlled by a few or many actors in the network? Is the product development concentrated within a few leading actors or does it proceed simultaneously in many different places. Who and how is the product development project managed?
3. The degree of integration and pattern of cooperation are of interest and to what extent are the product development activities integrated with each other? How much of the product development activity is carried out in cooperation between different actors?
4. The direction and content of the product development work. The emphasis is on:
   - process or product development
   - standardization or differentiation
   - incremental and radical product development
   - large-scale or small-scale product development
   - rationalization of existing or development of new techniques
   - revolutionary product development models and processes?

The results and experience from the special study on literature and cases, as well as other empirical studies carried out later on, had therefore been used to further develop the integrated network based product development model. The model is summarized in Figure 4.1. It consists of three main parts: the exchange process; the actors; and the environment. The relationship between the two actors A and B is created by the exchange process (i.e., the interaction), which is affected by the characteristics of the actors and the environment. The three parts of the model will now be described in more detail.

The Exchange Process in Network Based Product Development

The exchange which takes place within a relationship is a key concept in the model. It means that two actors consciously interact with each other in order to transfer resources or to perform some activity together. The most important

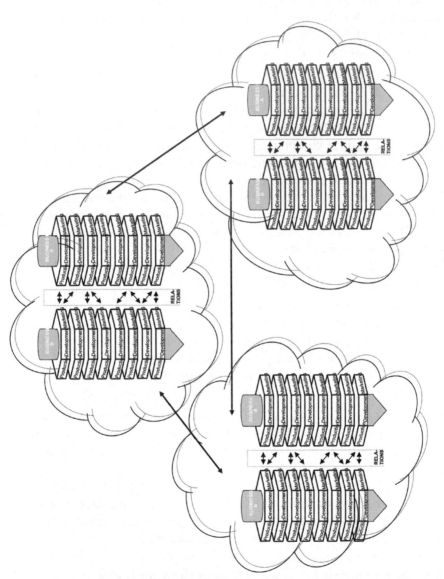

Figure 4.1   Several network partners.

categories of elements which are involved in the exchange between industrial product development actors are physical products, information, knowledge, proprietary rights, and financial assets.

The exchange process between industrial actors is complex and multidimensional in nature especially when more actors are involved to the product development project. Four main dimensions will be distinguished, viz. commercial, technological, social, and financial. These dimensions are not independent of each other, but constitute integrated aspects of the overall exchange process.

Commercial exchange, whereby products or services are transferred in exchange for money or some other form of payment, was the core of business relationships. Commercial exchange itself, in a narrow sense, may require a lot of exchange activities be carried out by the parties, e.g., sales calls, negotiations, technical discussions, delivery of goods, payments, adjustments of claims, and the performance of after-sales services. By definition, technological exchange comprises all those exchange activities which have a bearing on product development. The scope of technological exchange may therefore vary focus within broad limits. It can be anything from a one shot transfer of a small piece of information to an extensive joint R&D project in which both parties invest large resources. In some types of relationships, technological exchange itself can be the primary purpose of the relationship. This is the case when industrial businesses interact with universities, research institutes and development businesses.

Technological exchange can also lead to the establishment of relationships between firms producing similar or complementary products. It should be observed that even though the main focus was on development questions there could also be a commercial dimension in "R&D relationships" of these kinds. This was evident in connection with contract research, technology licensing and the commercialization of independent researchers' product ideas and inventions by other businesses. Furthermore, commercial goals were the underlying driving force when businesses interacted with competitors, complementary producers or end-users in technology matters.

To summarize, commercial, technological, financial, and social exchange constitute different aspects of the overall exchange process which takes place within a product development relationship. The different types of exchange are linked with and dependent on one another and cannot be seen as separate processes. To understand one type of exchange, for example technological, it is therefore not enough to analyze only that form of exchange. It is also necessary to examine how that exchange dimension is related to other aspects of the overall exchange process.

Product Development Exchange

Keeping in mind that product development exchange is just one of several aspects of the exchange process, the characteristics of the product development exchange are interesting.

Broadly speaking, product development exchange means that two actors interact with each other in a way which affects the product development in either or both of the actors. There are three basic mechanisms whereby this can happen:

1. The actors influence the each other's internal product development activities through exchange of information and knowledge or use of power.
2. The actors get involved in mutual exchange activities whereby R&D resources are transferred between the parties or put at each other's disposal.
3. The actors participate in a joint R&D project. Both parties allocate resources for a common purpose and perform activities together with the partner.

It was the hypothesis that information and knowledge transfer was a vital element – a catalyst to high speed product development.

Exchange Process and Degree of Bonding in NB PD Relationships

An important effect of exchange is that different kinds of ties or bonds are created between the interacting parties: technical, knowledge, organizational/administrative, economic, legal, and social. The degree of bonding in a relationship can be seen as a measure of its strength and long-term stability:

- Technical bonds
- Knowledge bonds
- Organizational/administrative bonds
- Legal bonds
- Social bonds

A bond may sometimes be created through exchange activities taking place within one major episode (e.g. the execution of a big product development project which could be radical to the actors). But more often bonds were created through a series of short episodes – incremental product development projects whereby the bond was gradually, and often relatively unconsciously, built up.

Product development exchange played an important role in the creation of bonds, especially the technical and knowledge-based bonds. Extensive product development cooperation presupposes a certain degree of bonding between the two partners.

Interacting Actors in NB PD

The process of exchange does not only depend on the elements exchanged and the degree of bonding, but also on the attributes of the actors themselves. Four main groups of influencing variables can be distinguished:

- the resources possessed by the actor,
- the transformation activities performed by the actor,
- the organizational structure by which the resources and activities are organized, and
- the objectives and strategies pursued by the actor.

Macro Environment

The interaction between two actors takes place within a wider context, which must be considered in analyzing the relationship. The interaction product development model describes this environment in terms of the market structure, the degrees of dynamism and internationalization, the position in the supply chain, and the wider social system which surrounds the relationship.

Technological exchange may be affected in different ways. For example, in an environment characterized by rapid market, technology, network and business based competence change reliance on a single or small number of development relationships can be risky. To obtain early access to new markets, new technologies, scientific advances, new networks and new competences may require a close and long-term collaboration which can only be maintained with a few counterparts. The degree of dynamism may thus affect the product development exchange within a relationship.

4.1.2 Network Interaction Model for HS PD

The actors (and the focal relationship) were seen as integrated parts of a product development network of relationships. This means that there is no distinct boundary which separates the actors from the environment. Actors are linked to other members of the network through direct and indirect exchange relationships.

As indicated, the network based interaction in the introduction of the product development model affects the attributes of the actors. In the model these are characterized in terms of resources, transformation activities, organizational structure and objectives/strategies. These characteristics are internal in nature. If the actor is regarded as an integrated part of the network, the position in the network also becomes an important attribute. The position describes how the actor is related to the network through external relationships. It can be defined in the following way:

1. The identity of those other actors with which the focal actor has direct or indirect relationships.
2. The characteristics of these relationships (the existence of different types of exchanges and bonds).

This means that all actors have some kind of unique position in the network. The position is a result of the actor's previous interaction with various counterparts. It is never static but changes continuously as a result of the exchange activities which take place, or do not take place, in the network.

Network position is an important concept for several reasons. Through the involved relationships, position gives the actor access to external resources which can be used as a complement to internal ones. These external resources can be seen as indirectly controlled, in contrast to those resources which are directly controlled through ownership or employment contracts. As a result, an actor's total resource strength, which is the base for its acting in the network, cannot be assessed without considering how it is linked up with other actors.

The network based product development model increase complexity. It makes it possible to include connections in the analysis of relationships and thereby achieve a better understanding of how actors interact with each other. Connection is an important concept.

Cooperation and Conflict in NB HS PD

The network based interaction model is based on the assumption that the two parties have mutual interests in the product development project. The exchange of goods, services, information, etc. which takes place within the relationship is perceived by both parties to be beneficial. Otherwise the exchange would cease.

It is therefore natural for the interaction model to emphasize the cooperative aspects of the interaction process. This is not to deny, however, that conflicts may occur between the parties. That a relationship is old and close docs not mean that the parties always act positively to each other and that everything is fine. The parties can have more or less conflicting goals, which may affect the exchange. Such conflicts can be solved, for example, by finding compromises which are acceptable for both parties.

Certain conflicts arise within the network relationship without any influence from the environment. By extending the model to include three or more actors, the possibility for other types of conflicts is opened up. Here it is the connections between the relationships in the triad which create tension and conflict among the actors.

As a first step, a simple framework for discussing the issue of conflict and cooperation in networks will be outlined below. It is based on the assumption that in all relationships there are both common and conflicting interests among actors. As illustrated in different research, relationships can be classified into more categories depending on the strength of the different interests.

The actors can act relatively independently without considering the consequences for the counterpart. The interests of the two actors are in strong conflict, for example because they are competing for the same customers or the same product development resources.

In a broad sense, cooperation means that two (or more) actors engage in mutually advantageous activities where both actors consciously take each other's interests into consideration.

The relationships are characterized by the existence of both strong common and strong conflicting interests. In this kind of relationship the actors have incentives to cooperate in certain fields and counteract each other in others. For example, within a particular product development relationship knowledge exchange may give rise to a conflict regarding the price or other terms of knowledge exchange. This may lead one partner to start negotiating with an alternative network partner in order to exert pressure on the partner. At the same time the two businesses may be engaged in long-term product development cooperation which is important for both of them.

The last situation is interesting. How does the interplay between the cooperative and conflicting aspects function? How is this interplay affected by connections to other relationships in the network? At the same time, the secondary cases provided by earlier indicated that this situation is not unusual in industrial networks. This does not only hold for supplier-customer relationships. Businesses that are competing in the same market frequently have strong common product development interests. They may, for example, be interested in keeping product development at a certain standard level, which may result in direct cooperation between competitors. Other common interests may have to do with standardization and technological development.

4.2 Computer Network Perspective

As our work became gradually more computer oriented up to 2003, the information technological need of the various specialisations of an organisation became increasingly accessible across functional barriers both internally and externally. Through the use of network IT it was possible to re-interpret specialist and fragmented information and to exchange such information freely and internally.

Computer networks, however, enable the entire business to receive feedback in the form of opinions and to add individual learning. Until now such knowledge had been stored in archives and in the heads of specific employees and had only been furtively discussed over coffee or had even forgotten. Computer networks encourage the employees to share and remember knowledge if it was designed right. Furthermore, employees/network partners were interested in transferring their knowledge.

Several researches showed that open computer network systems were instrumental in making network participants feel more involved.

A visible and accessible network organisation helped to increase the potential for improved idea and learning relation among network members. This was expected to bring about improved speed, reduced costs, and increased product performance and benefits e.g.:

- By allowing computer networks to expand time and compress space to develop products. Thus, it is not necessary for the network members to be in the same place at the same time in order to have a dynamic exchange. Each participant is able to contribute to the network when it suits him or her, and it is possible to establish relations across different internal and external computer networks independent of the geographical or physical placing.
- By abolishing competition for "transmission time" or "floor space" in computer networks. All network participants can contribute at the exact moment when they want to. Each network participant also has more time to consider his or her contribution to the network and thereby raise the quality of the outcome.
- By reducing the tendency among computer networks participants to keep their opinions to themselves. The influence of "powerful persons" and

"time-wasting persons" can be reduced by changing the facilities of the computer network in a smart and intelligent way. Computer networks can be used to advance learning in network cooperation, in particular "double loop learning" (Bout/Hover, 1996). Double loop learning allows for more flexible ways in which to organise the enterprise. Computer networks tend to loosen stagnate structures which can often be found in traditional organisation models based on the principle of control. Computer networks can also increase amount of data and information.

On the basis of a computer network specific software specifications designed specifically with the object of advancing learning could be established:

- Learning/problem solution databases
- Electronic learning packages
- Software systems designed for the support of individual and collective learning can be included in the network, e.g. personal development systems, catalogues on open learning, live resources etc.
- Product configuration

Practical Approach

The above-mentioned allowed for new network possibilities. The case research (Case No. 60 Holcase; Case No. 61 Dolle) showed that such possibilities were very important to high speed product development. Case 60 showed how the case business can develop for 24 hours a day with product development moving around the global market in a computer network. Furthermore, the case showed how the business can develop simultaneously on sub-tasks in the product development process. This meant that the computer network gave the business several, different high speed effects.

4.3 Virtual Network Perspective

An increasing number of networks were becoming virtual (Coldmann & Price, 1999) (Mackinsey, 2001) up to 2003. The definition of a virtual network is:

"A virtual network is a network that exist when there is a task for it
but does not exist when there is not a task for it"

The increasing number of virtual networks had not been lessened by the spreading of the Internet or by the tendency to work in networks. However, the virtual network was not a new phenomenon, but the virtual networks became more advantageous during the years up to 2003 and the advantages of running a network and being a participant in a virtual network had increased. The old and well-known networks and organisation forms – the physical networks – were often too inflexible, not sufficiently agile, and too burdensome in terms of costs. This aspect conflicted with increased market demands for increased speed, improved performance, and less costs as regards the industrial product development.

The physical network and the computer network were often too costly and resource consuming when running continuously. Therefore, the virtual network was interesting where the network was only activated when there was a task for the network.

A virtual network can be defined as follows:

> *"Customer-focused and opportunity based, and it must have a clear and agreed upon set of objectives. Through the combination of the core competencies of all its members, it must establish a set of world-class core competencies to meet each opportunity"* (Coldmann & Price, 1999)

Coldmann & Price focused on the core competences of the virtual network and on the drivers and benefits. They also argued that the above-mentioned items would have a major influence on customers' interest in buying products and achieving integration with the virtual network.

Coldmann & Price emphasized the focusing on customer and opportunities and stressed the importance of the virtual organisation having a common ultimate objective. On the basis of case examinations we were convinced that having a common ultimate objective of the product development task of the virtual organisation was particularly decisive for a virtual network which wished to improve time, costs, and performance. A discussion at the Salsa Group (www.salsa.de) stressed the importance of the tasks of the virtual networks. If the virtual network did not have a task, it did not exist and consequently, there was no need and want for the virtual network.

This imposes certain conditions on the virtual network:

> *"The relationship among the partners in a virtual organisation must include trust, open and honest communication, and compatible management styles. The organisation must be able to make decisions quickly and to disband relatively painlessly when the opportunity that occasioned its creation has passed. It must have been organized because no one member could have met the opportunity alone.*

It may be designed to be joined very easily to operate in dis-
tributed mode, to exploit concurrency, or to include competitors."
(Coldmann & Price, 1999)

When it came to the product development process, there were a number of
characteristics and advantages connected to the virtual network:

"The virtual organisation can appear to customers as if it were, and
it can act like, a big business because of the access to complementary
competencies that it has in a virtual organisation. Look bag and
retaining the entrepreneurial nature of being small."

"Regardless of size, the virtual organisation offers to its members
access to expanded markets, the ability to combine resources for
new markets, and the ability to cut the concept to cash time through
concurrency."

"The virtual organisation has the ability to combine a disparate
set of core competencies and offer advantages to customers in
terms of systems reliability and capability. [...] the combined set
of core competencies can exceed the capabilities of the members
organisations acting either alone or in a non-integrated network.
A virtual organisation of small businesses can collectively take on
systematic tasks and be responsible for the manufacturing function
of a customer firm." (Coldmann & Price, 1999)

At the same time, the virtual network provided the customer with a number
of advantages:

"Access to skills and experience of many different manufacturing
approaches (initially and over timer)"

"Focus on the system versus components"

- *higher reliability*
- *better quality and consistency*
- *lower cost for given functionality*
- *lower internal development costs*
- *Flexible access to resources*
- *variable production quantities*
- *pay only for services needed, with backup access as needed*
- *more robust partnering* (Coldmann & Price, 1999)

The major problem or drawback of virtual networks was:

"The protection of intellectual properties." (Coldmann & Price, 1999)

Many businesses deliberated whether to enter into or establish virtual networks these years. The strategic prerequisites for the functioning of a virtual organisation are as a minimum:

1. Sharing infrastructure and risk
2. Linking complementary core competencies
3. Reducing concept to cash time through sharing
4. Increasing facilities and apparent size
5. Gaining access to markets and sharing market or customer loyalty
6. Migrating from selling products to selling solutions

The backbone of a virtual network could be a mix of both a physical and a digital network (Case No. 1 Zara) (Case No. 19 UK Chemicals). Therefore, it was relevant to look into the existence of virtual networks.

4.4 Reflection on Network and High Speed Product Development

The importance of network types to high speed product development was of considerable interest to this project when considering the kinds of network used in product development and the intensity and influence of the network types on high speed product development. Furthermore, it was essential to this project to examine the kinds of network which were used and could be used in the early phases of a product development process to speed the process.

All our cases had therefore been carefully examined on the following points as shown in Table 4.2:

Table 4.2   Network analysis of cases

Network Type	Idea Phase	Concept Phase					
Participating Functions							
	Product Development	Marketing	Sale	Finance	Production	HR	Management
Traditional Network (Internal)							
Traditional Network (External)							

			Table 4.2 Continued				
Network Type	Idea Phase		Concept Phase				
Participating Functions							
	Product Development	Marketing	Sale	Finance	Production	HR	Management
Virtual network Computer Network (E-development)							

The objectives were to identify the involvement of networks and network partners in NB HS NPD.

4.5 Network Perspective and Analytical Framework of Research Project

The network perspective of this research project toke Håkonson and Håkonson's network concept as its starting point combined with the concept of the computer based and the virtual network. The network perspective presented network solutions to and tools for explaining and substantiating high speed in different product development courses.

The focus of this research project was to identify which internal and external players were performing in the actual product development networks, and to identify the roles such players were performing in order to produce high speed in the product development.

This research project also focused on the kinds of networks which were used in high speed product development.

It was of the utmost importance to examine the kinds of networks that were used and the extent to which they could contribute to the explanation of high speed in the product development.

In what way networks could help increase the speed of product development and thereby shorten the product development process?

It was the hypothesis of this project that HS NPD projects were generally characterized by two or more network partners as is illustrated in Figure 4.1.

NPD network partners were only relevant provided that the NB HS PD core prescribes or allows network partners to be included in the product development project. Consequently, this research project focused mainly on the empirical evidence containing and applying networks in HS product development.

It was the hypothesis of this project that a HS PD course could increase speed considerably when network partners were included. It was the hypothesis of this project that it was advantageous to the business to included network partners as early in the course of product development as possible according to Figure 4.2. Furthermore, this project puts forward the hypothesis that the involvement of network partners could influence all success criteria of a product development course.

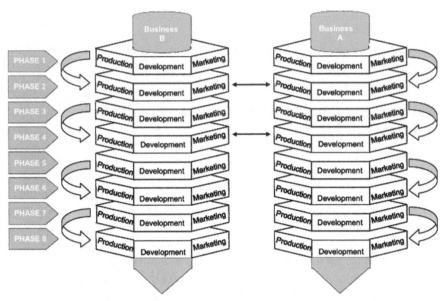

Figure 4.2   Network partners at identical levels.

It was the hypothesis of this project that the kind of network partners which the business asks to and should join the network and which were actually becoming involved differs and should differ widely depending on:

- product development project
- product development model and process
- the market
- the technology
- the competences of the business

It was the hypothesis of this research project that one of the characteristics of a NB HS PD course was that the individual network partners could be dragged in or dragged out at different stages and gates which meant that the partners did not necessarily have to be at identical stages as illustrated in Figure 4.3.

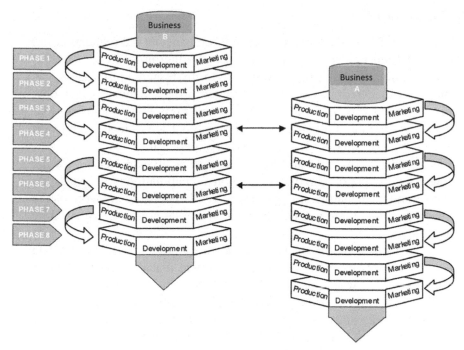

Figure 4.3 Network partners at different levels.

This project puts forward the hypothesis that it was critical to businesses product development and speed in product development that they were:

1. able to interfere with networks
2. able to find the right network partners for the situation in question
3. able to be fully aware of the individual differences of the partners in regard to stage and gate levels
4. able to attract and repel the right network partners throughout the project

The above-mentioned paragraphs have dealt with network based product development models and processes in relation to speed in product development. However, this subject also requires an additional discussion of the direction which a business or a network of businesses wish their product development to take, and of the success criteria according to which a product development project must be judged. Such a discussion was vital and necessary to manage high speed product development in 2003 and beyond. Consequently, the discussion is conducted in the following chapter.

4.6 Summary

The objectives of this chapter were to establish the theoretical foundation and the theoretical concepts for my research project. Furthermore, this chapter was meant to delimit and clarify the central areas of my research project.

Thus, it was determined that my research project defines the product concept as a total product in the business-to-business market consisting of tangible as well as intangible elements. Furthermore, I determined that I should take the marketing point of view as my point of departure.

Additionally, I toke my starting point in an analytical framework the fundamental structure of which was based on a combination of the stage and gate model and the department stage model. Yet, I also based my work on the hypothesis that not all product development processes could be explained by this model.

I had also taken the product development model – the stage and gate model – as my starting point and had determined that the central point of my research project was to be what Cooper 1986 calls the idea and concept phases together with the ensuing screening phases. My research project also deals with the horizontal and vertical processes which go on in network based product development – i.e. internally between the departments of the business and externally with suppliers and customers. My analysis also treated the use of different kinds of network in high speed product development as a means to increase the speed of the product development process.

The analysis of such processes had as their objective to advance a generic model explaining network based high speed product development. It had become apparent that when preparing such an analysis it was necessary also to address the iterative product development process and the parallel development of several elements and processes during the product development course.

To be able to evaluate the success of a product development course, I had chosen to address three central concept, viz. costs, time, and performance. Of these three, I centred my attention on time. The dependence of each success criteria had been discussed and alternative success criteria had also been deliberated.

The above definition of concepts had been necessary in order to carry on the empirical study of network based high speed product development as it would otherwise not be possible to describe, analyze, and satisfactorily explain the main phenomena of network based high speed product development.

5

High Speed Enablers to NB PD

This chapter presents an additional dimension of network based high speed product development as a result of a case study primarily on secondary business cases carried out under a previously defined analysis framework described in the article "Network product Development – Analysis Framework for Case Studies" (Bohn & Lindgren, 2000). The specific aim of the chapter is to determine the main enablers to high speed in network based high speed product development.

The chapter seeks to list these enablers and discuss their importance to high speed.

5.1 Introduction

Wheelwright & Clark (1999) emphasize three critical incentives for the product innovation process. The incentives were mentioned in connection with an account of the globalisation of the environment of the businessess:

- Intense international competition through an increase in the number of competitors competing at a certain performance level
- Fragmented market demand as customers and end-users demand high performance and reliability
- Alternative and speedily changing technologies as a result of increased knowledge of and access to new technology

My secondary case analysis confirmed that during the recent years up to 2003 businesses had centred their attention on the time dimension – high speed product development.

The focus on the time dimension during these recent years – high speed product development or the difference between success and failure in product development had in several surveys (Cooper, 1993) shown that it was primarily a question of the business being able at high speed to develop products for the market – time to market. Furthermore, it was essential that the new products enable the business to obtain advantages by implementing the products on the market before the competitors – the so-called "first mover advantage".

The questions was however

- what enablers explain the high speed at which new products were developed in specific businesses
- what was an high speed enabler. Such enablers are the focus of this chapter.

By way of introduction, an analysis of 74 case businesses had been used to identify the above-mentioned high speed enablers. The 74 businesses had proved their ability to increase the speed at which a product had been developed from idea to final product and market introduction. By means of the framework model of my research; see Figure 5.1.

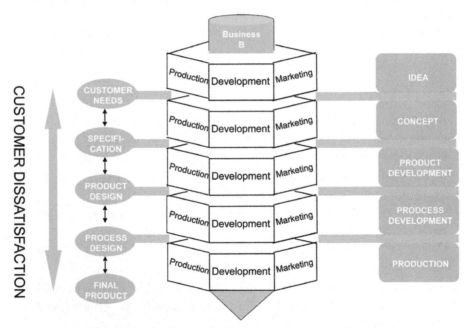

Figure 5.1   Analysis model for network product development.

In this chapter I wish to present, identify and define the enablers employed by the case businesses in order to increase speed and diminish product development time.

5.2 What Is a High Speed Enabler?

5.2.1 Theoretical Approach

"A high speed enabler is a catalyst put to the product development process that increase the speed and diminish the time at which a product development process can be completed".

The hypothesis of my research project was that high speed enablers used to speed up the product development process each have their own characteristics and central part to play in increasing the speed of the product development process.

5.2.2 Practical Approach

As a result of the research ten main enablers of highs speed product development were identified.

1. Use of information and communication
2. Customer satisfaction/customer focus
3. Optimization of PD processes
4. Network product development
5. Development of product development innovation
6. Human resource
7. Process optimization
8. From product to process
9. Product modelling
10. E-development

The main enablers could be described as catalysts which were included in and applied vertically and horizontally in the product development process. My hypothesis was that the enablers were in evidence in and applied in the businesses and in their product development processes with different intensity, frequency, and focus depending on:

- Business characteristics
- Product development situation
- Market characteristics (line of business, customers, competition)
- Network characteristics

Business characteristics were defined as specific and prevailing business characteristics such as size, business culture, interpersonal relationships, tradition, management philosophy, sales and marketing strategy, production strategy etc.

The product development situation could be defined as the task which the product development business face:

- incremental or radical product development
- idea or concept phase

The market characteristics were the conditions – trade and customer characteristics – defined as according to which the specific business had to act, i.e., type of industry, growth rate, legislation, customer segmentation, customer behaviour etc.

Competition characteristics were defined as the competitive conditions to which the specific business was subject. Such conditions include the number of competitors, the degree of rivalry, entry and exit barriers, threats from potential entrants or substitutes.

Network characteristics were defined as the network conditions to which the specific network was subject. Such conditions include the number of networks, the degree of network, the different types of networks etc.

Technology characteristics were defined as the technologies that were available, mainstream and coming up to the industry involved.

My hypothesis was that the above five characteristics influence to a considerable extent the kind of enablers employed in the businesses high speed product development. In the following each main phenomenon will be identified and explained in terms of content and sub-phenomena.

5.2.3 Information and Communication Utilization

Theoretical Approach

The high speed enabler which was first identified in the cases was the ability of the businesses to use the information and communication parameter and flow faster in the product development process to develop and create valuable information at an earlier point in the information creation.

Practical Approach

Firstly, the cases showed that the businesses had managed to streamline the information flow from customer to supplier and to sub-supplier. Primary information could be "reaped" directly with the customer and at an earlier point in time – real time consumption. Subsequently, the information was transported to the supplier and the sub-supplier often without any time delaying "filters". This meant faster access to information and consequently the possibility of swift use and swift analysis of market information. The advance of the e-business area was also pushing such development (BESTCOM Project, 2002).

An example of such use of information had been described in the case on the computer business Gateway 2000 (Case No. 16). Gateway 2000 allowed their sub-suppliers direct access to their customers real-time consumption.

In this way, production and product development could be based on such "real-time" data which allowed product development to be based not only on forecasts but rather on real-time consumption.

Similar examples of this were the space systems employed by the retail trade such as space management, category management, store management and others (Case No. 27) which were all based on real-time consumption. Information was gathered directly at the cash registers and is transferred back to the sub-suppliers for further product development.

This was also true for the e-business business DELL described in Case 44. In principle DELL and their sub-suppliers were able to monitor real-time consumption from one hour to the next or even from one minute to the next.

The main phenomenon was based on the prerequisite that access to primary information is open to all links in the supply chain. Subsequently, the information flow is streamlined e.g., by means of the Internet, high speed network, e-development tools, and data mining to ensure that information can be passed on at higher speed and with higher quality to relevant key players in the product development process.

One of the distinctive features of the businesses which work with the above-described enabler was that all players in the product development process had committed themselves to and had accepted the open access to primary information. Furthermore, they seek continuously to improve and make more effective the communication tools, the communication flow, and the quality of the information. The businesses in the product development process were confident that the above scenario was beneficial to the product development process and to the speed of the product development process (Price, 1995 and Cooper, 1996). In this connection, the trust factor was important (Price, 1995).

The result was that the speed at which new products were developed and the speed at which the decision processes in product development could be carried through, could be increased as the information and consequently the conditions for making such decisions were available at an early point in the decision making. Thus, information and the access to information no longer constituted limitations or "bottle necks". On the contrary, the decisive factor was now the speed at which creativity and the decisions of product development could be made.

5.2.4 The Customer Enabler

Theoretical Approach

Most businesses no doubt will claim that they aim for maximal customer satisfaction with due regard to the earnings of the business. At the same time many businesses will claim that an increase in the degree of customer satisfaction would delay the product development process and increase the costs of product development as the products and the development process were becoming more complicated.

However, the above scenario indicated a traditional and outdated assumption in 2003. The next main phenomenon – increased customer satisfaction – Enriching the customer (Price, 1990, Cooper, 1992, and others) verified this argument. Products could be developed at a higher speed at similar or even lower costs and with better performance, provided that they were developed "right" the first time.

This involved letting the customer join the product development process. The main phenomenon includes several dimensions i.a.:

1. Products are developed to satisfy the immediate wishes of the customers.
2. Customers and suppliers develop the products in cooperation. This will reduce the number of misunderstandings and result in applied product development.
3. The customer is put in a position to make product development process decisions which may increase speed and performance and reduce costs.
4. The customer will have the possibility of "freezing" the product development at a very late point in the product development phase.

Practical Approach

The Levis case (Case No. 3) presented a new product development concept – personalized clothing. In this case Levis develop trousers in cooperation with the individual customer in the shop. This results in increased customer satisfaction as the product was fully adapted to suit the customer's needs. As a consequence, the speed of Levis' product development process had increased and performance had improved.

By means of an "electronic stocking" The Customer Foot Shoe Business (Case No. 4) had been able to make customer adapted shoes. As was the case with Levis, The Customer Foot Shoe Business was able to develop exactly the shoe which the customer wanted while the customer was still in the shop. In the course of a few days the customer adapted shoe was sent directly from the shoe factory to the customer. The customer had taken part in the product development. The customer was allowed to make vital decision in the product development process and this information was sent directly to the supplier which helped ensure optimal performance within the briefest possible time.

Glunz and Jensen (Case No. 30) was developing advanced picture developing equipment particularly for the Japanese market. The Japanese customers were known to want the "best product". In terms of product development this often poses complicated problems as the customer may discover at a very late point of the product development phase that the product which was to be completed was not the "best".

Glunz and Jensen had solved this problem by enabling the customer to "freeze" the final version and performance of the product as late in the product development phase as at the time of product realization. In this way Glunz and Jensen evade a possible delay of the product in the final phases of the product development and at the same time succeed in bringing the product to market on time.

Case No. 37 and Case No. 38 confirmed that on their own initiative two major customers wanted to establish their own product development department at Lumonics and Lyngsø respectively. The motive for establishing their own product development departments was their wish to increase speed, improve performance, and reduce costs in two specific product development projects. Both customer and supplier had recognized that the product development process could be optimized by employing the above-described enabler.

Other businesses addressed the customer satisfaction phenomenon in another way. By developing customer products using the advantages and capacity of mass production the businesses were able to save time considerably This phenomenon stems i.a., from the concept of mass customization of which businesses such as Lego strived to make the most.

5.2.5 Optimization of PD Models

Theoretical Approach

In a previously published article (Product Development 2000) criticism of the Stage gate models was put forward. My information retrieval and preceding case analysis had shown that an increasing number of businesses tried to optimize existing product development models by moving away from traditional sequential product development models towards more simultaneously based product development models.

This also involved the theory and tools of flexible product development models and rapid prototyping as previously mentioned (Verganti, 2001).

Practical Approach

In particular Rossflex (Case No. 11), TC (Case No. 6), and TC2 (Case No. 8) had special focus on the processes of product development. The centre of attention was on the elimination of such processes which were not productive an on the optimization and speeding up of the remaining processes.

Toyota (Case No. 10) was another example of a business struggling to diminish the product development time of a new car model from 5–6 years to 2–3 years.

AKV (Case No. 39) deliberately seek to put pressure on their product development and product development department by forcing product development through the phases from idea, to concept, to prototype. AKV anticipate errors as a result of their approach but they believed that such errors and the pressure to overstep the bounds of their ability would eventually benefit their product development in terms of time, performance, and ultimately of costs.

Lumonics (Case No. 37) showed the way in which the business worked with three key product development tasks simultaneously. The case also showed that progress or new developments in one of these areas would result in decisive improvement in one or both of the other areas.

5.2.6 Network Enabler

Theoretical Approach

For many years the major part of product development had been based on internal product development. However, an increasing number of businesses realized that as a result of the demand for high speed product development they were no longer able to base their entire product development solely on internal product development. Consequently, they were called upon to consider network product development.

Network product development can typically be assigned to two main types:

1. The enterprise network – i.e. product development between equivalent businesses or businesses in the same line of business
2. Exterprise – partnership with non-traditional partners
3. A combination of enterprise and exterprise

Practical Approach

Canon/HP (Case No. 21) and UK Chemicals (Case No. 19) were good examples of enterprise network. Canon and HP who are traditionally competing on the printer market form a strategic product development relationship with the object of developing a colour laser printer. Subsequent to the development, the two businesses wish to market the product separately as two competing products. Canon had the motor and HP had the know-how on laser print. In this way both businesses were able to see the advantages of carrying out product development in an enterprise network.

UK Chemicals was a fusion of 11 chemical businesses which normally competed on the British market. However, they decided to form a strategic product development relationship with the object of developing and marketing products for the American market. Each business realised that when forming an enterprise network they would be able to take on and quickly meet large product development challenges in the American market. At the same time, they would be able to represent a wider selection of product development possibilities to the demanding American customers.

Apart from the fact that an increasing number of new networks were being established in order to satisfy the need for faster development, the nature of the network organisation was also changing. The networks were now organised on the basis virtual networks with dynamic relationships; this was illustrated by the UK Chemicals case (Case No. 19) and the Agile Web Network case (Case No. 20). The networks were organised as virtual networks or organisations (www.agileweb.com) which come up and became separated according to current needs. The actual product development task decided the extent and participation in the network intended to fulfil the task. The networks may also change in accordance with the time or the phase of the product development project. As a consequence, some players may participate in all phases whereas others may only take part in some of the phases.

5.2.7 Product Development Innovation

Theoretical Approach

The development of the PD innovation phenomenon focuses on the development process where product innovation is continuously developed e.g., directly with the customer either at the customer's address or at the supplier.

Practical Approach

The Rossflex case (Case No. 11), the Mayekawa case (Case No. 13), and the Lumonics case (Case No. 37) clearly show the way in which the product innovation process could change from being the business own product development or product development carried out exclusively with the customer and the supplier to being an integrated, joint product development cooperation between customer and supplier.

The two – or more – actors let their organisations merge in a joint attempt at and with the joint objective of developing product at the highest speed possible, with the best performance, and at the lowest cost possible.

5.2.8 Human Resource

Theoretical Approach

The main phenomenon which I toke the liberty to call "human resource" was the phenomenon which the major part of the businesses were presently addressing (18%). I believe that the reason could be found in the fact that this very phenomenon product valuable results in terms of speed, performance, and costs.

The main phenomenon contains several partial element such as:

- Empowering teams
- Flexible organisational structure
- Choice of optimal team size
- Clear rules governing the product development teams
- Control of distribution channels
- Raising of the level of team spirit

Practical Approach

Several cases (i.a., Case Nos. 34, 35, 36) show that it was indeed possible to increase the speed of product development by conferring power, defining a more exact framework of product development, and by empowering the product development teams.

At the same time, it appears that the speed at which the product development could be carried out depends on the way in which the business has organised the product development as well as on the number of participants involved in the product development. The size of the team seemed to be decisive for the efficiency and speed of product development (Jepser Larsen, 2001).

Thus, the Lyngsø case (Case No. 38), the Scanio case (Case No. 41), and the AKV Langholt case (Case No. 39) showed that advantages of speed can be gained by organising product development into minor teams, within a known framework of product development, and with a "strong" product development manager – most often the general manager or the sales manager. The "strong" product development manager was characterised by his thorough knowledge of the framework of product development in terms of profitability, finance, technology and organization. Furthermore, the "strong" product development manager was able to and dare take risks and face uncertainty. Consequently, the product development team was allowed to work in a "safe" setting and the project manager was allowed to act as a catalyst for high speed as the process was not constantly being halted because of non-acceptance of proceeding. The product development manager may even force an increase of speed in the product development team as his leadership allows him to see which participants and which decisions etc. were decisive to the increased speed of the product development; see the Langholt case (Case No. 39) for further reference.

Another aspect of the main phenomenon described here is the ability of the business to arrange for a flexible product development organisation. The Rossflex case show (Case No. 11) how the business was capable of constructing a flexible product development organisation (Verganti, 2001).

Additionally, the Langholt case (Case No. 39) showed a flexible structure of the product development organisation. The results of such a structure became immediately evident in performance, speed, and costs.

The third aspect of the present main phenomenon is the organisational structure and control of the distribution channels. The Zara case (Case No. 1) showed how the competitors viewed the practically inhuman product development speed at Zara. The competition report and analysis showed as far as they could see, Zara introducing a new collection each week which was impossible seen from the point of view of a traditional supplier of textiles. Thus, Zara exceeded according to their analysis all physical limits to product development in the textile trade and gained major first-mover advantages in the business model ecosystem of textile/clothing.

As a result of the way in which Zara organised their product development, an outsider would believe that a new collection was being developed each week. The business had decided that a new collection had to be introduced in the shops each week. Thus, the product development team defined the limits to the future product development and could act accordingly.

The scenario described above may quickly result in bottlenecks had Zara not ensured that they have full control of the distribution and of the distribution channels. Consequently, because of their distribution strategy which was a combination of franchising and full ownership, the Zara chain was able to remove four weeks "old" products from the shops when the products had completed the four weeks life cycle defined by the business. At the same time, new collections were introduced to the shops every week, making other manufacturers believe that the product development speed was just one week.

The Zara case also showed that the team behind the product development and the team behind the distribution control were important players when speeding up the product development process. When the players had defined the business, financial, technical, and organizational framework which allowed the participants to take large risks and to work under uncertainties which supported the possibility to speed product development. The product development team had been motivated to work at high speed and within the predefined "risky and uncertain" product development framework. Hereby they could develop new products at higher speed (Leifers, 2002). The effects of the concept "empowering the teams" thus became evident. At the same time, it became clear how important HRM including high motivation of internal and external actors was when focusing on high speed product development.

5.2.9 Optimization

Theoretical Approach

For many years researchers have tried to define the product and its core benefits. Researchers have tried to define which partial elements are important to a product and which can be removed without the customer experiencing inferior performance of the product. A determination of perceived value is important to find the performance, cost, and time which product development should match.

In recent years, researchers have focused intensely on the increased practical value of process optimization. The hypothesis has been that focusing on the definition of the parts of the core of a product which the business may remove, the business may help increase the speed of product development. Additionally, such focus may even result in increased performance of the finished product seen from the point of view of the customer and in improved performance in terms of costs and resources seen from the point of view of the business. This phenomenon is described on the basis of an understanding of and focus on process optimization and adding value by subtracting time (Price, 1995).

Practical Approach

The TC case (Case No. 8) describes the way in which the business will be able to reduce the classic product development process from 66 weeks to only 3 weeks by focusing on the processes of product development. This is done by introducing a quick response system.

Also AKV Langholt (Case No. 39) focus on this phenomenon when dividing their product development into a product part and a process part. To the customer, the process part is often the most important part. How fast can the product development process be completed and what production process improvements may the customer experience as a result of the new product?

The ability to "freeze" a product at as late a point in the product development process as possible, is another aspect where the process and – specifically – the product development process are in focus (Hein & Myrup 1986). The Glunz and Jensen case (Case No. 30) shows how the business is able to accommodate their Japanese customers by making changes to the product at a very late point in the product development phase without producing vital consequence to the product development.

A third aspect is to make the process more flexible (Verganti 2002) (Microsoft – Case No. 50).

5.2.10 Changing Focus from Product to Process

Theoretical Approach

Several of the cases used for this thesis describe how the businesses change their focus from looking narrow-mindedly at the product to looking at the product as a process. As will appear form the cases, the business may look at the current product or at the process in which the product is included at the customer's place. The businesses alter their perspective from considering the product development as having a beginning and an end. Instead, they think in terms of processes and they conceive of the product as a part of a process or maybe as a process which is continuously changing. Additionally, the businesses extend the product concept to include:

- Product families
- Product life cycle
- Project families

Practical Approach

Lyngsø (Case No. 38), Sony (Case No. 5), Mobilix (Case No. 40), GSI Lumonics (Case No. 37), and Nike (Case No. 14) show how the development of a product can be incorporated into product families and product life cycles already at the idea and concept phase. In this way, the strategy of the businesses, not least the marketing and production strategies, have already been incorporated and made ready when the product is introduced and settled on the market. The businesses have integrated their product in a "from the cradle to the grave" context. As a result, no need for alterations due to changes in the product family or in product life cycle needs will arise during the course of the product development Consequently, the product development process can take a faster course and the product which is already on the market can be developed faster. In this connection, the businesses focus partly on "new development" of a product, partly on a subsequent "variance creation" of a product.

Already in the idea and concept phase of the new product, the product is integrated and made ready for solving the problem of creating variants on the market, when market needs and product life cycle demands it. The product architecture is prepare to high speed product development.

The Mobilix case (Case No. 40) clearly shows how such an integration on the mobile telephone market have been achieved in that Mobilix consider the global needs of the different markets as well as the needs of the individual segments on the partial markets. VW, LY, and Martin Lys (Cases Nos. 36, 38, and 45) show how integration can be included in the platform or module strategy of a business. GSI Lumonics (Case No. 37) incorporate a modular service concept in their product by developing the product in order for it to be serviced as quickly as possible. This will increase performance in the eyes of the customer. A well-thought-out product architecture is characteristic and decisive of such a method.

Furthermore, product development projects are no longer seen as isolated development projects but rather as development project families. Case No. 37 describes how a break-through in one of the three product development projects help advance product development in another development project. Such advantages are fully realized by the business which is why they focus on project families in order to increase product development speed.

5.2.11 Product Modularisation

Theoretical Approach

The major problem of known product modelling techniques is that it is time and money consuming process to develop several prototypes and thus carry out product modelling. In addition, it was often difficult or even impossible to visualise the result of the product modelling. Concurrently with improvements in e-development tools and especially in product configurators, product modelling has become easier.

Practical Approach

The Levis case (Case No. 3) illustrates one of the aspect of the product modelling phenomenon. Levis obtain high speed by product modelling in cooperation with their customer either in the shop or on the market. In this way, the products become individualized according to the needs of the customers. By way of product modelling the customer and the supplier are taken through all phases from idea to product completion and shipment in the course of 1 to 2 hours.

Subsequently, an individualized product is delivered at the customer 2–3 weeks later. Thus, the speed of product development is decided by the amount of resources or time used for product modelling rather than by the speed at which product development decisions are taken by the customer and the supplier – in this case probably especially by the customer.

The opposite aspect is illustrated by the Lyngsø case (Case No. 38) in which the decision-making process at the manufacturer's place is crucial to the speed. In the Lyngsø case several financing models are employed concurrently with the on-going product development process in order to increase the speed of decision making.

As previously mentioned, another dimension of product modelling is the businesses increased focus on product modelling as an integral part of a platform or a modularisation outlook. Sony, VW, Lumonics, Lyngsø,

Martin Lys and Gateway 2000 (Cases Nos. 5, 36, 37, 38, 45, and 16) illustrate how the speed of product modelling is growing continuously with the help of platforms and modularization strategies in the product development of the businesses. The products have been prepared for such a strategy through their excellent product architecture. The speed at which product development is carried out is increased considerably because product modelling prerequisites are particularly advantageous once the product architecture has been decided on.

A third example of the employment of product modelling to increase speed and performance of the development of new products can be found in the Nike case (Case No. 14). On the basis of the customer's use of the product Nike attempts to individualize the products – the so-called "sneakerizing" phenomenon. In this way, Nike develop new products on the basis of a firm product architecture.

Finally, Gateway 2000 (Case No. 16) presents a fourth example of the product modularisation enabler when using product modelling to simplify the product even though they are competing in a fragmented market. The case demonstrates how customers are "lead" through a product modelling process which to the customer must seem complex and individualised when in fact Gateway 2000 have developed a simplified product modularisation process which helps them to increase the speed of product development.

Apart from reducing the time of product development, the above examples of product modularisation all help to increase performance and reduce costs.

5.2.12 E-Development

Theoretical Approach

The last identified enabler to high speed in product development to be discussed in this chapter is called e-development or electronic development. This enabler has been in focus for a long time and by virtue of increased and improved technology sue, it is now possible to increase the speed at which new products are developed. The constant and speedy development of the Internet and related software tools (DISPU please see Chapter 12) contribute to making the idea and concept development more precise and faster.

The hypothesis is that E-development plays a significantly important role at the very early stages of product development where time or finances do not allow the development of a physical prototype. By means of E-development tools it is possible to create an exact copy of the final product and explain to the customer all facilities of the product. In this way, the product development process is furthered. The supplier and the customer are able to make decisions as to the subsequent course of the product development at a very early stage and actually at any stage of the product development process. The previously encountered physical obstacles to prototype development are overcome.

Practical Approach

Glunz and Jensen (Case No. 30) use e-development tools for their internal product development carried out in cooperation with sub-suppliers. Likewise, Levis, The Shoe Maker Business, ODI, and Idémøbler (Cases Nos. 3, 4, 46, and 23) use E-development tools externally and in cooperation with the customers.

The Danish business Kellpo (Case No. 74) has developed their business in such a way that they only develop new products via advanced e-development software tools. This means that Kellpo cannot business economically develop new products with customers who are unable to develop together with Kellpo on e-development software platforms. Therefore, Kellpo has to reject customers who are only capable of developing products physically and not digitally.

On the basis of the cases included in this thesis it appears that suppliers' and sub-suppliers' ability to handle the tools will be the decisive factor when suppliers and sub-suppliers are chosen. Thus businesses begin to reject sub-suppliers and suppliers when such suppliers do not master the new tools. Additionally, some businesses are rejected because they do not master e-development.

5.2.13 High Speed Enablers in Future

Analysing these enablers to high speed product development gives the following hypothetical picture of future high speed product development as shown in Table 5.1.

Table 5.1   Enablers to HS PD

Enablers to High Speed Product Development	Today	In the Future
Information and communication	Physical and to some extent digital information and communication	Mix of all existing information and communication tools mixed with new and high speed communication tools

(Continued)

Table 5.1 Continued

Enablers to High Speed Product Development	Today	In the Future
Customer satisfaction/customer focus	Customer to some extent involved in product development process	Customer involved in all phases of the product development process
Optimization of PD processes		
Network product development	Few and narrow networks in PD	Network and all types of network in all product development projects
Development of product development innovation	Innovation ad hoc	Innovation continuously
Human resource	HRM not particularly involved. HRM not very important	HRM involved in all phases of the product development project – HRM very important in PD
Process optimization		
From product to process	Some businesses see the product as a process	All businesses see the product as a process
Product modularisation	Businesses try to use product modularisation but not with great success	All businesses use product modularisation
E-development	Some businesses use E-development	All businesses use e-development

5.3 Analysis Framework of PD

On the basis of 74 secondary case examinations the frequency with which each enabler appears in the secondary case examinations has been examined.

It is worth noticing that the main enablers appear with diverse frequency and that particularly the HRM enabler and the product modularisation enabler were the main phenomena which receive the most attention in the businesses or which are most frequently employed when exercising high speed product development.

On the basis of the above a need arises for a closer examination of the way in which product development processes and models are made adaptable to allow for the hypotheses and trends which we believe to be able to identify in the future network based high speed product development situations. Central aspects as to be included are shown in Table 5.2:

Table 5.2   Use of high speed enablers in secondary case businesses

Use of High Speed Enablers in Secondary Case Businesses	Total in %
ICT Communication Enabler	2
Customer Enabler	9
PD Model Enabler	11
Network Enabler	11
Innovation Enabler	6
HRM Enabler	27
Process Enabler	2
Product to Process Enabler	2
Modularisation Enabler	24
E-Development Enabler	8
Total	100

- How can the 10 enablers be integrated into the product development process?
- How do we achieve high speed in the internal and external networks to the product development to provoking first mover advantage?
- How do we ensure that high speed is integrated in all functional areas?

Thus, our future analyses and research must expose high speed product development processes in Danish productive enterprises. We define the process as a series of partial processes/activities in which internal connections are determined, in which the use of each separate enabler is determined, and in which their contribution in achieving the common goal – costs, performance and high speed – are defined. We believe, that the main focus in the initial phase should be on the following enablers:

- Customer satisfaction/customer focus
- Optimization of PD processes
- Product modelling

5.4 Summary

The literature and case studies of high speed enablers have shown that there are at least 10 enablers which may influence the speed of product development considerably. Consequently, in connection with ensuing empirical studies of current network based product development, it is necessary to extend the analytic foundation and selection of models which intercept, describe, and provide a better understanding of the enablers to high speed network based product development. Such an extension will be made in Chapter 8 of this thesis.

Before this elaboration a discussion of the success criteria to network based high speed product development is important to find measurements of NB HS NPD.

6

Success Criteria for PD

This chapter concentrates on a discussion of success criteria for network based high speed product development. The chapter introduces a general discussion of success criteria – cost, performance and time – for network based high speed product development and produces more observations and analysis on the issue of success criteria. Focus is both on short-term and long-term success criteria in order to reach the optimal NB HS NPD. This chapter discusses the definitions of short-term and long-term success criteria as well as their differences. Finally, the importance of short- and long-term success criteria seen in relation to NB HS NPD are discussed.

The chapter finalises the PhD project's discussion on concepts for NB HS NPD.

6.1 Introduction

Chapter 3 discussed Rauseneau's 1983 definition of the success criteria of product development was introduced. Focus at that time and up till now has to a large extent been on time, cost, and performance. The present PhD project, however, stresses the importance of looking at such success criteria in a strategic perspective and of focusing on other success criteria than time, cost and performance have been very poor. The reason for this can be seen in the businesses' lack of knowledge about other success criteria and the lack of need so far to focus on other success criteria. However, with added pressure on speed, network and fast changing conditions on the field of product development it seems as if the time has come to focus on other success criteria.

As we saw before, the definitions of the success criteria were:

As can be seen from the above, there is quiet a difference between the theoretical definition of success criteria to the practical definition of success criteria. This discussion can be intensified by discussing what short-term success criteria are as indicated in Table 6.1.

Table 6.1   Definition of success criteria

Success Criteria	Theoretical	Practical (Secondary Cases)
Time	Relative time – according to the view set for the PD task	Physical time – an working our e.g.
Cost	Direct and alternative cost	Direct costs
Performance	Perceived value	Value
Speed	Relative time it takes to move a PD idea from idea generation to encapsulation of a product – according to the view set for the PD task.	Physical time moving a product development project from idea to market implementation

6.2 Short-Term Success Criteria

6.2.1 Theoretical Approach

Matching time, cost, and performance in a given moment for a given product development project is essential to a business. Often time, cost, and performance have already been specified and can therefore be classified as the success criteria of a product development project.

However, a business matching only short-term success criteria matches the success criteria of only one product development project at a time often for only one customer. This means that the business matches the success criteria only in a vertical perspective as shown in Figure 6.1.

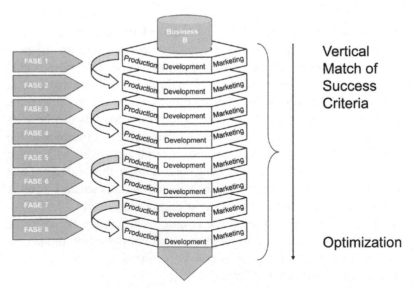

Figure 6.1   Vertical match of success criteria in NB HS NPD.

The business focuses on the challenge of getting the product from idea to market introduction as fast as possible, with a minimum of cost and with a performance to which the customer can here and now agree is basically very short-term oriented.

However, this match is a kind of sub-optimising as the business focuses only on optimising within the individual product development project. Hereby, the business focuses on a short-term optimisation without paying attention to horizontal optimisation as seen in Figure 6.2.

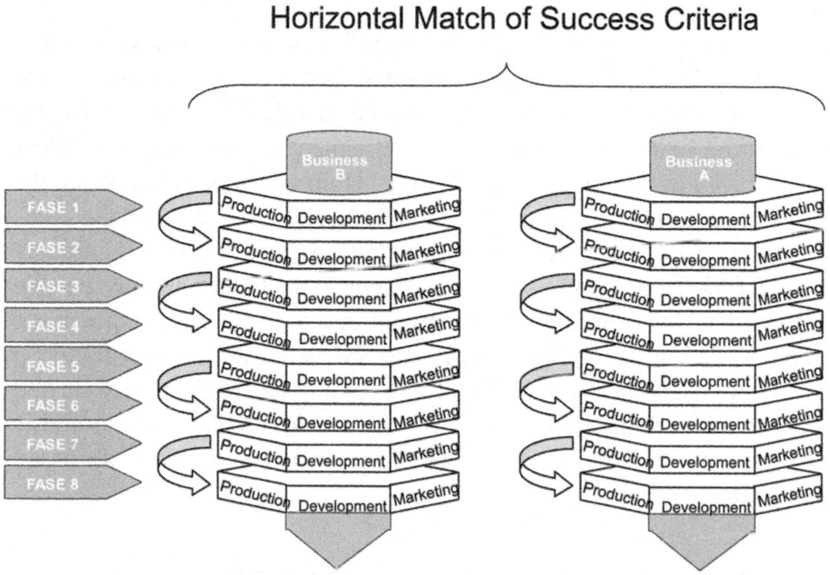

Horizontal Match of Success Criteria

Figure 6.2 Horizontal match of success criteria in NB HS NPD.

The business does not optimize the business overall product development activities seen in a horizontal perspective. This prevents the business from gaining an overall optimum of time, cost and performance across the business product development projects. Consequently, an optimization on a horizontal perspective has to be connected to an optimization seen in a long-term perspective because the different product development projects do not have the same beginning and end or the same match of time, cost and performance success criteria. A horizontal focus and perspective on long-term success criteria will teach the business how to gain a better match of time, cost and performance. However, learning demands knowledge transfer and learning across product development projects (Corso, 2001) (Gieskes, 2001).

The vertical and horizontal dimensions are still very much related to an internal view of the business product development activities. When seen in a NB HS NPD perspective, the optimisation of the success criteria becomes even more complex. Suddenly, success criteria cannot be defined solely to the individual product development project or to the individual business. It must instead be to optimize the success criteria for an entire network or more advanced for more networks as seen in Figure 6.3.

In this way, success criteria for NB HS NPD turn into a more complex perspective of optimisation of success criteria because network partners focus on different success criteria.

This new success criterion perspective forces the business product development managers to consider and reflect on strategic alliance, joint venture or even acquisition to match the success criteria across networks. It also forces the managers of product development to consider and reflect on the transfer of knowledge across networks. Both considerations demand a high degree of focus on long-term success criteria and long-term perspectives in the planning of product development. The management must continuously improve their product development management competence and must be able to visualise the business product development into a long-term perspective.

The hypothesis of the PhD project is that SMEs in 2002 mainly focus on short-term success criteria indicated in Table 6.1 and that they mainly carried out product development within a single or very few networks. The hypothesis of the PhD project is that hardly any SMEs focus on long-term network based success criteria.

6.3 Long-Term Success Criteria in NB HS PD

As can be seen from the above, long term success criteria have to deal with knowledge of product development and in our case particularly on NB HS NPD. Knowledge of best practice of NB HS NPD and of the performance of NB HS NPD in different situations has to be transformed between network partners.

A business which wants to develop knowledge about NB HS NPD has to develop a product development learning culture and to facilitate learning (Gieske, 2001) or a product development learning model (Leifers, 2002). If a business is not able to develop such conditions, it was my hypothesis that the business will remain in the same position and will not be able to make further developments on shortening time, cost or performance.

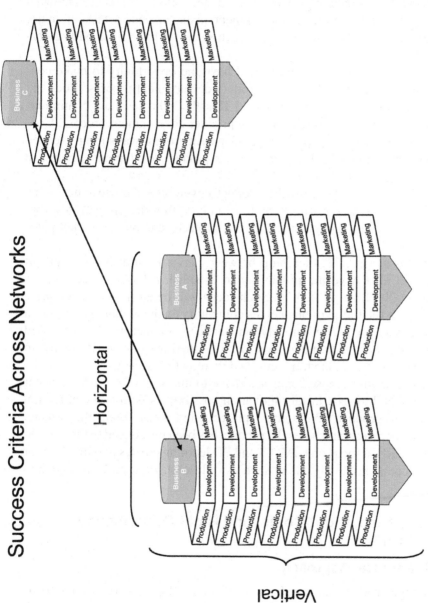

Figure 6.3 Optimization of NB PD success criteria across networks.

Learning in NB HS NPD means a business ability to transfer knowledge both vertically, horizontally, and between networks. This is a somewhat complicated task for product development managers but a task of major importance to gain competitive advantage in an increasing NB HS NPD global market.

SMEs focusing on PD learning will have the opportunity to make continuous improvement (CIM) of the product development process (Boer, H., 2000). Many of the cases (Case No. 11 Rossflex) showed how continuous improvement could be brought into the product development process. Continuous improvement in the product development process was strongly related to learning and knowledge transfer from one product development process to another.

CIM is important because the product development process has to be improved continuously in future both initially within the product development process, across product development projects, and on the market place (Corso, 2001).

SMEs focusing on PD learning between network product development projects will also have the possibility to establish a basis for continuous innovation (CI) (Boer, H., 2001). Such focus will be an important strategic competence in the future of businesses competing on the global market. For a long time, an ability to continuously innovate new products has been stressed as important (Cinet, 2002) and have been verified to be strongly related to knowledge and knowledge transfer – and thereby to learning.

CI is important because businesses have to innovate new products faster in the future and to seek faster innovation possibilities both at the attraction phase of new product development projects and ideas, along the product development process, and when the product has been introduced to the market.

Additionally, a focus on CIM, CI and PD learning creates a foundation for long-term competitive advantage such as right cost, right performance, and right time.

6.4 Right Time, Right Costs, and Right Performance in NB HS PD

6.4.1 Theoretical Approach

Speed or high speed in network based product development has been in focus for some time (Cooper, 1986) (Sanchez, 1996) (Bessant, 1999) (Verganti, 2001) but how can we define high speed in product development? Taking our point of departure in an analytical framework for product development (Bohn & Lindgren, 2000), a model for knowledge management in Product

Innovation (Corso, 2001), and a model for flexible design of product development models (MacCormark, Verganti and Iansiti, 2001) we are able to gain a deeper understanding of speed in product development.

Initially, I claim that high speed enablers (Chapter 5), management tools, technological tools, product development models, and product development processes have to be understood in a wider perspective to comprehend the ability to perform network based high speed product development. The hypothesis is that there are more types of speed in NB HS NPD. The research intends to find these and verify such types.

Until now there are only fragmented knowledge and research on the types of speed and tools of speed. We do not know which high speed enablers are available and appropriate in different situations of product development. Learning has to be established in all areas of high speed product development to find models of speed in NB NPD. The PhD project will try to increase this learning on HS enablers.

Nevertheless, our claim is that high speed in product development is not the issue and is not always advantageous. Our secondary case research has shown that it can even be advantageous to "hurry slowly". When characteristics in market, technology, network, and the competences of the businesses are in a certain position, a slow speed can be advantageous as learning of market, technology, network, and competences develop, proceed, and get ready for the new product. The opposite can of course also be the case (Case No. 1 Zara).

Even so, the question of speed is more complicated than outlined above. During the product development process it seems as if the speed sometimes has to be increased and sometimes has to be slowed down. Some of the main components in the field of product development can turn out to influence and make radical changes to "the game of product development". Therefore, businesses often have to change speed during the product development process (Verganti, 2001).

However, this is not always possible because of the businesses different choices of product development models.

6.5 Right Speed in NB PD

6.5.1 Theoretical Approach

Right speed in product development has to be learned and can be seen from different viewpoint as shown in Figure 6.4. The critical issue before talking about speed in product development is the ability of the management to analyse "the game of product development" and learn from one product

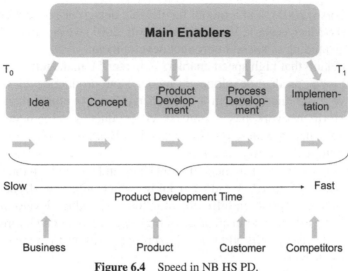

Figure 6.4   Speed in NB HS PD.

development project to another which speed is advantageous to this specific product development situation. Even more critical is the ability of the product development managers to learn throughout the product development process. The last learning area concerns the development process from idea to market introduction as well as the span of time after market introduction. Finally, the question of how to establish learning of speed tools and speed in product development across networks in the product development process is becoming important to our research project.

Several researchers have put forward models for speeding up NPD (Cooper, 1986) (MacCormark, Verganti, and Iansiti, 2001) but few of them have suggested which models to chose in different product development situations. Two main stream NPD models have been proposed. "The Stage-Gate" model and "The Waterfall" model have proved to be extremely efficient when market, technology, network, and the competences of the businesses are stable or to some extent evolving (Cooper, 1986) (Eppinger, 1996). The flexible models have proved to be efficient when the market is dynamic (Verganti and Iansiti 2001).

6.5.2 Practical Approach

Especially the car industry (Case No. FI, Case No. 36 VW), the furniture industry (Case No. 58 Tvilum), the pump industry (Case No. 54 Grundfos), the windmill industry (Case No. 59 NEG Micon), and several industries producing mainly hardware (Case No. 69 DAN and Case No. 70 BM) have shown until now to have profited from using stage-gate product development models.

As for the "flexible" product development models, especially the software industry and industries facing dynamic markets, technologies, networks, and competences as e.g. Telecase, Lycase, Metzacase (Microsoft Internetcase Macormarc, Verganti, Iansiti, 2001) have turned out to profit from using more "flexible" product development models.

The question is how and when is it appropriate to change product development model and process to gain right speed? Summing up on the case research shows some hypothetical proposals and mainstreams to the answer of which NPD models to chose and the appropriateness to high speed. This is indicated in Table 6.2.

Table 6.2 Product development models and HS PD appropriateness

	Stage Gate Model	Flexible Model
Characteristics		
Markets	Familiar markets	Unfamiliar markets
Technology	Familiar Technology	Unfamiliar Technology
Network	Physical networks and stabilised ICT networks	Dynamic networks, ICT – networks, Virtual and dynamic networks
Competences	Stable and physical competences	Dynamic and virtual competences
Product	Products are mainly hardware	Products are mainly processes Software, services,
Strength	When main components can be characterised as stable and in some case evolving on the product development field.	Flexible to sudden change in the main components on the product development field.
Weakness	Un flexible to sudden change on the product development field	When product development turns out to be stable for a long period.
Opportunities	When market, technology, network and competence turn to stabilise	When market, technology, network and competence turn to be dynamic and virtual
Threats	"Trapped in a dynamic process" either in market, technology, network or competence – performance does not match demand of market.	"Trapped in a stable process" either in market, technology, network or competence – too much cost.
Time for change of NPD – model and speed	Going from stabilised to dynamic PD – characteristics When products turn to processes	Going from dynamic to stabilised PD – characteristics When processes turn into products – standard modules

The tools proposed in the two types of models can only be considered as hypothetical guidelines as a result of the case research and the literature study. The empirical research will perform profound insight and verification into such questions.

6.6 Hypothetical Framework for Short- and Long-Term Success Criteria in NB HS NPD

When analysing the success criteria of NB HS NPD as shown in Table 6.3

Table 6.3   Short- and long-term success criteria

NB HS NPD Success Criteria Short-Term Perspective	NB HS NPD Success Criteria Long-Term Perspective
High Speed – Time	Right Time – Right Speed
Costs	Right Costs
Performance	Right Performance
	Continuous Improvement
	Learning
	Continuous Innovation

it seems as if the following framework of success criteria as shown in Figure 6.5 can be advanced:

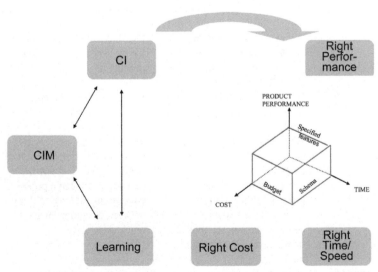

Figure 6.5   Relationship between long-term success criteria in network based product development.

Source: Lindgren & Bohn.

The hypothesis of this PhD project is that SMEs focus on the short-term success criteria and barely on the long term success criteria. My hypothesis was that this would prevent them from reaching sustainable competitive advantage via their product development activities on the global market.

NB HS PD knowledge has to be made available in an open form and transferred before it can be used to reach right speed in the businesses product development. When implemented in businesses and when learning interacts with CIM and CI, my hypothesis is that long-term success criteria such as right performance, right cost and right speed as shown in Figure 6.5 can be gained in product development.

The aim of this research project was to clarify how much focus the long-term product development success criteria have in SMEs today.

The framework finally recommends that businesses focus more on long-term success criteria to gain right performance, right cost, and – essentially – right speed because it seems as if this will be a major focus for the future as listed in Table 6.4.

Table 6.4   Hypotheses of network based success criteria

Hypotheses of Network Based Success Criteria	Today	In Future
Success criteria in the business	Short term primarily on cost Time Performance	Long term Right cost Right time Right performance (Perceived value) CIM CI Learning
Success criteria in the network	Vertical success criteria	Vertical, horizontal and network based success criteria
The match	Matching success criteria to the business or narrow network partners	Matching success criteria across businesses and a broader set of network partners

The discussion of concepts of network based high speed product development is now finished. Reflection on the concepts of NB HS NPD and the perspectives of NB HS NPD will be given in Chapter 7.

7

Reflection on and Perspectives of NB HS PD

High speed and network in product development will become major issues and major competitive weapons in future global competition. This is verified through the theoretical work in Chapters 3 to 6. Four main issues seem to be important to future product development.

1. The concepts of product development are changing
2. The network based product developments of the businesses are changing
3. The success criteria of product development are changing
4. The tools of product development are changing
5. The management of product development is changing

7.1 Concept of Product Development

In these years, both researchers and industry realise how the concepts of the products and product development models and processes are changing. The old concepts of product development do not match the demands made on product development on the global market. The pressure on speed in product development has the rules of concept for product development from clear cut encapsulated concepts definition to "fuzzy" process oriented dynamic definitions.

Speed in product development is, however, a rather new issue and therefore a rather new competitive weapon. Until now most businesses have thought about speed as cutting corners in order to save time and cost. The secondary cases show that this does not pay off. Speed is important, but it is only a valuable tool when used business economically optimal to gain competitive advantage on the global market and to get profitable, new products. Speed in NB NPD must be handled with strategic care and leadership in product development. Otherwise the business can be badly punished by the market as the secondary cases verified.

7.2 Network of Product Development

Networks and working in networks are also important and relative new concepts to the businesses. Many businesses know about physical networks and have joined these for many years. However, as was verified, new networks have turned up on the global scene – digital and vertical networks together with new ways of mixing networks and handling networks characterised by a high degree of openness, flexibility, and agility. Furthermore, the new network types demand trust and openness. It is in this context that businesses have to act with their product development for the future.

7.3 Success Criteria of Product Development

In recent years, many researchers have stressed the importance of changing the focus from short-term success criteria to long-term success criteria. However, it has been difficult for businesses to adapt this advice and especially to implement such guidelines. Many researchers have therefore tried to change their research to find the reasons why businesses do not change their focus. Furthermore, researchers have made specific guidelines for businesses to help implement the changes. However, many research results have shown that not much has been done.

I claim that this is mainly due to a lack of motivation in industry to adapt these guidelines to changes. When industry can develop without changing and when the global market is still in its initial phase then there is no need for changing. My suggestion is that this will soon cease to be the case because the pressure on product development will increase tremendously when new networks enter the global market. Especially the opening up from the far East (WTO) and Eastern Europe (EU) where new an often unknown business models enter the global market will add new pressure on the ability of western businesses to compete on product development.

7.4 Tools of Product Development

As previously mentioned, the use of HS enablers have until now only been fragmented and single-minded. New business and networks show new and quite effective ways to use and mix HS enablers. We have to learn form these businesses in order to improve.

7.5 Management of Product Development

The importance of managing a business product development activities must therefore be stressed. For a long time, management of the business product development activity has been in focus. However, most often focus in product development has been on the tactical level of product development rather than on the strategic level of product development.

Taking these above mentioned facts into account we will analyse and reflect on the challenge and issues of future product development that are presented in Chapters 3, 4, 5, and 6 and paint a picture of tomorrows product development.

This could be done by looking at the relationship and interaction between product development and the 4 issues commented on in Chapters 3, 4, 5, and 6 as shown in Figure 7.1.

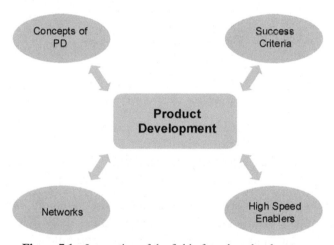

Figure 7.1   Interaction of the field of product development.

The aspect outlined in Figure 7.1 will be the issue of this chapter.

7.6 Change in Concept of Product and Product Development

From the analysis in Chapter 3 we saw major developments in our understanding of the concept of product development. In the following paragraph we will reflect on these.

7.6.1 Change of Product Concept

It is very clear that today's understanding of the product concept is not capable of defining and explaining the product. Until now we have understood the product as one physical item with a start and an end. In the future it will be characterised as a process with the following characteristics:

Table 7.1   Product characteristics – now and in future

Product Characteristics Before	Product Characteristics in the Future	Case Examples
A product	A total product A total process	
A Physical Product	From physical product to more immaterial products	
Single products	Multi-products (both physical, digital, immaterial and virtual products)	
One time encapsulation	Sequential encapsulation of a product – "never ending story"	
From product to a new product	The product changes to a process Dynamic products	
Focus on cost	Focus on direct and alternative cost	
The customer cannot change the product or is blamed if they change the product	The customer continuously changes the product together with the business	
The customers do not accept errors	The customer and the business accept trial and error	
Physical processes	Physical, digital and virtual process	

Table 7.1 shows that the core of the product becomes dynamic, rarely stable, and always under construction. The product becomes more dynamic and interactive in all areas and always in a process of continuous innovation and continuous improvement. The product becomes a total product or process and the business therefore has to "posses" more competences to fulfil the demands of the customers.

To live up to this, the business must attempt a process oriented product development model because the customer will not accept a one time delivery but will wish to receive more after the first part of the delivery has been fulfilled. The customer wants to have a relationship partner and to develop further on. The product development model and process must therefore have access to and involve more competences into the product development process. Hereby, product development must become relationship-based and long-term network based, because businesses – and SMEs in particular – cannot have all competences in house. Short-term product development thinking will kill businesses because it is just too expensive to change to a new product development process every minute. The only stable component of product development will be to build a strategic architecture of the product development process and of the product.

7.6.2 PD Models Become Process Models

When a product concept changes to process, the concepts of product development also have to change. Dynamic product development situations – with a high degree of dynamics and with all types of product development models and processes involved in a continuous process with many start and many ends – be the challenge for the future product development in businesses.

Businesses therefore have to change to a more agile style of choosing among stage-gate models, flexible and process oriented product development models. Hereby the businesses can gain a dynamic, flexible and agile product development processes.

Businesses have to implement more flexibility and agility into their product development process.

7.6.3 Focus on Advanced Use of High Speed Enablers

In my initial research I identified approximately 10 high speed enablers. The secondary case research showed that businesses were focusing on particularly the customer, the product development model, and the product modularisation enabler. However, my hypothesis is that this might not be optimal in all cases because every product development project is unique and demands a different set-up and a different choice of high speed enablers. These enablers will come into more focus in the future but will have to develop and be used in new ways as shown in Table 7.2.

Businesses will have to put a higher priority on high speed enablers in future according to the individual product development process. Especially the HRM enabler will be in focus because know-how of the location of existing competences and access to competences will be a major core competence to gain competitive advantage to product development. Some of the high speed enablers as e.g., modularisation and e-development will become standard high speed enablers used by all businesses because these enablers will allow the business to break the physical time limit of today's product development. When products can be visualised and modelled around without encapsulation

Table 7.2   New use of high speed enablers

High Speed Enablers	New Use of High Speed Enablers
Information and Communication	Mix of all Existing Information and Communication Tools Mixed with New and High Speed Communication Tools Will be the Case
Customer satisfaction/customer focus	Customers will be involved in all phases of the product development process. We will see more beta-versions or prototypes because customers are always on the look-out for new products and will accept beta-versions.
Optimization of PD processes	Businesses will always look for continuous optimisation of the product development process
Network product development	Network and all types of network in all product development projects will be used.
Development of product development innovation	Continuous innovation will be the case
Human resource	HRM will be involved in all phases of the product development project. HRM will be very important in PD because access and overview on competences will be essential
From product to process	All businesses will turn their view from product to process
Product modularisation	All businesses use product modularisation but in a more advance way than today
E-development	All businesses use e-development as a major tool in the product development process

of the product, the businesses can keep the products "floating" until the very last moment of customer acceptance.

7.6.4 PD in Networks

Product development will have to be done more in network in the future because the product development projects demand more competences. The challenge in network based product development will be to manage and lead the network in the product development process. The major challenge and conditions in this task will be as listed in Table 7.3.

Apparently, also the network aspect in product development will change. Therefore, businesses must improve on their network competences in several areas. Businesses have to learn more of networking and find best practice of networking competences to survive in future product development on a global market.

When this is the case the product development model and process for the future turns into network based product development models with a flexible,

Table 7.3 Network challenges

Network Challenges in Future Product Development	
Network characteristics	Multi network and global networks
Numbers of networks	Many and unknown actors because some actors will have to be brought in from places where the specific competences exist.
Network construction	Interactive in all areas of the product development process. Network actors can interact on all phases in the product development process.
The core of the network based product development project	Rarely static – but with a high degree of dynamic
Management	Not static and mostly joined management or fragmented management. Management can change over the product development process. Please see the new EU 6 program which are constructed like this.
Boundaries	More risky, un-formalised and fuzzy boundaries
Important network competences	Trust
	Competence to hub up on networks
	Language and communication competences
Success criteria for a network actor	Able to do continuous improve
	Able to be innovative
	Able to learn, learn fast when "walking around"
	Able to focus on value and cost at the same time

agile, and network oriented structures. Different functions and different phases in the product development model will suddenly involve a mixture of network actors and will be placed where appropriate to the common product development project or process.

Such new challenges stress the importance of changing today's belief in product development as a physical stage- and gate model placed inside the business. The product development process has to become digitalised at first sight and placed in the environment which is most appropriate to the network based product development process and which can also operate in a virtual environment. The product development process has to become independent of the physical environment to match the demand for flexibility and agility. This stresses the importance of the high speed enablers – the information and communication and the e-development enabler.

7.6.5 People and NB HS NPD

When the speed of product development increases and the market, technology and network begin to interact more and more with an increasing mixing of these components, the pressure on competences must increase. The network based product development projects will demand an increasing number of people and thus an increasing number of competences. However, the product development will need the participation of the employees in the product development in a more flexible and agile way where people will enter and leave the product development projects at all levels.

It is therefore interesting to examine the kind of people or functions which are involved in the product development project. Today mainly the sales, production and product development functions are said to be involved (Myrup & Hein, 1986) (Cooper, 1993). My hypothesis is that we will see more functions involved in the product development projects in the future.

7.6.6 PD Leadership and PD Management Become Central Issues

The major part of available literature on product development shows a prevailing focus on the management of the product development process within the process at a relatively tactical and operational level (Wind, 1973) (Wheelwright & Clark, 1992) (Cooper, 1993) (Ulrich & Eppinger, 1995) (Baker & Hart, 1999). This focus is merely due to the practical and theoretical challenges to manage the product development process through the product development process from idea to market introduction.

At the product development management level (PUM level) focus is on short-term success criteria such as time, cost, and performance. Furthermore, the focus is on the process within the development process. Very seldom are other product development projects inside or outside the business the centre of attention. The integration and knowledge transfer from other network based product development process is very seldom used.

PUM drifts among the four main components influencing NB HS NPD. PUM tends to be much involved in day-to-day product development management. Seen from the point of view of the initial product development management level.

PUM has however difficulties in developing and maintaining the objective strategic view of "the product development game". It is also difficult to elaborate a flexible strategic view and design to the "product development game" where all components are endogenous and exogenous variables played out and into the "product development field".

When a business solely performs PUM, the business lacks a higher strategic view of the business product development processes. The business therefore lacks e.g., the competence to choose from different and optimal product development models and processes to gain right speed in product development. Furthermore, the business ties itself to short-term success criteria and short-term management which will not ensure long-term success or long-term competitive advantage. The hypothesis of the PhD project is that this is the reason why the product development managers of the SME businesses focus mostly on short-term success criteria and on PUM. In addition, the major reason to why SMEs do not gain right speed in PD.

Consequently, the PhD project suggests that the business focus on product development leadership (PUL). PUL affects and is closely related to learning and knowledge management of product development. Learning and PUL of NB HS NPD creates knowledge about the interaction and development of the market, technology, network, and competence component in the product development field. PUL chooses the right product development model and process and thereby the right mix of main components to "the product development field" to obtain right speed.

PUL is elaborated along with product development management (PUM) as seen in Figure 7.2 with the aim to achieve right time product development.

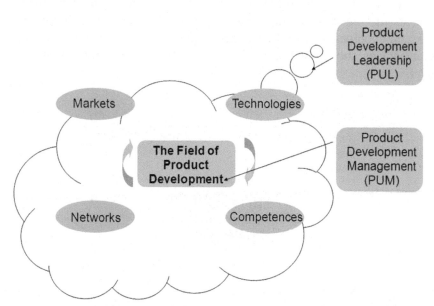

Figure 7.2 The field of product development.

PUL is the ability and the know-how of the "game of product development". PUL is to know **why** and **what** the "game of product development" is about to decide and perform; how the right speed in businesses product development activity should be.

On the PUL level focus is on CIM, CI, and learning to create long-term success criteria such as right cost, right performance, and right time. CIM is in focus because the product development process has to be improved continuously both initially within the product development process, across product development projects, and on the market place (Corso, 2001). CI is in focus because businesses have to innovate new products and seek innovation possibilities at the start of a product development project, along the product development process, and when the product has been introduced to the market.

Additionally, my hypothesis is that in order to reach their long-term success criteria the business must have a fundamental understanding of the field of product development. This understanding is closely related to PUL.

7.6.7 Understanding "Field of PD"

The main context and components in today's "product development game" can be characterised by the following "interactive picture" of the four main components in "the field of product development" – markets, technology, networks, and competences of the businesses playing in "the field of product development" as shown in Figure 7.3 on behalf of work and content in Table 7.4.

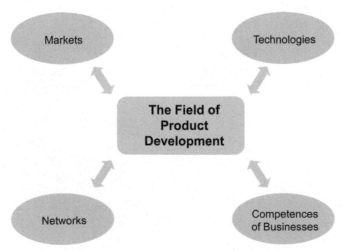

Figure 7.3　The contents and main components of the field of product development.

Table 7.4 Shape of main components in PD game

The Main Components Context	Characteristics	Example of Markets 2002
Market (Sanchez 1996)		
Stable markets	Stable market preferences	Food industry, Furniture industry
Evolving markets	Evolving market preferences	Agriculture industry, environment industry
Dynamic markets	Dynamic market preferences	Software industry, Bio and gene industry
Technology (Sanchez1996)		
Stable technology	Stable and known technologies	Audio and video technology
Evolving technologies	Evolving technologies	Bio – technologies
Dynamic technologies	Dynamic and mixed technologies	Nano-technology
Stable networks	Networks mainly based on physical and stable networks often internal and dominated network	Industrial groups, branch groups
Evolving networks	Networks based on a mix and evolving system of networks – Physical networks, ICT – networks, virtual networks	PUIN – network group, EU – community,
Dynamic network	Networks based on a mix of dynamic networks with high degree of dynamic where network partners constantly comes in and goes out. Often there is no formal network leader.	Virtual network groups, Ambias
Business competence context (Prahalad and Hammel, 1990)	Support competences Complementary competences Core competences	

An overview of the field of product development and the interaction of the 4 components to the field of product development is vital to decide further on in the product development process. A wrong analyse will perform a wrong decision of product development model and process along with HS enablers.

The interaction of market and technology in product development has been known for several years (Wind, 1973) (Myrup and Hein, 1986) (Eppinger, 1999).

What is new is the interaction that market and technology have with different types of networks and their relation to the competences of the businesses.

Furthermore, the main components can each be of different shapes both prior to the product development project and during the product development project when "the field of product development" is being analysed.

My claim is that until now many businesses have managed and developed a "blind" high speed product development strategy seeking speed, uniqueness, and innovativeness from a tactical point of view. Without reading the characteristics of the component in "the product development field". To a large extent, the management of product development has concentrated their efforts on the establishment of high speed product development and on being an innovative firm.

This was partly due to the tendency to regard such a strategy as the survival in future competition (Cooper, 1993) (Balwin et al, 1996). Unfortunately, the combination of high speed product development, uniqueness, and being the innovative firm has proved to be far more difficult to implement than originally expected (Bessant, 1999) (MacCormac, Verganti & Iansiti, 2002). Our initial case research showed businesses which have realised several failures and problems with a strong focus on high speed and cost reduction in NB PD.

> *"Our business has lost much money because of an expensive high speed product development approach. Our product was too early in the market and then the Internet came up and everything was turned around."* (Case No. 46 – ODI, UK)

> *"We developed several E-learning products with high speed, and we built up an international competence in multimedia production with high speed because all market and all expert signs indicated a future heavy demand for e-learning products. When the product development finished and our technology and competences were ready for the evolving market – the market did not evolve as we had expected and we had to close our e-learning activities."* (Case No. 52 – M2SIRE)

> *"We invested in a new particleboard surface machine for melamine production due to a strong market want and pressure from customers and sales. When we were ready for production, the market had turned to be minimal and covered by competitors."* (Case No. 53 – NOVDK)

My secondary case observations show that these businesses forgot to read "the field of product development". They pushed the product development too fast. They developed products which performed badly on the market because the main components were not in the right position for or had changed during the product development process. The speed of product development turned out to be either too slow or too high.

Therefore, I claim that it is necessary to focus on moving the product development activity in the business at a speed which matches the characteristics of the field of product development. I maintain that businesses should focus on right speed.

7.6.8 Management's Understanding of Incremental and Radical PD

My secondary case research showed that the case businesses were joining more unknown, dynamic and risky product development networks. The case businesses became very much dependent on and related to such types of networks and to the performance of such networks. Businesses act in this way because their product development competence often turns out to be too narrow and under a tremendous pressure for change, agility, and development. Therefore, because of "a perceived want from the product development field", many businesses force themselves into product development in uncertain and risky business areas where markets, technology, and networks are unfamiliar and where the product development task and innovation degree often exceed the competence of the businesses and turn out to be radical product development as indicated in Figure 7.4.

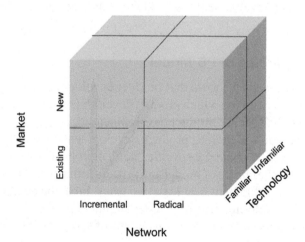

Figure 7.4 Incremental and radical PD.

Source: Lindgren & Bohn, 2002.

Therefore, in many cases the models and processes are not providing the businesses with right performance, right cost or, right speed because businesses and managers seem not to analyse and try to understand when product development are incremental and radical.

Furthermore, the learning gained by the businesses from several product development processes is rarely transferred to continuous improvement (CIM) or to continuous innovation (CI) in future product development processes of the businesses. This is because the product development processes are made radical or "one of a kind" every time, and because the knowledge transfer to other product development processes does not exist or is not formalised.

My case research shows that the main components of "the field of product development" "float" between being stable and dynamic. When markets, technology, network, and competences are changing constantly from stable to dynamic, which according to researchers will continue in the future (Fine, 1996) (Coldmann & Price, 1996) (Verganti, 2001), businessses are forced to look into new types of product development models and product development management tools. Businesses and researchers have realized that stage-gate models are effective for some product development tasks and situations but not when far more dynamic and flexible product development models seem to be more effective (MacCormac, Verganti, et al., 2001).

Therefore I claim that the management of PD must change the decision process of which PD model to use into a more agile and flexible decision process. Management must choose from many models. The choice must be related to the product development task and the characteristics on the field of product development.

7.6.9 Important Choice of PD Model

The secondary case research shows that product development managers face quite a difficult task of deciding which product development model and process are most suited for the specific product development task.

Before making such a choice, product development managers have to "read" and learn carefully about the specific "field of product development". It is our hypothesis that the choice of model and process influences the possibility of speed in product development in at least two ways. First, the height of speed which can be achieved during the product development project, secondly, the possibility to change speed and how much change of speed the business needs in the product development process.

It is verified that businesses can gain speed both in a stage-gate model and in a flexible product development model (Cooper, 1986) (McCormarck,

Verganti, Iansiti, 2001). However, the costs of changing speed differ in accordance with the chosen PD model and are influenced by the choice of product development model as indicated in Figure 7.5.

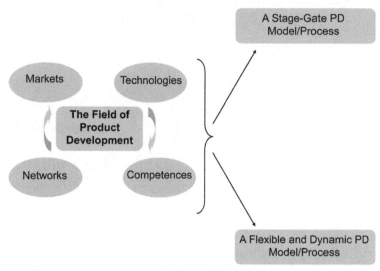

Figure 7.5   Choice of PD model and process.
Source: Lindgren & Bohn, 2002.

Firstly, the flexibility of speed possibilities is not the same in the two models. Secondly, the possibilities of performance of the final product differ from one model to the other in accordance with the point in time at which the change of speed in product development is required. Consequently, right performance, right cost, and right speed of " the field of product development" is very much related to the ability of the product development managements to "read the field of product development" both initially and as the product development progresses.

My hypothesis is that such a state of things is strongly related to learning. Additionally, managers of product development need to develop a strategic design of how to use NB HS NPD, to understand why NB HS NPD should be used, and to realise what NH HS NPD models and processes should be used. The strategic design of NB HS NPD has to be strongly related to product development knowledge and product development knowledge creation at the management level within the business. The managers of businesses have to learn about NB HS NPD and to develop and continuously improve product

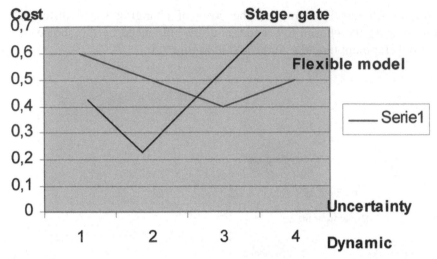

Figure 7.6   Costs of flexible and stage-gate PD models related to uncertainty and dynamic PD projects.
Source: Lindgren inspired by R. Verganti, 2002.

development leadership as indicated in Figure 7.6. Yet, learning in product development management is far from easy as my case research showed.

7.6.10 Change from Short-Term to Long-Term Success Criteria

It is my hypothese that businesses today acts on product development in a short-term way but will have to focus in the future more on management in a long term perspective – which means more focus on product development leadership".

Speed and time will be the issue for the future product development but seen in quite another perspective than today. In some product development cases, businesses will have time enough. It will be possible to do product development 24 hours a day in all places and with all types of networks. The limits will not be time and resources but the ability to find the right time and the right resources.

Different kinds of conditions on "the field of product development" will require different kinds of speed in network based product development. Strategies on speed in network based product development for different product development conditions will therefore be central. The ability to chose the right speed in which a product development process should run will strongly be related to learning and product development leadership. Successful businesses on the market will seem as acting with short-term success criteria but looking deeper into what is going on, a clear focus on long-term success criteria as shown in Table 7.5 will be the fact.

Table 7.5 NB HS NPD success criteria

NB HS NPD Success Criteria Short-Term Perspective	NB HS NPD Success Criteria Long-Term Perspective
High Speed – Time	Right Time – Right Speed
Cost	Right Cost
Performance	Right Performance
	Continuous Improvement
	Learning
	Continuous Innovation

The secondary case research has shown a rather "blind" focus on speed in product development. My hypothesis is that this forces sub-optimization and the development of informal product development models and processes to match the demand of speed. In future, product development managers must therefore focus more on product development leadership than on product development management. However, the hypothesis of this research is that this is still not the case in the major part of existing businesses. I want to verify this in my empirical research.

It is important for managers to realise that right speed is strongly related to learning in order to reach the long-term success criteria of the businesses – such criteria being right performance, right cost, and right time as shown in Figure 7.7.

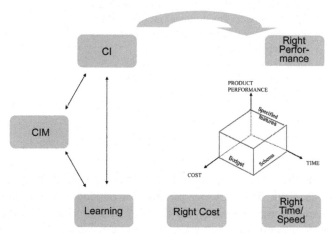

Figure 7.7 Relationship between long-term success criteria in network based product development.

Source: Lindgren & Bohn.

Learning demands knowledge and knowledge transfer. In this connection knowledge and knowledge transfer about possibilities, threats, strength and weaknesses of product development are required. This knowledge has to be made available in an open form and transferred before it can be used by other network partners to reach right speed in the businesses involved in the product development process. When product development leadership is implemented in businesses and learning interacts with CIM and CI, our hypothesis is that long term success criteria such as right performance, right cost and right speed as shown in Figure 7.7 can be gained in product development.

It is the hypothesis that the major part of businesses have not implemented learning and product development leadership. This will be verified in my empirical research.

I also claim that most businesses do not follow an analytical process of product development.

7.7 Important Issues in Future NB HS NPD

Taking all what has been mentioned above into our mind we are left with several important issues to be solved and to focus upon in future product development theory and praxis. These are as seen in Table 7.6.

Table 7.6 Important issues to future NB HS NPD

Important Issues of NB HS NPD in Future
The product development architecture
The product development process architecture
A new product development process model
Product development leadership (PUL)
Choice of high speed enablers
Choice and involvement in networks
Competence to "hub up on" networks
Trust and openness in network
Learning and dynamic learning
Long term success criteria
Analytical approach to NB HS NPD

On behalf of the above-mentioned important issues I propose new approaches to NB NPD. The existing practice of product development will not be able to match the demands on the field of product development. The product development models, product development management style and the existing focus on the success criteria of product development will not be able to give radical competitive advantage to SMEs. I claim this on the basis of Table 7.7 which shows what I believe is the major challenge to NB PD.

Table 7.7   Components and characteristics

The Main Components Context	Characteristics	Example of Markets 2002
Market (Sanchez 1996)		
Evolving dynamic markets	The markets becomes more evolving and dynamic in all markets	The textile line of business – Zara, Case No. 1, The mobile line of business The TDC case 57
Rapid cheap power technology	The technology gives us more possibilities and more technological power. The technology will be mixed faster and in a more dynamic and agile way.	Lyngsø Case No. 38, Mobilix, Case No. 40
Dynamic mixture of networks	Networks based on a mix of dynamic networks with high degree of dynamic where network partners constantly comes in and goes out. Often there is no formal network leader.	UK Chemical, Case No. 19
Increasing pressure on businesses competence	The pressure on businesses competences increase the businesses approach to perform more agile and flexible businesses where competences moves in and out according to the product development task	Rossflex, Case No. 11
The product becomes a process	The product turn into a process and the product becomes digital or virtual until "encapsulation"	Footshoe, Case No. 4; Scooter, Case No. 55

I therefore propose a new approach to NB NPD as elaborated in Chapter 8.

---

[i]Jianxin Jiao and Mitchell M. Tseng *A Requirement Management Database System for Product Definition*; Integrated Manufacturing Systems 10/3 [1999] 146–153.

PART III

Analysis Model

The analysis model part defines and elaborate the contents of the framework model for evaluating HS in NB PD. The analysis model finds the elements which the NB HS PD framework model must take into consideration. The analysis model also delimitates the focus of the research on network based high speed product development. This includes which main questions and main hypotheses for network based high speed product development should be finally verified.

This part sets the framework fundament for the empirical research.

8

Analysis Model of NB HS NPD

As demonstrated in the previous chapter, Network based HS in NPD processes is multi-dimensional and can be made operative in many different ways. A framework model for a Network Based HS NPD process must allow for this.

The first objective of this chapter is to create such a framework model for evaluating HS in NPD.

The chapter begins by drawing a picture of the elements which a framework model must take into consideration when discussing NB HS NPD.

From this point of view the elaboration of the framework model for NB HS NPD that can include the above-mentioned aspects will be made.

8.1 Elaborating Network Based HS to NPD

8.1.1 Introduction

The research takes its starting point in the analytical foundation of "the four Ps". According to this model the Process is defined, the Procedure is defined which means that the manual of the partial processes is uncovered. Subsequently, the People are described. This means that the actors taking part are described. Thirdly, Project management is described to determine who is actually leading or managing the process. Finally, the Point of entry is laid down in order to determine the player/players who initiate the partial process (Plat, 1999).

My hypothesis in 2003 continued to claim that the achievement of goals during PD processes and partial PD processes is only effective provided that decisions made during the process have been defined as depending on right speed, perceived value, cost both direct and alternative cost and thereby create the right performance.

As we have seen, the network based HS framework model had therefore to be defined and related to a number of main elements.

1. The product development task
2. The main components on the field of product development
3. The management of the product development project
4. The success criteria of the NB HS NPD
5. The product development model and process
6. The HS enablers to NB NPD

My proposal to the decision or analysis model for NB HS NPD was as follows:

The first two elements are generic and can be defined by different tools already known. The last four items are more dynamic and have to be defined and classified carefully to the specific situation. Different tools and frameworks can be chosen by product development management. It was my hypothesis that this choice would define and decide the speed of a product development project and would consequently define the top speed of a product development project. When PD management carefully choose the above, they would "hit" the right speed of PD.

The last four elements are elements which the management could choose to use and define differently from one product development project to another. The management could also choose to use the four elements in different combinations to different PD task.

For each element, the management could also choose to use sub-elements e.g. the HS enabler either alone or in combinations. This meant that the management of product development face the challenge of analysing the task and "the field of PD" and combine both the four main elements and also single sub-elements in different combinations.

The product development management thereby became more complex and must take into account factors as prevailing nature of NPD work (incremental/ radical NPD), time, cost, performance, culture, personnel, history, organisation, network etc.

The PD management became central to the definition the right architecture on behalf of the PD analysis of the PD task and the field of product development.

8.1.2 Framework Model for NB HS NPD

On behalf of the above-mentioned the overall framework model for NB HS NPD was elaborated as shown in Figure 8.1.

First, the manager must analyse and understand which type of product development task the business is facing. Secondly, managers must understand the position and characteristics of "the field of product development". Thirdly, the managers must decide on the success criteria for the specific product

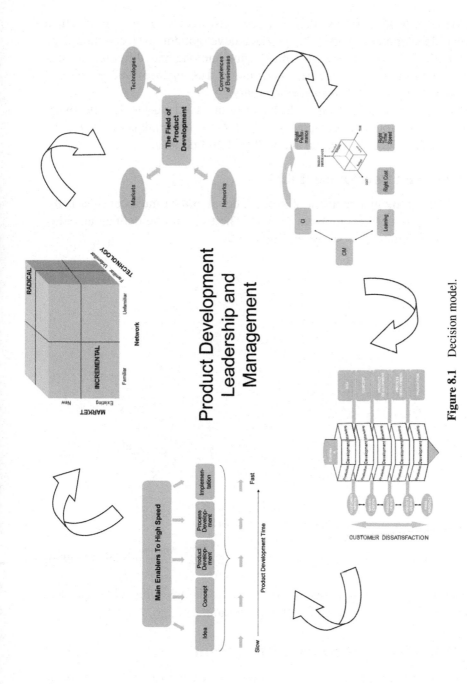

Figure 8.1 Decision model.

development task. Fourthly, the manager must decide according to which product development model the product development projects should be developed. Finally, the manager must chose among the different product development high speed enablers and mix of HS enablers and in the end implement the product development activity.

However, the management of the business involved must realise that this is not a one-time product development project. Once implemented, the product development process has begun and will seldom find an end.

8.1.3 Product Development Task

The first thing the management of product development must consider is the task of the product development project. The task can be defined as either radical or incremental as shown in Figure 8.2.

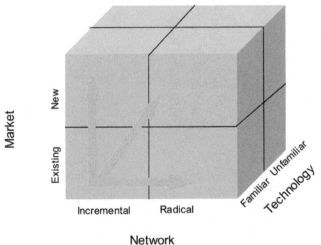

Figure 8.2   PD task model.

The empirical data will try to verify where to place the product development task.

The next thing the framework model tries to define is the field of product development explained in the following paragraph.

8.2 "The Field of Product Development"

The different main elements were considered as the basic minimum set-up to a PD project and require a careful analysis if network based HS is to be successfully managed, developed, and sustained to the extent where it can confer strategic competitive advantage to the business. "The field of product development" has to be analysed carefully by the management level. In the research a verification of the following questions will be done.

On the basis of the previous chapters I claimed that the main context and components in "product development game" up to 2003 could be characterised by the following "interactive picture" as shown in Figure 8.3 of the four main components on "the field of product development" – markets, technology, networks, and competences of the businesses playing in "the field".

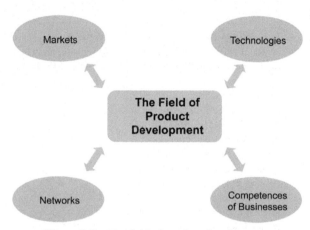

Figure 8.3 The field of product development.

Each of the main components can be of different shapes both prior to the product development project and during the product development project when "the field of product development" is being analysed. The main components interact continuously in the field of product development. In the empirical research verification were made in accordance with the following dimensions as shown in Table 8.1:

Table 8.1 The shape of the main components in the product development game

The Main Components Context	Characteristics	Analysis Tools
Market – (Sanchez 1996)		
Stable markets	Stable market preferences	Market analysis on behalf of Sanchez' characteristics
Evolving markets	Evolving market preferences	
Dynamic markets	Dynamic market preferences	

(Continued)

Table 8.1 Continued

The Main Components Context	Characteristics	Analysis Tools
Technology (Sanchez1996)		
Stable technology	Stable and known technologies	Technology analysis on behalf of Sanchez' characteristics
Evolving technologies	Evolving technologies	
Dynamic technologies	Dynamic and mixed technologies	
Network (Håkonson & Håkonson; Child and Faulkner, 1999)		
Stable networks	Networks mainly based on physical and stable networks often internal and dominated network	Network analysis on behalf of Håkonson and Håkonson and Child and Faulkner's characteristics.
Evolving networks	Networks based on a mix and evolving system of networks – Physical networks, ICT – networks, virtual networks	
Dynamic network	Networks based on a mix of dynamic networks with high degree of dynamic where network partners constantly comes in and goes out. Often there is no formal network leader.	
Business competence context (Prahalad and Hammel 1990)	Support competences Complementary competences Core competences	Analysis on behalf of Prahalad and Hammels characteristics

8.2.1 Success Criteria for PD Project

Successful businesses will have to decide on both short-term and long-term success criteria in their product development activities.

The research wanted to look into what was going on in businesses and gain a clear view on which short- and long-term success criteria businesses used. The results verified by my research are presented in the next part of this book.

As a minimum long-term success criteria and short-term success criteria as shown in Table 8.2 must be:

1. hierarchically
2. quantitatively
3. realistic
4. consistent

Table 8.2 Success criteria

NB HS NPD Success Criteria Short-Term Perspective	NB HS NPD Success Criteria Long-Term Perspective
High Speed – Time	Right Time – "Right Speed"
Costs	"Right Costs"
Performance	"Right Performance"
	Continuous Improvement
	Learning
	Continuous Innovation

Firstly, when businesses have analysed and decided which criteria should be in force, I assert that the product development model can be chosen.

8.2.2 NB HS NPD Framework Model

The NB HS NPD framework model was "built up" with different elements:

1. The point of entry
2. The core of the NB HS product development framework model
3. The internal framework model for NB HS NPD
4. Functions involved in the framework PD – model
5. The process in the NB HS NPD.

Point of Entry

In the framework model it was important to verify the point of entry for new ideas. Moreover, it should be determined how optimal this point of entry was to support the businesses and the markets' goals of speed of product development. Table 8.3 shows the different points of entry which the product development framework of a business can have.

Core of Product Development Task

The (ideal) specific NB HS PD project will consist of a PD core which in ideal situations and at the point of departure (t0) defines the overall strategic and tactical framework for the object of the product development.

The core also defines the strategic limits, the costs, the product architecture, the expected performance of the product, and the initial framework of the product development project. It must be realised that the core can be changed because of changes in the field of product development. The core of the product development project or task must therefore have a somewhat flexible design

Table 8.3 Points of entry

Sources to Product Development Ideas in General	In Per Cent Point of Entry
Customers	
Suppliers	
Marketing	
Finance	
Sales	
Leadership/Management	
Production	
Product Development	
Human Resources	
Competition	
Others	
Do not know	
Total	

but a strong product architecture to match the interaction, demands and change from all other main components in the field of product development.

It was the hypothesis of this research project that in the course of time, the core of the product development task will progress from c_0 to c_1. Consequently, my hypothesis was that the core of an NB HS PD was not necessarily static but can have a dynamic character which was influenced by all other elements in the frame work model; especially the elements on the "field of product development".

Furthermore, it was the hypothesis of this research project that the separate components of the core will develop differently over time so that some parts may at a given time be much developed whereas others were less developed or not at all developed at the beginning of the product development. It was also the hypothesis that occasionally, the core of the product development project cannot be defined because the product development project was so radical that the major part of the core could only be defined later in the project. Consequently, my hypothesis was that in some cases the mission of the product development project could be developed initially whereas in other cases it could not.

The strategic limits of the core define the limits to which the product development is subjected. Such limits are the limits of the business, the product, the technology, the market, the customers, and the competition. Additionally, the internal strategic and tactical limits as e.g. financial, organisational, and technical limits were defined. The network boundaries also define the boundaries allowed for the business's actors in the product development project. Such limits also determine the internal involvement and possibilities of the product development group as well as the external cooperation with network partners.

As previously mentioned my hypothesis was that the core of the product development project was often dynamic as e.g. resources can be drawn out or put in depending on developments in mission, goals, and strategy.

In this connection it was the hypothesis that it was crucial for the speed in NB PD:

- To carefully define the core of the pencil (mission, goals, strategy, resources, access to network partners) initially to the PD project
- To maintain, if possible, the main architecture of the core throughout the course of the development process
- To adapt the core to the specific situational position and characteristics in " the field of product development"
- To communicate and agree upon the core with all network actors; both internal and external actors should take part in the development of the product development
- To explain the core to all actors of the product development process. My hypothesis is that this is an important activity which will secure motivation, high involvement and thereby high speed in product development
- To "tailor" the core to suit the specific product development success criteria – in terms of time and right time

In the case of network based high speed product development it was my hypothesis that the core of NB HS NPD was crucial because it would decide how fast a product development process could go. The core will, however, also be influenced by the different cores of the network partners.

Subsequently, a common network core will have to be formulated in a product development project of NB PD. This core will be a common core to the network actors of the specific development project.

The formulation of the core may give synergic effects or conflicts as the network core can clash with the product development core of the individual network actors or businesses. Such clashes may be due to conflicts of:

- Mission
- Objectives
- Strategy
- Business resources
- Other actors or boundaries to network actors

It was the hypothesis that this was one of the major reasons to why it was difficult to perform high speed in network based product development.

This was tested by case, focus group, survey and by joining other research and product development projects. The following questions was central as listed in Table 8.4:

Table 8.4   Elements of core

Elements of the Core in NB HS NPD	Hypothesis to Be Tested
1. Mission	Does the core in HS NB NPD consist of
2. Objectives	the elements mentioned 1 to 5?
3. Strategy	When is the elements formulated in
4. Business resources	relation to the product development
5. Other actors boundaries to network actors	process?
	Can businesses which have formulated
	or formulate the core of NB HS NPD
	run with higher speed in PD than
	businesses with out a formulated core?

Furthermore, my hypothesis was that it was very important that each business has fully decided on their individual and their common core before they joined a NB HS NPD. It was also important that the network partners trust each other in this process and elaborate the core with trust, fullness, openness, and real input to the core. My hypothesis was that otherwise the product development project would face problems at a later moment, and speed will be diminished. However, this aspect was not of primary interest to this research project.

Internal Framework Model for PD

The framework model for NB HS NPD toke its starting point in the assumption that a product development project inside a business consists of various stages and gates. This hypothesis was supported and confirmed by both theory and by the majority of the secondary business cases analysed and examined. My hypothesis was that most businesses until 2003 had not adapted other types of models such as flexible models or rapid prototype models. The secondary case research of several authors confirmed this product development picture (Wind, 1973) (Clark and Wheelwright, 1992) (Myrup and Hein, 1986) (Cooper, 1993) (Eppinger, 2000). I wanted to verify this hypothesis empirically to verify the conditions of NB HS PD.

The hypothetical model for the NB HS NPD model and the analytical framework for the empirical research were therefore outlined in detail in the following Figure 8.4 and paragraph.

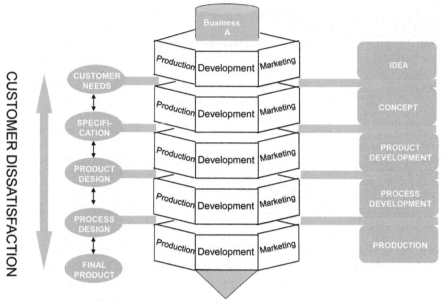

Figure 8.4 Product development pencil model.

Source: Bohn & Lindgren, 2000.

The PD framework model consists of a core surrounded by the potential functional areas which can be represented in the different stage and gates of the product development process. The product development model is shaped as a pencil and is called "the pencil model". The PD model consists of a process from before the formal idea stage to "on the market" stage. The PDD model consist of 7 stages and 6 gates.

The various stages and gates were defined as shown in Table 8.5.

Table 8.5 Definition of stages and gates

Stage and Gates	Definition	Activity
The pre idea stage	Actors network, discuss on an often informal basis	Networking
The idea stage	Actors find an idea and define it	Idea generation
The Idea gate	Actors screen the ideas	Idea Screening
The concept stage	The ideas are conceptualised	Concept generation
The concept gate	Actors screen the concepts	Concept screening
The Product development stage or the "prototype stage"	The concepts are transformed into prototypes	Prototyping

(Continued)

Table 8.5   Continued

Stage and Gates	Definition	Activity
The Prototype gate	Actors screen the prototypes	Proto type screening
The Process stage	The process for production of the new product is developed	Process development
The Process gate	The final product is tested against quality standards and the process is screened against success criteria	Process screening and quality screening
The Market implementation stage	The new product is introduced to the market	Market implementation
The Market evaluation and control gate	The market evaluate the product. The business control the performance of the product. Proposals for adjustments turns up.	Market evaluation and control

As previously mentioned one of the primary points of criticism of the stage and gate models up to 2003 was that such models give the impression that product development passes from one phase to another, and that product development cannot proceed until the previous phase has been completed. Several examinations had shown that when the businesses conform to the above-described principle, the speed of product development will be dramatically reduced.

The secondary case examinations showed that the businesses apparently operate in their formal NPD models with different stages and gates. However, several cases indicated that the stage-gate models were framework models which were merely used when the businesses presented their product development models to an audience. Therefore I proposed a more agile and flexible product development model with a structure that gave possibilities of dynamic and free choice of stages and gates related to the characteristics on the field of product development and the product development task. Yet, the above-mentioned scenario had the recurrent characteristic of the product development models found in the literature as well as in the case examinations which formed the basis of my hypothesis model.

This meant that all elements were variables and businesses can choose to:

- Attract ideas from all stage- and gates in the framework model
- diminish the product development model – jump stage and gates
- increase or diminish the stage and gates
- choose to develop through all phases or solely do "on the market" product development

- do network product development in all stage in the product development process

The above was a hypothetical model to NB HS product development. In my empirical research I wanted to verify the shape of this product development model.

Functions Involved in Framework PD Model

The framework model intended to establish an analytical framework which could identify the departments and functional areas which participate in the various stages and gates.

The framework model outlined above linked the classic stage-gate model by including several departments or functions of the business in the stages and gates. In this way the stage-gate model was combined with the department-stage model perception in terms of an analysis framework which toke into account the identification of the participation of various departments/ functional areas in the various stages and gates.

In the framework of the PD model marketing and sales were divided into two separate departments. The theoretical study verified that there was considerable difference between the sales and the marketing approach. Furthermore, it was the hypothesis that the marketing function becomes involved in the product development process at an earlier stage than the sales function.

Additionally, I had included three additional departmental and functional areas; namely finance and economy, and HR. These departments or functions had been included because my initial study of cases and literature emphasized the importance of addressing the involvement of such functions and departments at an early stage of the product development phase in order to achieve high speed in the course of the product development.

On the face of it, the management of the business was not represented in any of the above models. However, we had observed via the case analysis that management was a function which was represented either directly of indirectly via e.g. the core function or together with other functions. Consequently, the management function had been singled out as an independent function the result of which was that a total of seven functions were represented.

This meant that the analysis framework model contained seven potential functional areas which had been included in the analysis framework outlined in Figure 8.5.

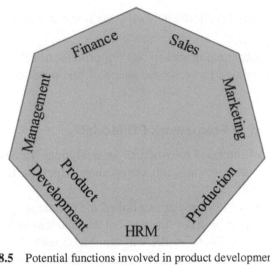

Figure 8.5   Potential functions involved in product development model.
Source: Lindgren, P., 2002.

The functions were defined as shown in Table 8.6.

Table 8.6   Internal functions of PD

Functions Involved in Product Development Model	Definition of Function	Function Activity in Product Development Process
Marketing function		-
Sales function		- ....
Product development function	-	
Finance function		- ....
Production function		- ....
HRM Function		
Management function		

As have appeared at a later point in this research the secondary case examinations have shown major differences from one business to another and from one product development course to another when examining which departments toke part in the various stage-gates. I wished to verify this observation further in my primary research. It was also worth noting that some functions can be represented in one participant, e.g. marketing and sales, production and product development, or economy and HR. All secondary cases were evaluated according to the tables shown in Table 8.7 and I wanted to evaluate the primary cases, focus group participants and the survey participants also on the basis of this framework.

Table 8.7 Evaluation of cases

Product Development Functions	Represented in One Person or Function								Represented Together with Another Function by One Person							
	Product Development Phase								Product Development Phase							
	Product Development		Process Development		Production				Idea Concept		Product Development		Process Development			
	Participant								Participant							
	Idea	Concept							Idea	Concept						
	I	IG	C	CG	PS	PG	PRC	M	I	IG	C	CG	PS	PG	PRC	M
Product Development																
Marketing																
Sales																
Economy/Finance																
Production																
Human Resource																
Management																

I   = Idea Stage
IG = Idea Gate
C = Concept Stage
CS = Concept Gate
PS = Product Development Stage or Prototype Stage
PG = Product Development Gate or Prototype Gate
PRC = Process Development
M = Market Introduction and Production
Shaded area is the focus of this PhD project.

The framework model intended to verify at the separate stages and gate which product development functions were involved related to our framework model stages and gates. This research project mainly focused on the very early stages and gates of the NPD model; i.e. typically the idea and concept phase as well as the screening phases belonging to these phases. The reason for this focus point was i.a. our hypothesis that speed in product development course was determined at the very early stages and gates, viz. in the idea and concept phase. Therefore, it was interesting to verify the participants in the initial phases and to verify their impact to speed in product development.

However, it had become apparent through the secondary case research that several businesses did not follow their stages and gates diligently. Furthermore, the stage and gate models were being criticised by the case businesses and researchers for not being sufficient when it came to describing the course of flexible and dynamic product development (Verganti, 2001). Thus, the product development models of several businesses were described as being integrated into and working simultaneously with each other (Sarens, 1984) (Hart et al., 2000) and also much more process oriented than envisioned until now.

Also the chain of events from idea to prototyping was questioned as several cases showed that the businesses employ a much higher degree of "prototyping" than previously imagined.

Therefore, the actual concept phase may be discontinued or distinctly shortened (Verganti, 2001). The empirical research will try to verify this hypothesis.

It is worth considering that the course of a product development was considered from a process oriented point of view (Corso, 2001).

The classic stage-gate models do not allow for "integration players" (network partners) in the NPD process as e.g. suppliers, customers and similar players (Biemans, 1992) (Hard, 2000) – in other words inter-organizational networking. This is mainly due to an inside out view of product development and not an outside in product development view. This lack was covered by the framework model outlined in the hypothesis model beneath.

The reason why product development models often do not follow the course of events prescribed by the theory up to 2003 may be found in the centre of the framework of the project – generally called the core of the product development model. The core sets out the framework and the environment which the product development project and its actors must address and with which it must comply.

If "the Core" allows for or even demands integration with network partners, the road is paved for interaction between network partners. Such interaction is illustrated in Figure 8.6.

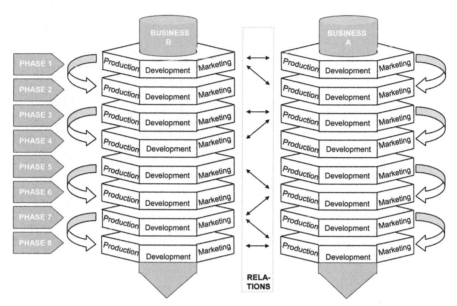

Figure 8.6 Framework model for network based product development.
Source: Bohn & Lindgren, 2000.

As the research project was focusing on network based NPD. I had developed the model further as can be seen in Figure 8.6 (Bohn & Lindgren, 2000).

The network perspective in product development had not been in particular focus and had not been particularly developed until now. On the basis of the above discussion of the NPD models, this research project puts forward an analysis framework based on an integration of network actors in the product development model and process. The above-mentioned model showed a very simplistic framework model with only few network actors. The hypothesis was that each network actor had their own product development model and process running inside their individual SME. These product development models and processes were linked and interact with each other via relations. Their interaction during the common product development model and process of the networks forms the final product. This is shown in Table 8.8.

By means of the aspects outlined in Table 8.8, my empirical research endeavoured to verify the network interaction and the network partners' participation in the SMEs' product development projects.

Table 8.8 Network partners' participation in SME's PD

	Customers			Suppliers			Competition			Other Network Partners		
	Y	N	ISC	DN	Y	N	ISC	DN	Y	N	ISC	DN	Y	N	ISC	DN
Idea Generation												
Concept Generation												
Product Development												
Process Development												
Phase												
Idea Screening												
Concept Screening												
Proto Type Test												
Process Test												

Y = Yes
N = No
ISC = In some cases
DN = Do not know

Which stages and gates are network actors involved in and which stages and gates are characterised by much interaction? What influence does this have on speed in product development? I want to examine the degree of network interaction in order to discover the potential of NB HS NPD and the impact on NB HS NPD. Furthermore, I wanted to learn who takes part in the management of the NB HS NPD projects.

Process in NB HS NPD

When dealing with the empirical data an identification of the four Ps in each separate process within the product development process was attempted in order to verify the processes employed by the businesses. The extent to which such processes were used in the idea and concept phase with the object of speeding product development was also examined.

In the secondary case research and literature study I observed dynamism in functions involved in the product development process. The hypothesis to be verified was whether some functions could change and change their involvement in the product development process from idea to market introduction. The empirical research wanted to verify how the above-mentioned circumstances influence speed in the product development.

In this way the empirical data would try to verify and comment on the following areas:

1. Interaction and change in involvement of functions during the PD process
2. Change in involvement of functions between phases in the product development phases
3. Functions involved between networks and change of functions within the network in the product development process
4. Impact on speed in PD by actors in the network

8.2.3 HS Enablers to NB PD

Which enablers are appropriate and in what form, will depend on the analysis of the characteristics of the field of product development and of the product development task. On the basis of the above a need arises for examining the high speed enablers of network based product development. Central aspects to involve were questions of:

1. What HS enablers are used by businesses
2. How is the ten high speed enablers used and incorporated in the product development process?

3. How do SME achieve high speed in the internal and external product development networks without causing first mover disadvantage?
4. Which types of speed in network based product development can be verified.

Thus, by empirical data the research wanted to verify the high speed product development process in industrial businesses. The process will be defined as a series of partial processes/activities in which internal connections were determined, in which each separate enabler is identified, and in which their contribution to the overall goal achievement – costs, performance, and speed – was uncovered.

PART IV

Empirical Results

This part presents the empirical results of the PhD project. The presentation will be divided into four chapters. Chapter 9 will present the result of the pilot case study in the primary case businesses. Chapter 10 will present the results of the network meetings in the PUIN group documented in the PUIN book on product development in network. Chapter 11 will present the results of the survey analysis made by the questionnaire tool *Question Mark* on SME businesses. Chapter 12 will present the results and observations of other research activities on NB HS NPD.

9

Pilot Case Study

This chapter presents the empirical results of the pilot case studies in the research project. The presentation will be divided into four parts. First, the case businesses will be briefly presented in general terms. Secondly, an analysis of each business's NB HS NPD activities will be elaborated on behalf of the NB HS NPD framework model presented in Chapter 8 and the research questions set up in Chapter 1. Thirdly, a specific NB HS NPD project for each case business will be presented and analysed also within the above-mentioned framework model for NB HS NPD. Finally, a cross case business analysis focusing on NB HS NPD in the first stage and gates of the product development process will be offered to summarise on the empirical results of the chapter.

9.1 Empirical Results of Primary Case Research Evaluating HS in NB PD

9.1.1 Introduction

On the basis of 10 exploratory semi-structured case studies carried out in 5 Danish and British businesses, the empirical case results can be presented. The case businesses were illustrated in Table 9.1.

Table 9.1   Business facts on case businesses

Business	Line of Industry	Employees	Turnover
Lyngsø Industries	Software Industry	220	120 mill. DKK
The Language Centre	Multimedia – E-learning industry	6	4 mill. DKK
AKV Langholt	Food Industry	120	70 mill. DKK
Lindholst	Machine Industry	55	120 mill. DKK
GSI Lumonics	Welding machine industry	180	250 mill. DKK

All businesses were business-to-business businesses operating on global markets.

The businesses represent a variety of industries and a variety in numbers of employees, turnover, range of products etc.

The businesses were interviewed with a semi-structured questionnaire framework an example of which can be found in appendix to this book. In Chapter 2 the methodology of this case research was described in details. The case analysis was supported by questions filled out by the businesses on an Internet-based survey tool and additionally supported by telephone interviews on specific questions. These questions can be seen in appendix.

The businesses' general profiles can be found in appendix together with their websites. The individual cases will be documented and analysed within the framework model of the network based high speed product development prepared in Chapter 8.

The aim of the chapter is:

- to verify, test and give answers to the research hypotheses and questions set up earlier in Chapter 1
- to show and verify different NB HS NPD models and processes carried out under different characteristics on the field of product development
- to show different SME businesses solutions to NB HS NPD
- to reflect on which consequences high speed and right speed would have on different parameters as shown in Table 9.2

Table 9.2   Consequences of highs speed on parameters

Consequences	High Speed	Right Speed
Time		
Cost/Value		
Performance		
Market Fit		
Risk		
Security		

In Table 9.3 the contributions to the research questions of each cases are shown.

9.1.2 The High Tech Industry Business

LYNGSØ Industri (LYNGSO Industries) is a leading software manufacturer and systems integrator of a wide range of logistics solutions based on automatic

Table 9.3 Hypotheses to be verified in Chapter 9

Empirical Results – Pilot Case Studies	Hypothesis to be Verified and Tested	Lyngsø Industries A/S (Lyngsø)		The Language Centre (TLC)		AKV Langholt (AKV)		Lindholst (Linco)		GSI Lumonics (GSI)	
		General PD	Specific PD	General PD	Specific PD	General PD	Specific PD	General PD	Specific PD	General PD	Specific PD
Overall Research Questions to be Verified											
What is network based high speed NPD?	NPD can be seen from different views (macro environment, business, product, market, customer, technology, competitive, and network view)	X	X	X	X	X	X	X	X	X	X
	HS NPD is a matter of right speed and not high speed		X	X	X		X		X		X
What enablers to NB HS PD can be identified?	Businesses use different HS enablers	X	X	X	X	X	X	X	X	X	X

(*Continued*)

Table 9.3 Continued

Empirical Results – Pilot Case Studies	Hypothesis to be Verified and Tested	Lyngsø Industries A/S (Lyngsø) General PD	Lyngsø Industries A/S (Lyngsø) Specific PD	The Language Centre (TLC) General PD	The Language Centre (TLC) Specific PD	AKV Langholt (AKV) General PD	AKV Langholt (AKV) Specific PD	Lindholst (Linco) General PD	Lindholst (Linco) Specific PD	GSI Lumonics (GSI) General PD	GSI Lumonics (GSI) Specific PD
Overall Research Questions to be Verified	HS enablers are identical to the 10 enablers – 1–10	X	X	X	X	X	X	X	X	X	X
	There can be more than these 10 enablers to HS PD	X					X		X		
	The enablers will play a different role according to the PD situation and project (Secondary focus)		X		X		X		X		X
	The customer enabler, the network enabler, and the PD model enabler plays an important role in the upper phase of the HS PD + phase	X	X	X	X		X	X	X		X

Question	Statement								
What framework models and processes in the idea and concept stage/gate of HS PD based on networks can be measured?	The HS PD projects can be divided into radical and incremental PD projects	X	X	X	X	X	X	X	X
	The radical and the incremental PD projects follow different generic HS PD models and processes and can thereby be described by different generic frameworks	X	X		X	X	X		
What success criteria can be used for measuring HS PD based on networks?	The success criteria for HS PD are dependent on the specific PD project – radical or incremental.	X	X		X	X	X		
	HS PD success criteria can be formulated as short term and long term success criteria	X	X		X	X	X		

(Continued)

Table 9.3 Continued

Empirical Results – Pilot Case Studies		Lyngsø Industries A/S (Lyngsø)		The Language Centre (TLC)		AKV Langholt (AKV)		Lindholst (Linco)		GSI Lumonics (GSI)	
Overall Research Questions to be Verified	Hypothesis to be Verified and Tested	General PD	Specific PD	General PD	Specific PD	General PD	Specific PD	General PD	Specific PD	General PD	Specific PD
	Time, costs, and performance are central success criteria in a short-term perspective	X		X	X	X	X	X	X	X	X
	Continuous improvement (CIM), continuous innovation (CI), and learning are central success criteria in a long term perspective so reach right time, right cost and right performance in NB HS PD.	X		X		X		X		X	

identification equipment. LYNGSØ Industries supplies solutions for track & trace, automation, planning and visualization of processes and assets in logistics and supply chains. For more details please see appendix. and website www.lyngsoe-industri.dk

The Lyngsø product portfolio was strongly focused on physical and service products as shown in Table 9.4. However, on the new product introduction a strong focus on knowledge and consultancy products were seen.

Table 9.4   Focus on product types

	Physical Products	Service Products	Knowledge and Consultancy
Existing Product Portfolio	40	50	10
New Products	10	20	70

This could be seen as a reaction to the market needs and wants but also as a penetration of the existing product potential within the business. When looking at Lyngsø's products 40% of the products were physical products, 60% were digital products, and there were no virtual products as shown in Table 9.5. The business had 20% on physical processes, 70% on digital processes and 10% on virtual processes.

Table 9.5   Focus on products and processes

	Physical Products	Digital Products	Virtual Products
Existing Product Portfolio	40	60	0
	Physical Processes	Digital Processes	Virtual processes
	20	70	10

Lyngsø's customers were a wide range of businesses within the transport sector, public and private service sector, as well as the sector of trade and industry. These were all customers with varied, unique requirements and wishes for their particular logistics solutions.

Lyngsø used special system technology to streamline the daily efforts of more than 1,100 businesses worldwide.

Lyngsø had a wide range of networks both on customer, supplier and other organisational institutes side. Lyngsø was very network oriented.

Lyngsø had competences within solution that ensure the customers the optimum flow of goods and utilisation of materials, information and human resources.

Product Development Tasks

On the basis of the case research, Lyngsø's task of product development could be defined as seen in Table 9.6.

Table 9.6 Lyngsø's product development task

	Physical Products	Service Products	Knowledge and Consultancy
Existing Product Portfolio	40	50	10
New Products younger than 1 year	10	20	70
Product Development	40	30	30

Of the business's product development tasks 40% could be related to hardware or physical products whereas 60% could be related to service and knowledge products. Obviously, Lyngsø's product portfolio mainly focused on physical products and service products as seen in Figure 9.1. However, the

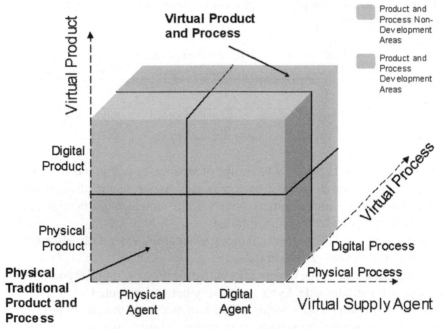

Figure 9.1 Lyngsø's product and process development matrix.

introduction of new products and the product development efforts were now strongly focused on knowledge and consultancy products.

The Lyngsø business claimed that product development projects could in general be divided into 40% strategic known and old areas and 60% unknown and new areas as seen in Table 9.7.

Table 9.7 Product development projects in relation to strategy

	Strategic Areas	
	Known and Old Areas	Unknown and New Areas
Total Average	40	60

When applying the product/market model to the product development projects of Lyngsø, it appeared that product development projects at Lyngsø could generally be characterized as radical product development as seen in Figure 9.2.

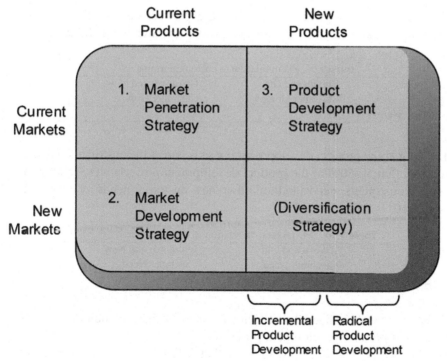

Figure 9.2 PD task at Lyngsø.

Additionally, Lyngø's product development projects were mainly development of old products more than three years old.

The case research also showed that the major part of product development projects on old products which were older than 3 years need big adjustments as indicated in Table 9.8. The data showed evidence of previous statements about a diminishing product lifecycle. In an overall perspective, 55% of the business's products needed big redevelopment after 1 year's lifetime. This indicated high pressure on radical product development at Lyngsø Industri.

Table 9.8   Product development in relation to product

	Old Products More than 3 Years with a Need for Small Adjustments	Old Products More than 3 Years with a Need for Big Adjustments	New Products Older than 1 Year with a Need for Small Adjustments	New Products Older than 1 Year with a Need for Big Adjustments
Total Average (%)	25	35	20	20

The survey also showed that 85% of the product development in the business was on known and old customer groups as indicated in Table 9.9.

Table 9.9   PD in relation to customer groups

	Known and Old Customer Groups	Unknown and New Customer Groups
Total Average	85	15

Looking at the product development projects and at the customers' needs, we realized that 50/50% of the product development projects were related to known/old customers' needs and unknown/new customer needs as indicated in Table 9.10.

Table 9.10   PD projects in relation to customer needs

	Known and Old Customer Needs	Unknown and New Customer Needs
Total Average	50	50

This once again indicated that Lyngsø was a business which dealt with rather radical product development projects.

On the technical level Lyngsø claimed that 30% of their product development projects involved new technology – radical technology areas and 70% of their projects were carried out in known areas or in development areas in which small, incremental technology adjustments were necessary as shown in Table 9.11.

	Known Technology	Known Technology with Small Adjustments (Incremental Technology)	Completely New Technology (Radical Technology)
	Table 9.11 PD projects in relation to technology		
Total Average (%)	40	30	30

The case research also showed that 50% of the product development project were related to market areas with fierce and rival competition as seen in Table 9.12. However, a large amount of projects were carried out in low competition areas. This was due to Lyngsø's high degree of radical product development.

	Table 9.12 PD projects in relation to competition		
	Markets with Low or No Competition	Markets with Medium or Intensive Competition	Markets with Fierce and Rival Competition
Total Average (%)	30	20	50

Lyngsø consider 50% of their product development projects as having a high element of innovation as shown in Table 9.13. This means high pressure on Lyngsø's competences. This also designates the radical element of the product development projects at Lyngsø.

	Table 9.13 PD projects in relation to degree of innovation		
	No Degree of Innovation – Routine NPD Project	Medium Degree of Innovation – Modified Product Development with Minor Demands on Adjustment (Incremental)	High Degree of Innovation – with Many Elements of Innovation (Radical Innovation)
Total Average (%)	25	25	50

Consequently, on the basis of the case research carried out at Lyngsø, the following general framework picture of the field of product development for Lyngsø Industri A/S could be verified.

Lyngsø were generally in interaction with all components on the field of product development from and outside in perspective. The field of product development was not under high pressure by any of the components in the field of product development. Many of the product development projects were in projects rather radical and new to the market. Both the market, the technology and the network components offered Lyngsø many opportunities as indicated in Figure 9.3. Lyngsø's challenge was to find the right PD projects and develop them within the right time and to meet and develop Lyngsø's competences to the opportunities in the field of product development.

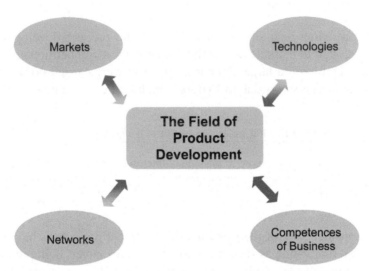

Figure 9.3 The interaction on field of product development at Lyngsø Industri.

General Product Development at Lyngsø

The case research showed the following general characteristics of the product development at Lyngsø as shown in Table 9.14.

Table 9.14 Sources of PD ideas in general

Sources to Product Development Ideas in General	In Percent
Customers	36, 4
Suppliers	9,1
Sales	27,3
Leadership/Management	9,1
Production	4,5
Product Development	9,1
Competition	4,5
Total	100

The product development ideas came mainly from the customers (36,4) and from the sales function (27,3%). This indicated that the product development of the business was generally strongly based on identified needs and wants in the market, and the business was therefore strongly customer/sales oriented.

The Core – Goals and Limits for NPD Projects

The core of Lyngsø's product development project was generally formulated at the strategic level inside the business. The reason why the core of the product development project in general was specified in Lyngsø can be related to the business's ISO9001 standard, which demanded such a specification. This was further supported by the fact that formal goals and limits (goals, cost, resources etc.) for the product development project were always specified. This was illustrated in Table 9.15 below.

Table 9.15 Goals and limits to product development

Definition of Goals and Limits to Product Development Project	
Mission	Yes
Goals	Yes
Strategy	Yes
Economic Resources	Yes
Personnel/Organisational Resources	Yes
Contact Limits to Network Partners	Yes

The goals and limits for the product development projects in Lyngsø were always defined in details in the areas as shown in Table 9.15.

Lyngsø maintained that these specifications practically always helped the business to reach the success criteria for the product development project. Generally, Lyngsø could be characterized as a planning oriented business.

External Networks Involved in PD

To a large extent as shown in Table 9.16 the product development projects at Lyngsø were managed by the customers (60%) with Lyngsø's sales department in the second place (25%). This gave a strong indication that the product development at Lyngsø was highly market oriented and based on a network consisting of the customers and Lyngsø's sales department.

Table 9.16   Management of projects at Lyngsø

Management of Project (%)	
Customer	60
Supplier	5
Marketing	10
Sales	25
Total	100

Lyngsø claimed that the network HS enabler was a major catalyst to high speed product development especially in the initial part of the product development process. This argument was analogous to what was found in the secondary cases (Case No. 13 Mayekawa, Case No. 11 Rossflex, Case No. 19 UK Chemicals).

Moreover, the argument was supported by the nature of the networks partners involved in the product development process.

Neither the suppliers or the competition were involved in Lyngsø's product development process. The customers were mainly involved at the initial stage and small gates of the product development process.

General Lyngsø Product Development Model

Formal Stages and Gates

In the case research Lyngsø claimed that they had a formal stage gate model. The model can be seen in Table 9.17. In addition, the case research could verify

Table 9.17   Network partners involved in PD process

	Customers	Suppliers	Competition	Other Network Partners
Idea Generation	Y	N	N	N
Concept Generation	Y	N	N	N
Product Development	N	N	N	N
Process Development Phase	N	N	N	N
Idea Screening	Y	N	N	N
Concept Screening	Y	N	N	N
Proto Type Test	Y	N	N	N
Process Test	Y	N	N	N

Y = Yes
N = No
ISC = In some cases
DN = Do not know

Lyngsø's classification of the PD model. The model was defined very much in accordance with the ISO 9001 standard. The stage and gates of Lyngsø's formal product development model are shown in Table 9.18.

Table 9.18 Stages and gates of Lyngsø's formal PD model

Idea	Concept	PD Phase	Process Development Phase	Idea Screening	Concept Screening	Proto Type	Process Testing
N	Y	Y	Y	N	Y	Y	Y

The case research showed that Lyngsø's formal product development model had three stages – a concept stage, a product development stage, and a process development stage. The business claimed that it had no formal idea stage but the initial phase of the product development process started with the concept stage.

In the screening area, the picture showed that Lyngsø had three gates – a concept screening gate, a prototype test gate, and a process test gate.

The above case research results proved that the stage gate model existed at Lyngsø. Nevertheless, the results also proved that the stage gate model was slightly different from our research hypothesis model because the formal idea stage and idea screening gate did not exist. This was due to informal stages and gates carried out beforehand.

Informal Stage Gate

The case research showed that there was also an informal model running parallel to the formal model. Lyngsø confirmed the existence of such an informal product development model. The existence and the importance of the informal product development model in different areas at Lyngsø were interesting as seen in Table 9.19.

Table 9.19 Stages and gates of Lyngsø's informal PD model

Idea	Concept	PD Phase	Process Development?	Idea Screening	Concept Screening	Proto Type	Process Testing
Y	Y	Y	Y	Y	Y	Y	Y

It was very interesting to see that Lyngsø's informal product development model contained all stage and gates as defined in the research framework model. The idea stage and gate existed in Lyngsø but only on an informal basis. The Sales Manager of Lyngsø claimed that the reason for this was that the idea stage and gate could not "live" in a strict ISO9000 model with formal procedures etc. Furthermore, he claimed that because of demands of ISO 9000 when ready for conceptualising the idea was "put" into the formal stage-gate product development model.

The Sales Manager of Lyngsø claimed that the informal product development model was important for all listed success criteria at Lyngsø. This state of affairs is illustrated in Table 9.20.

Table 9.20   Importance of informal PD model in relation to success criteria

	Time	Costs	Performance	CIM	CI	Learning
Yes	Y	Y	Y	Y	Y	Y

Focusing on time in product development Lyngsø claimed that the informal product development model influenced in particular the idea and concept stage and gate at Lyngsø when Lyngsø wanted to achieve high speed. It was verified that the informal model was used to speed the time of product development in Lyngsø. The research discovered a high speed enabler that was not registered before in the secondary case studies.

The case research showed that an informal product development process existed and that it carried an impact on all success criteria of Lyngsø's product development projects as seen in Table 9.20 and Table 9.21.

Table 9.21   Influence of informal PD process on success criteria

Time	Costs	Performance	CIM	CI	Learning
Y	Y	Y	Y	Y	Y

The case research at Lyngsø gave more details on the running of informal processes at the idea and concept stage as well as on the influence on time and speed in the product development process. This will be verified later in the specific Lyngsø case.

Internal Functions Involved in PD Process

In the case research carried out at Lyngsø, the following functions were involved at the different stages and gates of the product development process.

Lyngsø had a rather traditional involvement of functions at the product development stage and gates. The business was very focused on the involvement of sales, management and the product development department at the initial idea and concept stage and gates as seen in Table 9.22.

Table 9.22 Functions participating in PD idea stage

	Marketing	Finance	Sales	Management	Production	Product Development	HRM
Idea Generation	ISC	N	Y	Y	ISC	Y	N
Concept Generation	N	N	Y	Y	ISC	Y	N
Product Development	N	N	Y	Y	ISC	Y	N
Process Development Phase	N	ISC	ISC	Y	Y	Y	N
Idea Screening	ISC	N	Y	Y	N	Y	N
Concept Screening	ISC	N	Y	Y	ISC	Y	N
Proto Type Test	N	N	ISC	ISC	Y	Y	N
Process Test	N	N	N	ISC	Y	Y	N

Y = Yes
N = No
ISC = In some cases
DN = Do not know

Sales, management, product development and marketing were the main actors at the idea stage of the product development process. The sales department was very much the initiator of the informal product development models and processes and thereby the initiator to speed the PD process at the idea stage. HRM and finance were practically not involved in the product development process; production comes in occasionally in the initial phases of the product development stage and gates but were very much involved in the final part of the product development stages and gates. Management played a major role at the initial stages and gates of the product development process.

Focus of Success Criteria of PD Process at Lyngsø

Lyngsø Industry was very much focused on performance and continuous innovation as both received priority 1 on a scale from 1 to 5 as seen in Tables 9.23 and 9.24.

Table 9.23 Priorities of success criteria at Lyngsø

Priorities	
Time	2
Cost	2
Performance	1
CIM	2
CI	1
L	2

Table 9.24   Lyngsø's focus on success criteria

	Time	Costs	Performance	CIM	CI	Learning
Idea stage				▓		
Concept stage					▓	
PD Stage	▓					
Process Stage		▓				
Idea Screening	▓					
Concept Screening	▓					
Proto Type Test			▓			
Process Test			▓			
Implementation	▓					

The reason why Lyngsø did not focus very much on time was that much of their product development could be characterised as radical product development. Nevertheless, Lyngsø's priorities of success criteria changed both during the product development stage- and gates and from PD project to PD project.

The focusing on time dominated at the product development stage, at all initial screening gates, and at the implementation stage. Lyngsø claimed that focus on time was less central when product development was very radical. Performance was then very much in focus an particular very much in focus in the last gates of the product development process. Lyngsø put a high priority on CIM at the idea stage and on CI at the concept stage. This verified that the pressure on time was reduced at Lyngsø because focus on maximal innovation, continuous improvement at the initial stages.

The HS Enabler

The use of high speed enablers showed that all enablers were considered at Lyngsø but especially enablers Nos. 1, 2, 4, 8, and 9 were in focus as seen in Table 9.25.

Table 9.25   HS enablers in use at Lyngsø industri

	Very Much	In Some Cases	No	Do Not Know
ICT Communication Enabler	Yes			
Customer Enabler	Yes			
PD Model Enabler		Yes		
Network Enabler	Yes			
Innovation Enabler		Yes		
HRM Enabler		Yes		
Process Enabler		Yes		
Product to Process Enabler	Yes			
Modularisation Enabler	Yes			
E-Development Enabler		Yes		

However, it was verified that Lyngsø changed enablers from project to project and also during a product development project. The customer and the network enabler were very much used in the initial product development phase but not so much in the middle of the product development phase.

The consequences for Lyngsø Industri performing high or right speed in general were reflected as seen in Table 9.26:

Table 9.26   Consequences of high or right speed

Consequences at Lyngsø Reflected on Their General Product Development Model and Processes Related to Their Characteristic on the Field of PD	High Speed	Right Speed
Time	High speed will not give Lyngsøs product development any benefit on time because they will loose in alternative time	Right speed will give Lyngsø a competitive advantage because they will have time to develop and implement their product at the right time for the market.
Cost/Value	The direct and alternative cost will be too high and the value both direct and alternative will not match the markets demand for value.	The direct and alternative cost will diminish as both the value curve both and the cost curve both direct and alternative will match the markets demand for value. The business gains a possibility to find the right value and the right cost.
Performance	The performance on the products will either come out with a too high performance related to the market demand or a product that cannot match the performance of the market because of marketing and technological "bugs".	The performance on the products will come out with the right performance related to the market demand or a product that can match the demand for performance on the market. It will be possible to use play with perceived value because of marketing, design technology, production are matching each other.

(Continued)

Table 9.26   Continued

Consequences at Lyngsø Reflected on Their General Product Development Model and Processes Related to Their Characteristic on the Field of PD	High Speed	Right Speed
Market fit	Will not fit the market – "over valued" and customer demand are out of fit.	Will fit the market – "valued" and customer demand are in fit
Risk	The profit will diminish and the span between value and cost will diminish. Too many cost to product development because technology are not stable.	The profit will increase diminish and the span between value and cost will diminish. Too many cost to product development because technology are not stable.
Threat	Lyngsø implement a product too early and the products will either fail to be adapted to the market or be very early copied by competitors.	Lyngsø implement a product too early and the products will either fail to be adapted to the market or be very early copied by competitors.
ROI	No return on investment or late ROI	ROI and earlier ROI
Proposal for improvement	Increase the innovation capacity – use more PD enabler, Innovation enabler and modularisation enabler Be careful about the product architecture and choice of product development model. A right product architecture will give the possibility to use the Modularisation HS enabler later in the product development process and in the future, which can make Lyngsø speed product development further seen from a competitive and customer perspective.	

The PD Case – "A New High Tech Airport Promotion Tool"

The sales director of Lyngsø Industri met a supplier at the airport in Singapore while the sales director was investigating another new product development project at the airport in Singapore. The supplier initiated the possibility of a new product development project for Lyngsø. The product development idea was only possible to realise when the two network partners joined each other in the development, and the competences of the network partners would strongly complement each other.

The product development task could be characterised as radical on the technological side because the product development component was not previously known in Lyngsø, neither were the side effects known. The PD project was radical on market dimensions because until now competitors had not had this product feature in their products and the customers had not been presented to this product before. It was new to the market. The technology suddenly offered the possibility of doing what the supplier explained as an idea and concept. On the market side the product development would be an incremental "add on" to some other physical existent products. The customers were known and familiar. On the whole, the result of the product development task could be a radical breakthrough product if success was achieved. It would give the customer a possibility to have added value to their existing products whereby they could gain a big increase in turnover. The product development project could in the research frame work be characterised as shown in Table 9.27.

Table 9.27 Perspectivising the "Lyngsø Singapore case"

Dimension	Incremental	Radical
Where was the idea discovered?	On the field of product development by a network partner – a supplier.	
Initiator of idea	Network partner – a supplier	
Product type	Hardware 40% software 60% Physical product 20% Digital product 40% Virtual product 40%.	
Consequences for product core		new core
Placement in product development stage		idea stage
Innovation degree		High
Market		New
Customer needs		New
Customer group	Old and known	
Technology		New
Network	Old	
Innovation degree and challenge to competence	Middle	

(Continued)

Table 9.27 Continued

Dimension	Incremental	Radical
Product management	Loin Management	
Competition	Low not existent	
Strategic importance	low and long term not critical	
Success criteria	Performance – high	
	Cost – middel	
	Speed – middel	
	CI – none	
	CIM – none	
	Learning – none	
Product development task	Incremental Radical	
PD model – formal	Network based stage gate model	
Functions involved in initial phase	Sales	
Partners involved in initial phase	Supplier and in concept customer	
Enablers involved	Network enabler, Innovation enabler in the initial phase.	

As can be seen the product development project was very much radical to Lyngsø and was in the area of what Lyngsø normally dealt with in relation to product development projects. The case showed that the pressure on time was hardly existent because neither the customer nor the competitors knew about the radical idea.

Therefore, focus was more on performance and continuous innovation as usually in Lyngsø Industri's product development. However, both network partners knew that competitors could come up with a similar product in a short time. Therefore, some pressure on time did exist.

Related to the framework model of NB HS NPD and the Lyngsø PD model the case can be analysed and reflected upon with nearly the same contents as those of the general product development at Lyngsø. This is due to the fact that the Singapore case has nearly the same characteristics as Lyngsø's general product development model and product development task. The NB PD model does not change because of the new project.

9.1.3 The Language Centre – "The Multimedia Learning Business"

The Language Centre were specialists in the publication of software for use in language learning, both CBT (Computer-Based Training) and WBT (Web-Based Training). The Language Centre expertise included authoring, product design, storyboard writing, production, marketing and distribution. The Language Centre competence was the area of languages, but the businesses technological expertise could be used for all types of training courses.

Until 2003 the TLC product portfolio had been strongly focused on physical products and on the new product introduction as seen in Table 9.28.

Table 9.28 Focus on product types

	Physical Products	Service Products	Knowledge and Consultancy
Existing Product Portfolio	100	0	0
New Products	100	0	0

When looking at TLC's products the business claimed that 100% of the products were physical products and 100% of the products dealt with physical processes.

TLC had a very high focus on physical products and processes but some of these products had potential or were to some extent what could be called digital products as seen in Table 9.29.

Table 9.29 Focus on products and processes

	Physical Products	Digital Products	Virtual Products
Existing Product Portfolio	80	20	0
	Physical Processes	Digital Processes	Virtual processes
	100	0	0

Product Development in General at TLC

On the basis of the case research, TLC's task of product development could be defined as seen in Table 9.30.

Table 9.30 TLC's PD task

	Physical Products	Service Products	Knowledge and Consultancy
Existing Product Portfolio	100	0	0
New Products (3 years)	100	0	0
Product Development	80	10	10

Of the business's product development tasks 80% could be related to hardware or physical products, whereas 20% can be related to service and knowledge products. As can be seen in Figure 9.4, TLC's product portfolio mainly focused on physical products but the product development efforts were now also to some extent on service and knowledge/consultancy products.

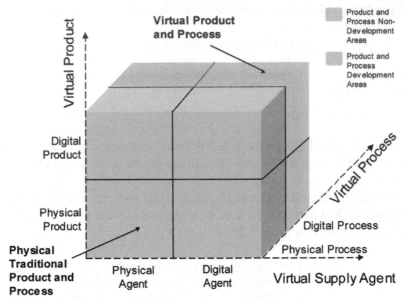

Figure 9.4   TLC's product and process development matrix.

TLC also claim that product development projects could in general be categorised into 100% strategic known and old areas as seen in Table 9.31.

Table 9.31   PD projects in relation to strategy

	Strategic Areas	
	Known and Old Areas	Unknown and New Areas
Total Average	100	0

The product/market model showed that TLC's product development projects were characterised as incremental product development as seen in Figure 9.5.

The product development projects of TLC were mainly on new products older than 1 year (100%) with needs of small adjustments. This was due to a market with very short lifecycles and many introductions of incremental product developments. In other words, the core of the products live for a longer

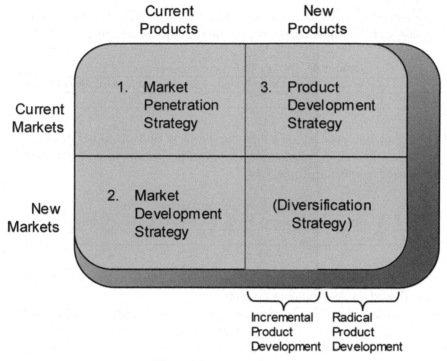

Figure 9.5 PD task at TLC.

time, but the "rings" around the product were continuously under pressure for development. They were rather dynamic.

The case research showed that out of the product development projects the main product development projects were on products that need small adjustments as seen in Table 9.32. The data showed evidence of what had been said about a diminishing product lifecycle. This indicated a high pressure on incremental product development at TLC.

Table 9.32 PD in relation to product

	Old Products More than 3 Years with a Need for Small Adjustments	Old Products More than 3 Years with a Need for Big Adjustments	New Products Older than 1 Year with a Need for Small Adjustments	New Products Older than 1 Year with a Need for Big Adjustments
Total Average (%)	0,0	0,0	100,0	0

Moreover, the survey showed that 100% of the product development in the business were on known and old customer groups as seen in Table 9.33.

Table 9.33   PD in relation to customer groups

	Known and Old Customer Groups	Unknown and New Customer Groups
Total Average	100	0

Looking at the product development projects and customer needs it appeared that 100% of the product development projects were on known/old customers needs as seen in Table 9.34.

Table 9.34   PD projects in relation to customer needs

	Known and Old Customer Needs	Unknown and New Customer Needs
Total Average	100	0

Obviously, this was another indication of a business who dealt with rather incremental product development projects.

On the technical level TLC claimed that 80% of their product development projects were into known technology – and 10% into incremental technology areas as seen in Table 9.35. Another 10% were in areas of radical technology areas where big technology adjustments were necessary. This indicates that the product development at TLC at this dimension was very incremental.

Table 9.35   PD projects in relation to technology

	Known Technology	Known Technology with Small Adjustments (Incremental Technology)	Completely New Technology (Radical Technology)
Total Average (%)	80	10	10

The case research showed that 100% of the product development projects were related to market areas with high and rival competition as seen in Table 9.36. This indicates that TLC was under a very intense pressure from the market by competitors; especially by global competitors – illustrated in Figure 9.6.

Table 9.36   PD projects in relation to competition

	Markets with Low or No Competition	Markets with Medium or Intensive Competition	Markets with Fierce and Rival Competition
Total Average (%)	0	0	100

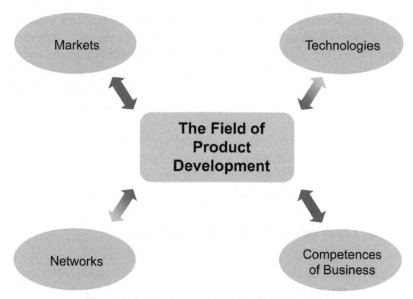

Figure 9.6   The interaction on TLC's field of PD.

TLC considered 20% of their product development projects as having a high element of innovation and another 80% in areas with medium innovation degree as seen in Table 9.37. This indicated that the product development projects at TLC at this area was between incremental and radical innovation.

Table 9.37   PD projects in relation to degree of innovation

	No Degree of Innovation – Routine NPD Project	Medium Degree of Innovation – Modified Product Development with Minor Demands on Adjustment (Incremental)	High Degree of Innovation – with Many Elements of Innovation (Radical Innovation)
Total Average (%)	0	80	20

The case research showed the following general characteristics of the product development in TLC.

TLC was generally in interaction with all components on the field of product development from an inside out perspective. The field of product development was under high pressure from all components in the field. Many of the product development projects at TLC were into rather incremental areas of product development. Both the market, the technology and the network components offered TLC many opportunities but the cost of product development were very high because there were high costs on developing contents to the e-learning products. TLC's challenge was to find the right speed for several PD projects and develop them within right time to harvest the market at the optimal time. However, a continuous innovation activity was necessary to keep the business in business.

As a network business with a big outsourcing activity TLC had no problems finding competence to develop the new products. Instead the problem was to find the right competences at the right costs and at the optimum time for product implementation on the market.

The General TLC Product Development Model

Sources to product development ideas in general were formed as follows.

The product development ideas at TLC came from different sources but mainly from the management (30%), the sales function (20%), and the competitors (20%) as seen in Table 9.38. This indicated that the business's product development in general was based on identified needs and wants in the market. Compared to the other case businesses more functions were involved in the idea generation, as can be seen in Table 9.38. This was due to a strong management effort to involve and make all functions responsible of product development and also because the business was so small that everybody had to be involved in or at least informed of product development projects going on in the business.

Table 9.38   Sources of PD ideas in general

Sources to Product Development Ideas in General	In Per Cent
Customers	15,0
Sales	20,0
Leadership/Management	30,0
Product Development	15,0
Competition	20,0
Total	100

The Core – Goals and Limits for NPD

The core of a product development project was in general formulated in the business. The reason why the core of the product development project in general was specified in TLC, was the management's strong focus on the strategic critical issue for the business to formulate the core of a product development project at the start of a project. Product development projects was vital for TLC and can mean life or death for the business. This was further supported as the formal mission, goals, and strategy for the product development project was always specified together with the economic resources as seen in Table 9.39.

Table 9.39 Goals and limits to PD

Definition of Goals and Limits to Product Development Project	
Mission	Yes
Goals	Yes
Strategy	Yes
Economic Resources	Yes
Personnel/Organisational Resources	No
Contact Limits to Network Partners	No

Personnel/organisational resources and contact limit to network partners were not specified. TLC focus strongly on the strategic limits of product development projects whereas the rest was up to the product development team behind the product development project.

Surprisingly TLC said that these overall specifications were not important for the business to reach the success criteria of the product development project. This meant that TLC had a clear feeling that although efforts were made to carry out the strategic planning behind the product development projects, such efforts were not always tantamount to the product development project meeting the success criteria.

Formal Stages and Gates

In the case research TLC claimed that they had a formal stage- gate model. The case research showed that the formal stage- and gates model has the same stage and gates as our hypothesis model.

The case research showed that TLC's product development model had 4 stages – an idea, concept, product development, and process development stage. On the screening area the picture showed that TLC had 4 gates – an idea screening, a concept screening, a proto type test, and a process test gate as seen in Table 9.40.

The above case research results proved that the stage gate model existed at TLC and that the stage-gate model was equal to our research hypothesis model.

Table 9.40 Stages and gates of TLC's formal PD model

Idea	Concept	PD Phase	Process Development Phase	Idea Screening	Concept Screening	Proto Type	Process Testing
Y	Y	Y	Y	Y	Y	Y	Y

TLC confirmed that an informal product development model did not exist in the business. The lack of existence of an informal product development model in TLC was interesting as this indicates that small businesses may not need to establish informal product development models and processes. This may explain higher speed and lower cost for SMEs in product development because there are no informal model and processes to consume alternative time and cost.

Internal Functions Involved in PD Process

In the case research at TLC the following functions were involved in the different stages and gates in the product development process.

TLC had a characteristic involvement of functions involved in the product development stages and gates. All functions were involved except HRM. The reason for this was the very small size of TLC where all function were concentrated on a single or on very few employees as seen in Table 9.41. HRM did not really exist. The function was integrated in the manager function.

Table 9.41 Functions participating in PD idea stage

	Marketing	Finance	Sales	Management	Production	Product Development	HRM
Idea Generation	Y	Y	Y	Y	Y	Y	N
Concept Generatino	Y	Y	Y	Y	Y	Y	N
Product Development	Y	Y	Y	Y	Y	Y	N
Process Development	Y	Y	Y	Y	Y		
Phase Idea						Y	N
Screening	Y	Y	Y	Y	Y	Y	N
Concept Screening	NA	Y	Y	Y	Y	Y	N
Proto Type Test	Y	Y	Y	Y	Y	Y	N
Process Test	Y	Y	Y	Y	Y	Y	N

Y = Yes
N = No
ISC = In some cases
DN = Do not know

External Networks Involved in PD

To a large extent, the product development projects at TLC were managed internally and only by the management of the product development department (100%). Management played a major role in teambuilding in the product development projects.

This gave a strong indication that the management of product development at TLC was strongly concentrated internally and based on strong management by the manager of TLC. This was quite another management model compared to the other case businesses except Lindholst case.

The argument was further supported when looking at the network partners involved in the product development process as seen in Table 9.42.

Table 9.42 Network partners involved in PD process

	Customers	Suppliers	Competition	Other Network Partners
Idea Generation	Y	N	N	N
Concept Generation	N	N	N	N
Product Development	N	Y	N	Y
Process Development	N	N	N	Y
Phase Idea				
Screening	N	N	N	Y
Concept Screening	N	Y	N	Y
Proto Type Test	Y	N	N	Y
Process Test	Y	Y	N	Y

Y = Yes
N = No
ISC = In some cases
DN = Do not know

The competitors were not involved in the product development process at all. This was due to the very fierce competition in the market (Manager TLC). The customers were mainly involved in the idea phase and the prototype and process test phase. The suppliers were mainly involved from the middle to the late part of the product development.

Focus on Success Criteria of PD Process

TLC was very much focused on nearly all success criteria; both short term and long term success criteria as seen in Table 9.43.

Table 9.43 Priorities of success criteria at TLC

Priorities	
Time	1
Cost	2
Performance	1
CIM	1
CI	1
L	1

TLC's high focus on time often resulted in higher costs than expected. Although TLC tried to learn from previous product development projects to continuously diminish costs. However, so far this approach had not been successful. This was due to the fact that TLC started all PD projects as radical development projects. Furthermore, TLC had not gained a learning curve on product development. Yet, the prioritising changed during the product development stages and gates as seen in Table 9.44.

Table 9.44 TLC's focus on success criteria

	Time	Costs	Performance	CIM	CI	Learning
Idea				▓		
Concept				▓		
PD Stage		▓				
Process Stage		▓				
PD Test			▓			
Idea Screening	▓					
Concept Screening	▓					
Proto Type Test			▓			
Process Test			▓			
Implementation	▓					

The time focus was very much concentrated on the screening gates when CIM was prioritised on the idea and concept stage. This was something to do with the focus on incremental adjustment and high pressure from competition where small adjustments could result in a competitive advantage (Manager TLC). At the product development and process stage, focus was very much on cost and not much on value.

The HS Enablers

The use of high speed enablers showed that all enablers were considered at TLC at seen in Table 9.45, whereas the HRM and the modularization enablers were not so much considered.

Table 9.45 HS enablers used at TLC

	Very Much	In Some Cases	No	Do Not Know
ICT Communication Enabler	Yes			
Customer Enabler	Yes			
PD Model Enabler	Yes			
Network Enabler	Yes			
Innovation Enabler	Yes			
HRM Enabler		Yes		
Process Enabler	Yes			
Product to Process Enabler	Yes			
Modularisation Enabler		Yes		
E-Development Enabler	Yes			

TLC claimed to focus on many HS enablers. However, it was verified through the case discussion that TLC changed enablers from project to project and also during a product development project.

The consequences for TLC performing high or right speed in general were reflected as seen in Table 9.46:

Table 9.46 Consequences reflected on general PD model and processes

Consequences at TLC Reflected on their General Product Development Model and Processes Related to their Characteristics on the Field of PD.	High Speed	Right Speed
Time	TLC is under high pressure on time and cost from "the field of product development". Especially the market and the technology press TLC to continuously lower cost and increasing introduction on new products. The product life cycle of the products are diminishing continuously incremental new product must be introduced to the market at high speed	Right speed in TLC means high speed but with a focus to competitive advantage TLC will in this market not have time to develop and implement their product at a slow speed A market oriented out side in management style is necessary at TLC. The field of product development" has to be analysed carefully to find the right time for product introduction.

(Continued)

Table 9.46   Continued

Consequences at TLC Reflected on their General Product Development Model and Processes Related to their Characteristics on the Field of PD.	High Speed	Right Speed
	Diminishing time in product development is a must for TLC. However, high speed focused only on cost will not give TLC benefit or competitive advantage on the market because competitors can develop at the same speed and with often lower cost. TLC come either too fast or too slow to the market and loose on alternative time and cost	TLC has to focus on a strong product architecture. TLC must find the right time for market introduction.
Cost/Value	The direct and alternative cost are too high and the value both direct and alternative do not match the markets demand for value and perceived value.	The direct and alternative cost will diminish as both the value curve and the cost curve directly and alternatively will match the market's demand for value when TLC focus on perceived value it will gain bigger profitability. The business gains a possibility to find the right value and the right cost.
Performance	The product either comes out with a too high performance related to the market demand or the product cannot match the performance of the market because of marketing and technological "bugs".	The product comes out with the right performance related to the market demand or the product can match the demand for performance on the market. It will be possible to use and play with perceived value because of marketing, design technology, production are matching each other.

Table 9.46 Continued

Market fit	Will not fit the market – "over valued" or "under valued" and customer demand are out of fit with the product.	Will fit the market – "value" and customer demand are in fit
Risk	The profit will diminish and the span between value and cost will diminish. Too many costs to product development because technology is not stable. And its very expensive to develop content.	The profit will increase because the span between value and cost will increase. Cost to product development because technology are not stable.
Threat	TLC implement a product too early and the products will either fail to be adapted to the market or be very early copied by competitors.	TLC implement a product at the right time continuously and the products are adapted immediately to the market. It will be very difficult for the competitors to copy the product and if so the competitors will be too late before a new incremental product development has taken place.
ROI	No return on investment or late ROI	ROI and earlier ROI
Proposal for improvement	Increase the customer enabler – use more the modularisation enabler and formulate the product architecture very carefully at the very initial idea and concept stage. Secure the possibility of fast variation and diminish cost continuously. Focus on perceived value and right time.	

Product Development Case – "A New Multimedia Learning Product"

The managing director of TLC met the new buying manager of Bogpa to discuss how TLC could sell more of their new language series. At the same meeting the managing director asked the buying manager how the competitor's product was selling. The buying manager told him that the competitor's product was sold at a much higher price and was – to some extent – of a better quality; for instance it had a few more features, a smarter package that the customers liked etc. However, the market for the language products and multimedia products in general was stagnating mainly because consumers were illegally copying the products. A pressure from substitute products was also very strong. The buying manager wanted a new product from TLC which could match the competitor's product and play "the game of perceived value" and prevent illegal copying.

At first sight, the product development task could be characterised as incremental on the technological side because the component as well as the side effects were known in the business before. The PD project could be characterized as incremental on the innovative side because everybody would have the possibility to include this product feature in their products. In terms of market, the product development could be characterised as radical to TLC because TLC had to come up with some new perceived value to the customers – however, this was mainly to the end user. The customer dimension was incremental because the business customer was known and familiar. Altogether, on the face of it the result of the product development task could be considered an incremental breakthrough on the market, if success was reached. It would give TLC a possibility to differentiate the businesses' products and thereby gain a major increase in turnover. However, when carefully analysing the product development project the core of the product was changing very dramatically because TLC now had to move the business from thinking inside out to an outside in thinking. The TLC business had to play the game of perceived value which was quite different from previously when high quality and best performance had been the game. Furthermore, the business had to develop an "anti copy" product.

Special characteristics of the development task is seen in Table 9.47:

Table 9.47   Characteristics of development task

Dimension	Incremental	Radical
Where was the idea discovered	On the marketplace	
Initiator of idea	Customer	
Product type	Hardware 5% Software 95%	
Consequences for product core		Modified to radical modification of the core
Placement in product development stage		Concept stage
Innovation degree		High to TLC

Table 9.47   Continued

Market		Old and mature market
Customer needs		Stable interest
Customer group	Old	
Technology		Old and stable
Network		Old
Competence's		To some extent new – "perceived value".
Product management	the business	
Competition		High and radical
Strategic importance		high, important, short term and critical – survival
Success criteria	CI – none CIM – some Learning – none	Performance – high on perceived performance Cost – high on diminishing the cost Speed – very high – pressure from competitors
Product development task		
PU model – formal	Stage gate model	
Functions involved in initial phase	Sales and production	
Partners involved in initial phase	Business and customer	
Enablers involved	Customer enabler	

Perspective on "TLC Bogpa Case"

As can be seen, the product development project was to some extent radical to TLC. The case shows that the pressure on time was very much existent because customer and competitors pressed for new products. Therefore, focus was on time, speed and perceived performance. Usually in TLC product development focus was on performance and cost.

Related to the framework model of NB HS NPD and the TLC development model the case could be analysed and reflected upon with another content as of the general product development at TLC. This was due to the fact that the BOGPA case www.bogpa.dk was different from the characteristics of TLC's general product development model and task.

Table 9.48   Perspectives and reflection on High Speed related to TLC and the Bogpa case

Consequences at TLC Reflected on their General Product Development Model and Processes Related to their Characteristic in the Field of PD.	High Speed	Right Speed
Time	High speed will not give TLC product development any benefit on time because they will loose in alternative time	Right speed will give TLC a competitive advantage because they have time to develop and implement their new product at the right time. However right speed in this case is both focused on high speed and perceived value because of the characteristics on the field of product development.
Cost/Value	The direct and alternative cost will be too high and the value both direct and alternative will not match the markets demand for value and perceived value.	The direct and alternative cost will diminish as both the value curve will increase and the cost curve diminish direct alternative will match the markets demand for value. The business gains a possibility to find the right value and the right cost.
Performance	The performance on the products will either come out with a too high performance related to the market demand or a product that cannot match the performance of the market because of marketing and technological "bugs" or the technology is out of date.	The performance on the products will come out with the right performance related to the market demand or a product that can match the demand for performance on the market. It will be possible to use play with perceived value because of marketing, design technology, production are matching each other.
Market fit	Will not fit the market – either "over valued" or "under valued" and customer demand are out of fit.	Will fit the market – "valued" and customer demand are in fit

Table 9.48 Continued

Risk	The profit will diminish and the span between value and cost will diminish. Too many cost to product development because technology are not always stable.	The profit will increase because the span between value and cost will increase. Cost to product development will diminish because the market is ready to the product. Implementation and because TLC will choose when technology are stable.
Threat	TLC implement a product too early and the products will either fail to be adapted to the market or be very early copied by competitors.	TLC implement a product at right time and the products will be adapted to the market or be very early copied by competitors.
ROI	No return on investment or late ROI	ROI and earlier ROI
Proposal for improvement	Increase the innovation capacity – use more PD enabler, innovation enabler and modularisation enabler with focus on product architecture – use the e-development enabler more intensively.	

9.1.4 AKV Case – "The Biochemical Business"

Business Description

The AKV Langholt business was a modern potato flour plant situated in Northern Jutland approximately 15 km north of Aalborg. The business was owned by approximately 300 potato growers. The business was a co-operative and the owners had committed themselves to supplying the business with the required raw materials. The business also produced seed potatoes, various sorts of potato starch and various potato sorts for cooking. Potato juice is a by-product resulting from the production of potato flour. Potato juice is the cell sap which has been thinned with water. The product is distributed with an organic fertilizer distributor or with a tanker bearing trailing pipes. Another by-product from the production of potato flour is potato pulp which is used for cattle fodder

The AKV Langholt business can be seen on www.akv-langholt.dk.

The AKV product portfolio was strongly focused on 100% physical products as seen in Table 9.49. There was no change on the new product introduction.

Table 9.49   Focus on product types

	Physical Products	Service Products	Knowledge and Consultancy
Existing Product Portfolio	100	0	0
New Products	100	0	0

This could be seen as a strategy from the business to focus only on physical products and fulfil the needs and wants of the markets in this area. The business had also 100% focus on physical processes and no focus on digital or virtual processes as seen in Table 9.50.

Table 9.50   Focus on products and processes

	Physical Products	Digital Products	Virtual Products
Existing Product Portfolio	100	0	0
	Physical Processes	Digital Processes	Virtual processes
	100	0	0

The market for AKV was very stable but with a very intense rivalry and pressure on price. The technology was strongly dynamic and evolving in these years and offering new opportunities. The network was very stable and narrow minded. However, AKV had a strong network cooperation with the French business Cerestar Pharma www.cerestarpharma.com

Product Development in General at AKV

The case research showed the following general characteristics of the product development at AKV.

Sources to product development ideas in general were formed as seen in Table 9.51:

Table 9.51   Sources of product development ideas in general

Sources to Product Development Ideas in General	In Per Cent
Customers	15,0
Suppliers	5,0
Sales	50,0
Production	25,0
Competition	5,0
Total	100

The product development ideas came mainly from the sales function (50%), the production department (25%), and the customers (15%). This indicated that the business's product development in general was based on identified needs and wants in the market but that the production department played a central role in generating new ideas. The new products were mainly discovered by internal research by the network partner Cerestar Pharma – who were an important product development network partner.

The Core – Goals and Limits for NPD Projects

In most cases, the core of a product development project was formulated in the business. The reason why the core of the product development project in general was specified in AKV could be related to the business's ISO9001 standard, which demands such a specification. This was further supported as the formal goals and limits (goals, cost, resources etc.) for the product development project were in most cases specified as seen in Table 9.52.

The goals and limits for the product development projects in AKV always included detailed definitions in the areas as shown in Table 9.52.

Table 9.52 Goals and limits to product development

Definition of Goals and Limits to Product Development Project	
Mission	Yes
Goals	Yes
Strategy	Do not know
Economic Resources	Yes
Personnel/Organisational Resources	Yes
Contact Limits to Network Partners	Yes

As can be seen from Table 9.52, AKV had always specified missions, goals, economic resources, limits of personal and organisational resources and which contact limits to network partners existed for the specific product development project. According to the managing director, the strategy for the product development project was not known initially. The strategy depended on the idea and was often formulated in the course of the PD process. The business said that such initial goal and mission helped the business to reach the success criteria of the product development project.

Product Development Tasks

On the basis of the case research, AKV's task of product development could be defined as seen in Table 9.53.

Table 9.53   AKV product development tasks

	Physical Products	Service Products	Knowledge and Consultancy
Existing Product Portfolio	100	0	0
New Products	100	0	0
Product Development	100	0	0

Of the business's product development tasks 100% could be related to physical products. As can be seen in Figure 9.7, AKV was strongly focused on physical products and the introduction of new products and the product development efforts were strongly related in the same line of business.

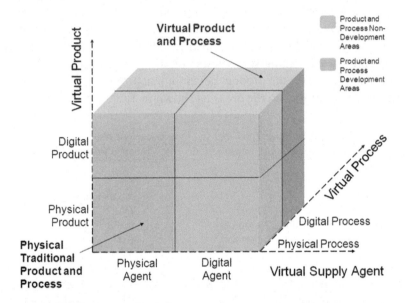

Figure 9.7   AKV's product and process development matrix.

AKV also claimed as seen in Table 9.54 that in general, product development projects could be divided into 95% strategic known and old areas and 5% unknown and new areas.

Table 9.54   PD projects in relation to strategy

	Strategic Areas	
	Known and Old Areas	Unknown and New Areas
Total Average	95	5

The product/market model gave the following picture as shown in Figure 9.8 of AKV's product development project.

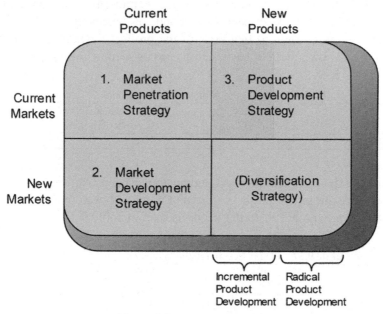

Figure 9.8   PD tasks at AKV.

The AKV product development projects were mainly situated in areas of a rather well known and incremental product development area as seen in Table 9.55.

Table 9.55   Product development in relation to product

Old Products		New Products	
Old Products More than 3 Years with a Need for Small Adjustments	Old Products More than 3 Years with a Need for Big Adjustments	New Products Older than 1 Year with a Need for Small Adjustments	New Products Older than 1 Year with a Need for Big Adjustments
Total Average (%) 40	10	30	20

The case research showed that of all the product development projects, the major part were Product development projects on products that needed small adjustments. The data showed evidence of what had been said about a diminishing product lifecycle. In an overall perspective 50% of the business's products had to have redevelopment after 1 year's lifetime. This indicated a pressure on product development at AKV. AKV did not have a large need for major adjustment on product development on their products. This could either indicate a stable industry concerning market and technology and/or a product development ability at AKV to meet the demands of the customers.

The survey further showed that 90% of the product development in the business was on known and old customer groups as seen in Table 9.56.

Table 9.56   PD in relation to customer groups

	Known and Old Customer Groups	Unknown and New Customer Groups
Total Average	90	10

The product development projects and customer needs showed that 95% of the product development projects were on known/old customer needs and 5% were on unknown/new customer needs as seen in Table 9.57.

Table 9.57   PD projects in relation to customer needs

	Known and Old Customer Needs	Unknown and New Customer Needs
Total Average	95	5

Table 9.58 showed yet another indication of a business which dealt with rather incremental product development projects.

Table 9.58   PD projects in relation to technology

	Known Technology	Known Technology with Small Adjustments (Incremental Technology)	Completely New Technology (Radical Technology)
Total Average (%)	80	15	5

On the technical level AKV claimed that 5% of their product development projects were into new technology – radical technology areas and 95% were in areas of known or development areas where small incremental technology adjustments were necessary.

The case research also showed that 45% of the product development projects were related to market areas with high and rival competition and another 45% to markets with middle to intensive competition. This indicated that AKV was in a mature and rival industry which could be characterised with high competition; especially on price as seen in Table 9.59.

Table 9.59 PD projects in relation to competition

	Markets with Low or No Competition	Markets with Medium or Intensive Competition	Markets with Fierce and Rival Competition
Total Average (%)	30	20	50

AKV saw 50% of their product development projects as having no element of innovation as seen in Table 9.60.

Table 9.60 Innovation degree at AKV Langholt

	No Degree of Innovation – Routine NPD Project	Medium Degree of Innovation – Modified Product Development with Minor Demands on Adjustment (Incremental)	High Degree of Innovation – with Many Elements of Innovation (Radical Innovation)
Total Average (%)	25	25	50

However, the innovative part was solved by the businesses' cooperation with the network partner Cerestar Pharma. Cerestar Pharma had all the necessary experts and they continuously develop new products to the market. This was another example which showed that product development projects were mainly incremental at AKV.

The AKV Product Development Model

Formal Stages and Gates

In the case research AKV claimed that they did not have a formal stage- gate model. However, the case research showed there was an informal product development model running in the business which had nearly the same structure as the stage- and gates of a formal product development model. This model is shown in Table 9.61.

Table 9.61 Stages and Gates of AKV's formal PD model

Idea	Concept	PD Phase	Process Development Phase	Idea Screening	Concept Screening	Proto Type	Process Testing
N	Y	Y	Y	N	Y	Y	Y

The case research showed that AKV's informal product development model had 4 stages – an idea stage, a concept stage, a product development stage, and a process development stage. The business claimed that it had no idea and concept screening which meant that their product development model used direct prototyping very much.

On the screening area the picture showed that AKV could identify 2 gates – a proto type gate, and a process test gate. The business could not identify an idea or a concept screening phase. This could be related to the high degree of incremental product development which may not need the considerable effort of screening at the idea and concept gates.

The above case research results proved that the stage gate model existed in AKV but that the stage- gate model was slightly different from our research hypothesis model because the idea gate and concept gate could not be identified or did not exist. Further the AKV business had a informal PD model very much related to the network partner Cerestar Pharma.

Informal Stages and Gates

AKV confirmed that an informal product development model existed in all stage and gates in the business. In this area the informal model should be understood as a model that was existing besides the normal product development model the business practice as seen in Table 9.62.

Table 9.62   Informal PD model at AKV Langholt

Idea	Concept	PU Phase	Process Development?	Idea Screening	Concept Screening	Prototype	Process Testing
Y	Y	Y	Y	Y	Y	Y	Y

The informal product development model was important for especially the time and cost success criteria at AKV as seen in Table 9.63.

Table 9.63   Importance of informal PD model in relation to success criteria

	Time	Costs	Performance	CIM	CI	Learning
Yes	Y	Y	0	0	0	0
To some extent	0	0	Y	Y	0	Y
No	0	0	0	0	Y	0
Do not know	0	0	0	0	0	0
Not answered	0	0	0	0	0	0

Focusing on time in product development the informal product development model had influence, particularly in the idea and concept stage and gate at AKV. The case research showed explicitly that the informal product development model influenced the success criteria of product development within AKV.

Internal Functions Involved in PD Process

In the case research at AKV the following functions as shown in Table 9.64 were involved in the different stage- and gates in the product development process. AKV had a special characteristic in involvement of functions at the idea stage as all functions except HRM were involved at this stage.

Table 9.64 Functions participating in PD idea stage

	Marketing	Finance	Sales	Management	Production	Product Development	HRM
Idea Generation	Y	ISC	Y	Y	Y	Y	N
Concept Generation	Y	N	Y	ISC	Y	ISC	N
Product Development	ISC	N	ISC	ISC	Y	Y	N
Process Development	N	N	N	N	Y	ISC	N
Phase Idea Screening	Y	Y	ISC	ISC	Y	ISC	Y
Concept Screening	Y	N	Y	Y	Y	ISC	N
Proto Type Test	N	ISC	ISC	N	Y	ISC	N
Process Test	N	N	N	N	Y	NA	N

Y = Yes
N = No
ISC = In some cases
DN = Do not know

Sales, marketing and the product development department were the main actors in the idea stage of the product development process. This also indicated that to a large extent the sales department was the initiator of the product development models and processes. It seemed as if AKV seek a very early commitment and screening to the product development as most functions were involved in the idea and concept screening stages and gates. Subsequently, AKV seemed to rely on production. HRM and economics were practically not involved in the product development at all. Management played a major role in the initial stages and gates of the product development process but was barely involved until just before product implementation in the market as seen in Table 9.64.

External Networks Involved in PD

PU Management

To a large extent, the product development projects at AKV were managed by the customers (85%) as seen in Table 9.65. This strongly indicates that product development at AKV was exceptionally market oriented and based on a network between customers and AKV.

The argument was furthermore supported when looking at the network partners involved in the product development process as seen in Table 9.65.

Table 9.65   Management of projects at AKV

Management of Project (%)	
Customer	85
Do not Know	15
Total	100

The competitors and other network partners were not involved in the product development process. The customers were involved in nearly all phases of the product development process. The supplier was strongly involved subsequent to the generation of the idea as seen Table 9.66.

Table 9.66   Network partners involved in PD process

	Customers	Suppliers	Competition	Other Network Partners
Idea Generation	Y	N	N	N
Concept Generation	Y	Y	N	N
Product Development	Y	Y	N	N
Process Development Phase	N	N	N	N
Idea Screening	Y	N	N	N
Concept Screening	N	N	N	N
Proto Type Test	Y	Y	N	N
Process Test	Y	Y	N	N

Y = Yes
N = No
ISC = In some cases
DN = Do not know

Focus of Success Criteria of PD Process

AKV Industry was very much focused on performance and time as these criteria had received 1st and 2nd priority on a scale from 1 to 5 as seen in Table 9.67.

Table 9.67   Priorities of success criteria at AKV

Priorities	
Time	2
Cost	3
Performance	1
CIM	3
CI	4
L	4

The long-term success criteria did not receive high priority at AKV as seen in Table 9.67 and Table 9.68. Prioritising changed during the product development stages and gates.

Table 9.68   AKV's focus on success criteria

	Time	Costs	Performance	CIM	CI	Learning
Idea	▓					
Concept	▓					
PD Stage			▓			
Process Stage	▓					
PD Test						
Idea Screening		▓				
Concept Screening		▓				
Proto Type Test			▓			
Process Test			▓			
Implementation						

The focus on time was very pronounced at the initial product development stage. At all initial screening gates the focus was on cost. Performance came very much into focus at the last gates of the product development process.

HS Enablers

The use of high speed enablers showed that especially enabler number 2 and 9 were in focus as seen in Table 9.69. Three enablers, viz. nos. 1, 3, and 10 were not considered.

Table 9.69   AKV's use of HS enablers

	Very Much	In Some Cases	No	Do Not Know
ICT Communication Enabler	Yes		Yes	
Customer Enabler				
PD Model Enabler			Yes	
Network Enabler		Yes		
Innovation Enabler		Yes		
HRM Enabler		Yes		
Process Enabler		Yes		
Product to Process Enabler		Yes		
Modularisation Enabler	Yes			
E-Development Enabler			Yes	

Product Development Case – "A New Chemical Ingredient..."

The managing director of AKV received a re-complaint from one of AKV's essential customers. The quality of the latest delivery was not within the defined and agreed tolerance areas. The managing director ordered the production manager to examine the problem and come up with the explanation and reason for the complaint. The production manager found new ingredients in the product which made the process react radically different at the customer's production line. The management team held a meeting and after some discussion they agreed that they had discovered a new product idea which would be radical to the market – new to the market. They agreed on speeding the product development process to get a first mover advantage.

Product Development Task

The product development task could be characterised as radical on the technological side because the component was not known before in the business; neither were the side effects known. The task could be characterised as radical on the innovative side because no competitive products offered this feature. The product would be new to the market. On the market side the product development was incremental because the customer was known and familiar. All in all, the result of the product development task could be a radical breakthrough product if success was reached. It would give the customer a possibility to run their production faster whereby they could achieve a big cost reduction in the production. The question was how much would the customer pay for the product? Some further special characteristics can be seen in Table 9.70.

Table 9.70 Special characteristics of the development task

Dimension	Incremental	Radical
Where were the idea discovered	Internal the business – production – due to a reclamation from a customer	
Initiator of idea	Director and production manager	
Product type	Hardware 50% Software (knowledge of the process and combination of chemicals) 50%	
Consequences for product core		modification of core and ad to functions
Placement in product development stage		concept stage
Innovation degree		high
Market		old and mature
Customer needs		new
Customer group	Old	
Customer technology		new
Technology		new
Network		old
Competences		new and unknown
Product management	the business manager	
Competition		high
Strategic importance	high, important, middle term and critical – competitive advantage	
Success criteria	Cost – middle CI – none CIM – some Learning – none	Performance – high Speed – very high
Product development task	Incremental Radical ⬅━━━➤	
PD model – formal	Stage gate	
Functions involved in initial phase	Management and production	
Partners involved in initial phase	business alone firstly – internal product development project afterwards prototyping with one or two main customers.	
Enablers involved	HRM – enabler and a new enabler to the PhD research – the Management enabler. The product development enabler – Rapid prototyping.	

9.1.5 Lindholst – "The Food Machine Business"

Lindholst was a medium-sized business which produce machinery for the food industry, particularly poultry slaughterhouse machinery. During the last 10–15 years up to 2003, the business had become a market leader of machinery for this line of business.

The Lindholst product portfolio was strongly focused on physical and service products. However, on the new product introduction a small implementation on new products on knowledge and consultancy products were seen.

This could be seen as a reaction to the market needs and wants but also as a penetration of the existing product potential within the business. When looking at Lindholst's products Table 9.71 – 80% of the products were physical products and 20% were digital products.

Table 9.71   Focus on product types

	Physical Products	Service Products	Knowledge and Consultancy
Existing Product Portfolio	75	25	0
New Products	80	15	5

75% concern physical processes and 25% digital processes as seen in Table 9.72.

Table 9.72   Focus on products and processes

	Physical Products	Digital Products	Virtual Products
Existing Product Portfolio	80	20	0
	Physical Processes	Digital Processes	Virtual processes
Existing Product Portfolio	75	25	0

Lindholst had a very high focus on physical products and processes but it seemed as if there was a minute tendency to shift focus from physical to digital and service/knowledge based products. This could be a reaction or a response to a market need.

The market was experiencing a huge pressure on price. There were large customers in the market who were focusing on cost and efficiency. New technology was constantly being introduced to the market, and new and often unknown networks were introduced. A high pressure on the competences of the businesses was felt.

PD in General at Lindholst

The case research showed the following general characteristics of the product development at Lindholst.

Product development ideas at Lindholst came from different sources but mainly from the customers (20%), the sales function (20%), and the product development department as seen in Table 9.73. This indicates that in general the business's product development was based on identified needs and wants in the market and consequently on the product development department. Compared to the other case businesses, more departments were involved at Lindholst during idea generation. This was due to strong management efforts to involve and make all functions responsible of product development.

Table 9.73 Sources of PD ideas in general

Sources to Product Development Ideas in General	In Per Cent
Customers	20,0
Suppliers	5,0
Marketing	5,0
Sales	15,0
Leadership/Management	10,0
Production	10,0
Product Development	15,0
Competition	20,0
Total	100

Core Goals and Limits for NPD Projects

As a general rule, the core of a product development project was formulated in the business. The reason why the core of the product development project was specified at Lindholst was their strong focus on the formulation of the product development project core at the start of a project. This was also supported by the fact that formal mission, goals, and strategy for the product development project were always specified.

The economic resources, personal/organisational resources and contact limit to network partners were not specified. Here, Lindholst had a strong focus on the strategic limits of their product development projects whereas the rest was up to the product development team behind the project in question.

As can be seen from Table 9.74, Lindholst said that such overall specifications were important for the business to reach the success criteria of the product development project.

Table 9.74   Goals and limits to PD

Definition of goals and Limits to Product Development Project	
Mission	Yes
Goals	Yes
Strategy	Yes
Economic Resources	No
Personnel/Organisational Resources	No
Contact Limits to Network Partners	No

Product Development Tasks

On the basis of the case research, Lindholst's product development task could be defined as shown in Table 9.75.

Table 9.75   Lindholst's PD task

	Physical Products	Service Products	Knowledge and Consultancy
Existing Product Portfolio	75	25	0
New Products	80	15	5
Product Development	80	15	5

Of the business's product development tasks 80% could be related to hardware or physical products whereas 20% could be related to service and knowledge products.

As can be seen Lindholst's product portfolio was mainly focused on physical products and service products. However, the introduction of new products and the product development efforts were now also slightly on knowledge and consultancy products.

Lindholst also claim that in general product development projects could be divided into 80% strategic known and old areas and 20% unknown and new areas as seen in Table 9.76.

Table 9.76   PD projects in relation to strategy

	Strategic Areas	
	Known and Old Areas	Unknown and New Areas
Total Average	80	20

The product/market model gave the following a picture seen in Figure 9.9 of Lindholst's product development project situated in areas of rather incremental product development.

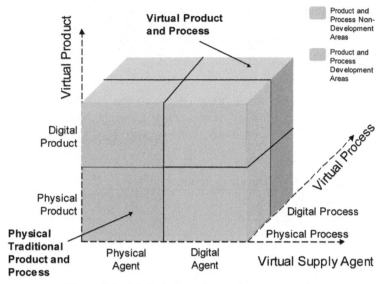

Figure 9.9 Lindholst's product and process matrix.

Lindholst's product development projects were mainly on new products older than 1 year (60%) and old products older than 3 years (40%) as seen in Table 9.77.

Table 9.77 PD in relation to product

	Old Products More than 3 Years with a Need for Small Adjustments	Old Products More than 3 Years with a Need for Big Adjustments	New Products Older than 1 Year with a Need for Small Adjustments	New Products Older than 1 Year with a Need for Big Adjustments
Total Average (%)	0	40	0	60

The case research showed that out of all product development projects the main projects were on products that need big adjustments. The data showed evidence of what had been said about a diminishing product lifecycle. In an overall perspective, the product development at Lindholst concerned major adjustments and major redevelopment after 1 year's lifetime. This indicated a high pressure on radical product development at Lindholst.

The survey also showed that 90% of product development in the business was on known and old customer groups as seen in Table 9.78.

Table 9.78   PD in relation to customer groups

	Known and Old Customer Groups	Unknown and New Customer Groups
Total Average	90	10

Looking at the product development projects and customer needs we observe that 80% of the product development projects were on known/ old customer needs and 20% on unknown/new customers needs as seen in Table 9.79.

Table 9.79   PD projects in relation to customer needs

	Known and Old Customer Needs	Unknown and New Customer Needs
Total Average	80	20

This is yet another indication of a business which deals with relatively incremental product development projects as indicated in Figure 9.10.

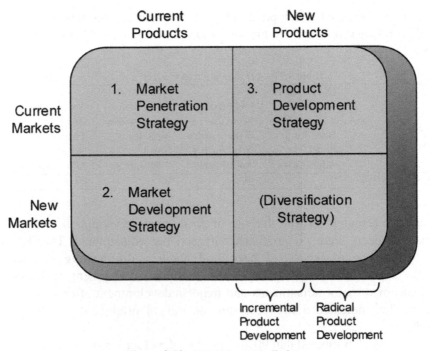

Figure 9.10   PD tasks at Lindholst.

On the technical level, Lindholst claimed that 70% of their product development projects were on new technology – radical technology areas and 30% were in areas of known technology or development areas where small incremental technology adjustments were necessary as seen in Table 9.80.

Table 9.80 PD projects in relation to technology

	Known Technology	Known Technology with Small Adjustments (Incremental Technology)	Completely New Technology (Radical Technology)
Total Average (%)	15	15	70

This indicated that product development at Lindholst was very radical on the technological side.

The case research showed that 50% of the product development projects were related to market areas with high and rival competition. There were no product development projects in low competition areas as seen in Table 9.81.

Table 9.81 PD projects in relation to competition

	Markets with Low or No Competition	Markets with Medium or Intensive Competition	Markets with Fierce and Rival Competition
Total Average (%)	0	50	50

Lindholst considered 35% of their product development projects with a high element of innovation and another 35% in areas with medium innovation degree as seen in Table 9.82. This indicated that the product development projects in this area at Lindholst were rather radical.

Table 9.82 PD projects in relation to degree of innovation

	No Degree of Innovation – Routine NPD Project	Medium Degree of Innovation – Modified Product Development with Minor Demands on Adjustment (Incremental)	High Degree of Innovation – with Many Elements of Innovation (Radical Innovation)
Total Average (%)	30	35	35

The Lindholst PD Model

Formal Stages and Gates

In the case research Lindholst claimed that they had a formal stage gate model. The case research showed that the stages and gates of Lindholst's formal model were identical to the stages and gates of our hypothesis model.

The case research showed that Lindholst's product development model had 4 stages – a concept stage, a product development stage, and a process development stage. In the screening area the picture showed that Lindholst had 4 gates – an idea screening gate, a concept screening gate, a proto type test gate, and a process test gate.

The above case research results proved that the stage gate model exists at Lindholst and that the stage gate model was identical to our research hypothesis model.

Informal Stage and Gate

Lindholst confirmed that an informal product development model also existed in the business. The existence and the importance of the informal product development model at Lindholst on different areas were interesting; see below in Table 9.83.

Table 9.83   Stages and gates of Lindholst's formal PD model

Idea	Concept	PD Phase	Process Development Phase	Idea Screening	Concept Screening	Proto Type	Process Testing
Y	Y	Y	Y	Y	Y	Y	Y

It is very interesting to see that the informal product development model at Lindholst mainly played a role in the lower levels of the product development process. The idea and concept stage and gate hold no informal stages or gates as seen in Table 9.84. My observation was that this was due to

Table 9.84   Stages and gates of Lindholst's informal PD model

Idea	Concept	PU Phase	Process Development	Idea Screening	Concept Screening	Prototype	Process Testing
N	Do not know	Y	Y	N	N	Y	Y

immense involvement and openness at the initial stages and gates at Lindholst. Furthermore, pressure on speed in the middle and lower part of the product development phase caused a need for informal models.

As previously explicated, the informal product development model was important to the time and speed success criteria. Moreover, Lindholst claimed that the informal product development model influences all long term success criteria – CIM, CI, and learning. This scenario differed very much from those of the other businesses examined.

Internal Functions Involved in Product Development Process

In the case research at Lindholst the following functions were involved in the different stages and gates of the product development process. The Lindholst business had a quite characteristic involvement of functions taking part in the product development stages and gates.

Although it seemed as if the business was very focused on involvement of sales, management and the product development department in the initial idea- and concept stage and gates.

Sales, marketing, management and the product development department were the main actors at the idea stage of the product development process. HRM and finance were not involved in the product development as seen in Table 9.85. Production enters the development in the middle stage of

Table 9.85 Functions participating in PD idea stage

	Marketing	Finance	Sales	Management	Production	Product Development	HRM
Idea Generation	ISC	N	ISC	ISC	N	Y	N
Concept Generation	N	N	ISC	ISC	N	Y	N
Product Development	N	N	N	ISC	Y	Y	N
Process Development Phase	N	N	ISC	ISC	ISC	Y	N
Idea Screening	ISC	N	ISC	ISC	N	Y	N
Concept Screening	ISC	N	ISC	ISC	N	Y	N
Proto Type Test	N	N	N	N	ISC	Y	N
Process Test	N	N	N	N	ISC	NA	N

Y = Yes
N = No
ISC = In some cases
DN = Do not know

the product development process. Management played a major role in the teambuilding in the product development projects.

External Networks Involved in Product Development
PU Management

To a large extent, the product development projects at Lindholst were managed internally by the product development department (70%) as seen in Table 9.86. This gave a strong indication that the management of product development at Lindholst was strongly concentrated internally and based on a strong leadership by Lindholst. This was quite another management model as the ones seen in the other businesses of this research.

Table 9.86   Management of projects at Lindholst

Management of Project (%)	
Sales	10
Production	20
Product Development	70
Total	100

The argument was further supported when looking at the network partners involved in the product development process as seen in Table 9.87.

Table 9.87   Network partners involved in PD process

	Customers	Suppliers	Competition	Other Network Partners
Idea Generation	Y	N	N	N
Concept Generation	Y	N	N	N
Product Development	N	N	N	N
Process Development Phase	N	N	N	N
Idea Screening	Y	N	N	N
Concept Screening	Y	N	N	N
Proto Type Test	Y	N	N	N
Process Test	Y	N	N	N

Y = Yes
N = No
ISC = In some cases
DN = Do not know

The suppliers and the competitors were not involved in the product development process; moreover the customers were mainly involved in the initial phase of the product development process. Other network partners were not involved.

Focus of Success Criteria of PD Process

Lindholst was very much focused on performance and continuous innovation as these received priority 1 on a scale from 1 to 5 as seen in Table 9.88.

Table 9.88 Priorities of success criteria at Lindholst

Priorities	
Time	2
Cost	2
Performance	1
CIM	2
CI	2
L	2

The priorities changed during the product development stages and gates as seen in Table 9.89. The focus on time was very predominant at the product development stage, at all initial screening gates, and at implementation. Performance was very much in focus at the final gates of the product development process. Lindholst put a high priority on CIM, on the idea stage, and CI in the concept stage and this meant that the pressure on time was lower because Lindholst wanted to have maximal innovation and continuous improvement at these stages.

Table 9.89 Lindholst's focus on success criteria

	Time	Costs	Performance	CIM	CI	Learning
Idea				■		
Concept					■	
PD Stage	■					
Process Stage		■				
PD Test			■			
Idea Screening	■					
Concept Screening	■					
Proto Type Test			■			
Process Test				■		
Implementation	■					

HS Enablers

The use of high speed enablers showed that all enablers were considered at Lindholst but especially enablers nos. 1, 2, 4, 8 and 9 were in focus as seen in Table 9.90. Lindholst tried to mix the enablers to the product development task.

Table 9.90   Use of HS enablers at Lindholst

	Very Much	In Some Cases	No	Do Not Know
ICT Communication Enabler	Yes			
Customer Enabler	Yes			
PD Model Enabler		Yes		
Network Enabler	Yes			
Innovation Enabler		Yes		
HRM Enabler		Yes		
Process Enabler		Yes		
Product to Process Enabler	Yes			
Modularisation Enabler	Yes			
E-Development Enabler		Yes		

Product Development Case – "A New Chicken Slaughter Machine"

The product development team was gathered around the managing director to discuss an idea for a new machine for hanging up chickens for slaughtering at one of the business's major and main customers. If Lindholst could develop the new product it would be a new to the market product and Lindholst would have gain a first mover advantage. However there were high pressure on time from the customers side because of some environmental new regulations.

Product Development Task

The product development task could be characterised as radical on the technological side. On the innovative side, the task could also be characterised as radical because no competitor had this product feature in their products. This would be new to the market. However, the task could be characterised as incremental on the market side because the customer was known and familiar. Yet, altogether the result of the product development task could be a radical breakthrough product if success was reached as the component would give the customer the possibility to run his production faster and thereby gain a big cost reduction in the production. Furthermore it would give the customer environmental benefits regarding their working environment. Finally, it would improve the quality of customers final product. In Table 9.91 the special characteristics of the PD task are seen.

Table 9.91   Special characteristics of the PD task

Dimension	Incremental	Radical
Where was the idea discovered	External the business – customer and sales – due to a strong customer need to solve a working environmental problem and a need for faster production in the chicken slaughter	
Initiator of idea	Customer	
Product type	Hardware 50% Software (knowledge of the technology to solve the hanging up process of chicken for slaughtering) 50%	
Consequences for product core		Modification of core and ad to functions
Placement in product development stage		Concept stage
Innovation degree		High
Market		Old and mature
Customer needs		New
Customer group	Old	
Customer technology		New
Technology		New
Network		Old
Competence's		New and unknown
Product management	The business – manager	
Competition		High
Strategic importance		High, important, short term and critical – competitive advantage
Success criteria	Cost – middle Speed – middle CI – none CIM – some Learning – none	Performance – high
Product development task	⬅➡	
PU model – formal	Stage gate	
Functions involved in initial phase	Management, product development	
Partners involved in initial phase	Customer and the business	

9.1.6 GSI Lumonics – "The Welding Machine Business"

GSI Lumonics was one of the largest and most resourceful worldwide providers of laser-based manufacturing systems and components.

GSI Lumonics helped customers to create and enhance the value of their products through the use of lasers.

The GSI product portfolio was strongly focused on physical and service products. However, in connection with a new product introduction a small implementation on service together with consultancy products and new products on knowledge were seen.

This could be seen as a reaction to the market needs and wants but also as a penetration of the existing product potential within the business. An observation of GSI's products showed that 80% of the products are physical products and 20% were digital products. 65% concern physical processes and 35% digital processes as seen in Tables 9.92, 9.93 and illustrated on behalf of the data in Figure 9.11.

Table 9.92 Focus on product types

	Physical Products	Service Products	Knowledge and Consultancy
Existing Product Portfolio	80	15	5
New Products	70	25	5

Table 9.93 Focus on products and processes

	Physical Products	Digital Products	Virtual Products
Existing Product Portfolio	80	20	0
	Physical Processes	Digital Processes	Virtual processes
Existing Product Portfolio	65	35	0
Product Development	55	45	0

GSI had a very high focus on physical products and processes but it seemed as if there was a small tendency to shift focus from physical to digital and service/knowledge based products and to digital processes. This could be a reaction or a response to a market need and hence a change of products.

The market for GSI Lumonics' products was characterised by larger customers and with fierce competition, yet with a strong relationship between customer and supplier. The technological development was very intense and dynamic and new network partners were constantly entering the market.

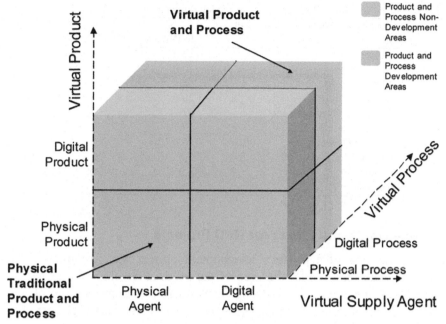

Figure 9.11 GSI's product and process development matrix.

The pressure on the competences of the businesses was very high and as a consequence, GSI Lumonics had many network partners and ad hoc consultants.

PD in General GSI

The case research showed the following general characteristics of the product development in GSI.

The product development ideas at GSI came from different sources but mainly from the customers (25%), the sales function (20%), and the product development department (20%) as seen in Table 9.94. This indicated that the business's product development in general was based on identified needs and wants in the market and at product development department. Compared to the other case businesses more departments at Lumonics were involved in the idea generation. This was due to a strong management effort to involve and make all functions responsible for product development. Furthermore, GSI used other sources such as universities and research centres to generate new ideas for products.

Table 9.94   Sources of PD ideas in general

Sources to Product Development Ideas in General	In Percent
Customers	25,0
Suppliers	10,0
Marketing	5,0
Sales	20,0
Leadership/Management	5,0
Production	5,0
Product Development	20,0
Competition	5,0
Others	5,0
Total	100

The Core – Goals and Limits for NPD Projects

In general, the core of a product development project was formulated in the business. The reason why the core of the product development project in general was specified at GSI strong focus on formulating the core of a product development project at the start of a project and the very high cost of developing a new product. This was further supported as the formal mission, goals, and strategy for the product development project were always specified as seen in Table 9.95.

Table 9.95   Goals and limits to product development

Definition of Goals and Limits to Product Development Project	
Mission	Yes
Goals	Yes
Strategy	Yes
Economic Resources	Yes
Personnel/Organisational Resources	Yes
Contact Limits to Network Partners	Yes

The financial resources, personal/organisational resources and contact limit to network partners were always specified. GSI had here a strong focus on the strategic limits of the product development projects. Particular to GSI a strong development together with the main customers often with prototype development of big plants in house GSI was important. These plant were afterwards destroyed and built up again at the customer's production line.

Because of very fierce and intense competition there was huge focus on which network partners could and should be involved in the product development.

As can be seen from Table 9.95 GSI said that such overall specifications were important for the business to reach the success criteria of the product development project.

Product Development Tasks

On the basis of the case research GSI's task of product development could be defined as seen in Table 9.96.

Table 9.96   GSI's product development task

	Physical Products	Service Products	Knowledge and Consultancy
Existing Product Portfolio	80	15	5
New Products (3 years)	70	25	5
Product Development	70	25	5

Of the business's product development tasks 70% could be related to hardware or physical products whereas 30% could be related to service and knowledge products. As appears from Table 9.96 GSI's product portfolio mainly focused on physical products and service products. However, the introduction of new products and the present product development efforts now also moved towards knowledge and consultancy products as can be seen in Table 9.97.

Table 9.97   PD projects in relation to strategy

	Strategic Areas	
	Known and Old Areas	Unknown and New Areas
Total Average	85	15

GSI also claimed that in general product development projects could be divided into 85% strategic known and old areas and 15% unknown and new areas.

The product/market model gave the following picture as seen in Figure 9.12 of GSI's product development project situated in the area of rather incremental product development.

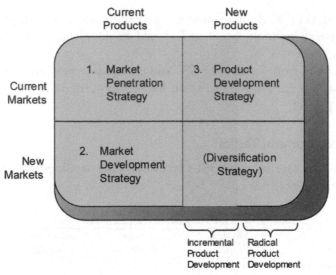

Figure 9.12 PD tasks at GSI.

GSI's product development projects were mainly on old products older than 3 years (70%) and only 30% on products older than 1 years (30%).

Table 9.98 PD in relation to product

	Old Products		New Products	
	Old Products More than 3 Years with a Need for Small Adjustments	Old Products More than 3 Years with a Need for Big Adjustments	New Products Older than 1 Year with a Need for Small Adjustments	New Products Older than 1 Year with a Need for Big Adjustments
Total Average (%)	20,0	50,0	25,0	5,0

The case research showed that out of the product development projects the main product development projects were products that need big adjustments (55%). The data showed evidence of GSI's deviation from the theory of diminishing product lifecycle. This was due to the lifetime of the products which were characterised by large investments both directly and indirectly concerning installation costs etc. In an overall perspective the product development at GSI were major adjustments and major redevelopment after 3 years of lifetime. This indicated a high pressure on radical product development at GSI.

The survey also showed that 90% of the product development in the business was on known and old customer groups as seen in Table 9.99.

Table 9.99   PD in relation to customer groups

	Known and Old Customer Groups	Unknown and New Customer Groups
Total Average	90	10

Looking at the product development projects and the customer needs we could see that 80% of the product development projects were on known/old customer needs and 20% on unknown/new customer needs as seen in Table 9.100.

Table 9.100   PD projects in relation to customer needs

	Known and Old Customer Needs	Unknown and New Customer Needs
Total Average	80	20

This designates a business which dealt with rather incremental product development projects on this dimension.

On the technical level GSI claimed that 65% of their product development projects were in new technology – radical technology areas and 35% in areas of known or development areas where small incremental technology adjustments were necessary. This indicates that product development at GSI was very radical on the technological side as seen in Table 9.101.

Table 9.101   PD projects in relation to technology

	Known Technology	Known Technology with Small Adjustments (Incremental Technology)	Completely New Technology (Radical Technology)
Total Average (%)	10	25	65

The case research showed that 50% of the product development project were related to market areas with high and rival competition. Only 10% of the product development projects were in low competition areas as seen in Table 9.102.

Table 9.102   PD projects in relation to competition

	Markets with Low or No Competition	Markets with Medium or Intensive Competition	Markets with Fierce and Rival Competition
Total Average (%)	10	35	55

GSI considered 55% of their product development projects to be in high competition areas.

GSI considered 40% of their product development projects to have a high element of innovation and another 45% to be in areas with medium innovation degree as seen in Table 9.103. This indicates that the product development projects at GSI were rather radical.

Table 9.103 PD projects in relation to degree of innovation

	No Degree of Innovation – Routine NPD Project	Medium Degree of Innovation – Modified Product Development with Minor Demands on Adjustment (Incremental)	High Degree of Innovation – with Many Elements of Innovation (Radical Innovation)
Total Average (%)	15	45	40

The GSI Product Development Model

Formal Stages and Gates

In the case research GSI claims that they had a formal stage- gate model. The case research showed that the formal model had the same stages and gates as our hypothesis model.

The case research showed that GSI's product development model had 4 stages – a concept stage, a product development stage, and a process development stage. In the screening area the picture showed that GSI has 4 gates – an idea screening gate, a concept screening gate, a prototype test gate, and a process test gate as seen in Table 9.104.

Table 9.104 Stages and gates of GSI's formal PD model

Idea	Concept	PD Phase	Process Development Phase	Idea Screening	Concept Screening	Proto Type	Process Testing
Y	Y	Y	Y	Y	Y	Y	Y

The above case research results proved that the stage gate model existed at GSI and that the stage gate model was identical to our research hypothesis model.

Informal Stages and Gates

GSI confirmed that an informal product development model also existed in the business. The existence and importance of the informal product development model at GSI were interesting as can be seen in Table 9.105.

			Table 9.105	Stages and gates of GSI's informal PD model			
Idea	Concept	PD Phase	Process Development Phase	Idea Screening	Concept Screening	Proto Type	Process Testing
Y	Y	Y	Y	Y	Y	Y	Y

It was very interesting to see that the informal product development model of GSI plays a role in all parts of the product development process. My observation was that this was due to the necessity of major investments in both the formal and informal product development models and processes. Furthermore, the pressure on speed in the middle and lower part of the product development phase caused a need for informal models.

As previously stated, the informal product development model was important to the time and speed success criteria. Moreover, GSI claimed that the informal product development model influenced cost and performance success criteria but also CI and learning as seen in Table 9.106.

Table 9.106	Importance of informal PD model in relation to success criteria					
	Time	Costs	Performance	CIM	CI	Learning
Yes	Y	Y	Y	0	Y	Y
To some extent	0	0	0	Y	0	0
No	0	0	0	0	0	0
Do not know	0	0	0	0	0	0
Not answered	0	0	0	0	0	0

Focusing on CI and Learning in product development the informal product development model influenced particularly the upper and middle part of the product development phase.

Internal Functions Involved in Product Development Process

The case research at GSI showed that the following functions as shown in Table 9.105 were involved in the various stages and gates of the product development process.

Although it seemed as if the business was very focused on involvement of the product development department at the initial idea and concept stage and gates, all departments except finance were involved in the initial stages and gates of the product development as seen in Table 9.107.

Table 9.107 Functions participating in PD idea stage

	Marketing	Finance	Sales	Management	Production	Product Development	HRM
Idea Generation	Y	N	Y	Y	Y	Y	N
Concept Generation	Y	Y	Y	Y	Y	Y	N
Product Development	Y	Y	Y	Y	Y	Y	N
Process Development Phase	ISC	Y	ISC	Y	Y	Y	N
Idea Screening	Y	Y	Y	Y	Y	Y	N
Concept Screening	Y	Y	Y	Y	Y	Y	N
Proto Type Test	Y	Y	Y	Y	Y	Y	N
Process Test	Y	Y	Y	Y	Y	NA	N

Y = Yes
N = No
ISC = In some cases
DN = Do not know

Our case research showed a business based very much on team management where all levels and functions had responsibilities to product development as seen in Table 9.107. Sales, marketing, management and the product development department were the main actors in the idea stage of the product development process but HRM and production were also involved in the product development as the very radical product development projects demand also their skills and competences. The financial department entered the product development project very early in the product development process because the product development project at GSI toke a lot of resources and demanded a lot of financial calculation beforehand. Management played a major role in the teambuilding in the product development projects and as can be seen later in the case analysis, management supports the product development group with external consultancy because there was a strong need for external management competence in the heavy weight product development projects.

External Networks Involved in Product Development

PD Management

The product development projects in GSI were very much managed as a combination of internal and external product development management. However, the management of product development at GSI as seen in

Table 9.108 was strongly concentrated internally and based on a strong leadership by GSI's management group. This was quite another management model compared to the other case businesses as seen in Table 9.108.

Table 9.108 Management of projects at GSI Lumonics

Management of Project (%)	
Customer	10,0
Supplier	10,0
Marketing	5,0
Sales	10,0
Management	10,0
Production	10,0
Product Development	30,0
Common Leadership	15,0
Total	100

The argument was additionally supported when looking at the network partners involved in the product development process.

The suppliers and the customers were involved in the product development process and external consultancies were also placed with management responsibilities. The management of the product development project was a combination of team management and functions or external partners with management responsibilities of particular parts of the product development project. The customers were involved in essentially all product development phases especially in radical product development projects.

Focus of Success Criteria of Product Development Process

GSI was very much focused on time, performance, continuous innovation and learning as these criteria received priority 1 on a scale from 1 to 5 as seen in Table 9.109.

Table 9.109 Priorities of success criteria at GSI

Priorities	
Time	1
Cost	3
Performance	1
CIM	3
CI	1
L	1

However, the priorities changed during the product development stages and gates as seen in Table 9.110.

Table 9.110   GSI Lumonics' focus on succes criteria

	Time	Costs	Performance	CIM	CI	Learning
Idea	X		X		X	X
Concept	X		X		X	X
PD Stage			X		X	X
Process Stage	X		X			
PD Test			X			
Idea Screening		X	X			
Concept Screening		X	X			
Proto Type Test		X	X			
Process Test		X	X	X		
Implementation			X			

The focus on time was predominant at the process development stage and at all initial generation stages. Performance was very much in focus at all gates of the product development process. GSI put a high priority on CI and learning in the generation stage because new technology, market needs, and networks had to be learned and innovated to develop the competence of GSI. Cost was very much in focus at the concept stage and gate as product development projects were very costly for GSI to implement.

HS Enablers

The use of high speed enablers showed that all enablers were considered at GSI and more enablers than we had seen in other case businesses as seen in Table 9.111.

Table 9.111   HS product enablers used at GSI

	Very Much	In Some Cases	No	Do Not Know
ICT Communication Enabler	Yes			
Customer Enabler	Yes			
PD Model Enabler		Yes		
Network Enabler	Yes			
Innovation Enabler	Yes			
HRM Enabler	Yes			
Process Enabler		Yes		
Product to Process Enabler	Yes			
Modularisation Enabler	Yes			
E-Development Enabler	Yes	Yes		

Product Development Case – a New Welding Machine

The management team was gathered around 4 new product development tasks. The discussion centred on the way in which to develop 4 central product development projects parallel and at same time. One product development task were considered as rather incremental but however placed inside one of GSI Lumonics new products sold to one of the strategic important customers.

However, the product development management considered the product development task to be of a character that it would be potential to outsource to a sub supplier. It was decided to do so also because the existent product development department at GSI Lumonics did not have the time to do the development task.

Product Development Task

The product development task was characterised by the management group as somewhere between radical and incremental on the technological side because the technology was known – power supply, beamer, and steering was known technology. The product development project also seems to be incremental on the innovative side but turned out to be radical. The product development task was incremental on the market side as the customers were known and familiar. However, altogether the result of the product development task could be a radical breakthrough product if success was reached because it would leave the competitors behind for some span of time.

Special characteristics of the development task are shown in Table 9.112.

Table 9.112   Special characteristics of the development task

Dimension	Incremental	Radical
Where was the idea discovered	Internal the business – product development, management – due to a strong business need to gain an competitive advantage in the laser welding market specifically the automobile production	
Initiator of idea	The business	
Product type	Hardware 50% Software(knowledge of the technology to solve the laser welding process)	

(Continued)

Table 9.112   Continued

Dimension	Incremental	Radical
Consequences for product core		Modification of core and ad to functions
Placement in product development stage		Concept stage
Innovation degree		Regarded as middle but turned out high
Market		Old and mature
Customer needs		New
Customer group	Old	
Customer technology		New
Technology		New
Network		Old
Competence's		New and unknown
Product management		The sub-supplier - product development manager
Competition		High
Strategic importance		High, important, short term and critical – competitive advantage
Success criteria	Cost – middle advantage CI – none CIM – none Learning – none	Performance – high Speed – high – competitive first mover
Product development task		
PU model – formal	Stage gate	
Functions involved in initial phase	Management, product development	
Partners involved in initial phase	Sub-supplier and the business	
Enablers involved	The network enabler	

9.2 Cross-Section of Case Results

The case examination proved partly a general status for the product development situation of the businesses, partly a specific status for a specific product development course in the businesses. In this paragraph the main emphasis is on the general, empirical observations made during the case examination.

9.2.1 General Results
Product and Process Portfolio of Case Businesses

The case examination showed that the existing product portfolio differed from one business to another. Nevertheless, there was a clear focus on hardware and physical products. Similarly, the products most recently developed by the businesses were primarily hardware and physical products. However, it was evident that the share of service, knowledge, and consultancy was growing. Especially new product development was increasing in these areas.

Birth of Product Development Ideas

An examination of the case businesses showed that the ideas to the product development of the business may be borne by many different sources. However, the primary sources seem to be the customers, the sales department, the management, and the product development department. The idea creation turned out to be rather more subtle than indicated by my immediate hypothesis as the ideas are created before, during, and after the course of a product development. In particular, many small, incremental product development ideas were created after the implementation of the product on the market.

Product and Product Development Task

Additionally, the case examination demonstrated that product development was not concerned primarily with changing the actual core of the product but rather with adapting certain partial elements of the product to the market, i.e. related product development.

Thus, the core of the product could be relatively stable and fixed in its product development phase, while the "rings" surrounding the products may be under pressure and undergoing continuous development. The question was when the development in the outer rings will bear an impact on the core of the product.

Goals and Framework

It was evident from the case examination that the businesses would generally specify goals and budgets for at product development project prior to the establishment of the project. Some of the businesses were forced by their ISO 9000 standard to make such specifications whereas other businesses did it in order to set up milestones to steer by.

The overall mission of a product development project was generally less specifically formulated. On the other hand, personnel and organisational resources were most often formulated in all case businesses.

The drawing-up of a strategy and the interface to the networks partners was seldom prepared at the beginning of a product development project.

Apparently, these two areas were much more dynamic than others and were the object of continuous definition throughout the product development course concurrently with the emergence of needs and demands.

It was characteristic that the strategic importance of a product development project may influence the wording of the goals and framework of the product development project. In other words, there was a tendency to draw up all parts of the goals and frameworks of the product development in the case of important product development projects.

Product Development Task

The actual product development tasks in the case businesses were characterised by being primarily what was previously known as incremental product development tasks. The product development task were mainly found within known and old strategic areas, i.e. product market areas. Similarly, the businesses characterized the innovation degree of their project as existing mainly in the low innovation area. Thus, only very few of the case businesses worked deliberately and mainly in the high innovation area. However, major variations in the product development profile were found in the case businesses.

Similarly, the product development tasks of the businesses were characterised by being primarily placed in old markets with known customer need and known customer groups. Likewise, the customer technology – i.e. production technology – which the businesses contemplated using in order to fulfil the needs of their customers was known by the business.

On the other hand, the businesses quite often had to include several basic technologies in order to solve the product development task, and my observations generally showed an growing degree of technological mergers and mix in the solving of product development tasks. In this aspect my case research demonstrated a general pressure on the competences of the businesses in terms of mastering and acquiring the competence to mix and employ several technologies in their product development. This was also the "root" of a growing use and involvement of external network partners with special competences required in order to solve the product development task. However, the businesses were still committed to the use of known and geographically close network partners.

The product development tasks were generally aimed at competitive areas with medium to high competitive intensity. In other words, product development was used to provide the businesses with a competitive advantage or "first mover advantage" and to evade a unilateral competitive situation in which focus was solely on price and quality.

Thus, the radicality and incrementalness of a product development task can be worded in many different ways.

Product Development Life Cycle

One of the characteristics of product development were that the products of the businesses were generally in need of re-development and major adjustments after a life of three years in the market. Furthermore, there is a tendency among the businesses to carry out product development on products with only one year and less than three years of life on the market.

This indicates that the life cycles of the products were diminishing but at the same time it seemed to signify that the modularisation outlook had not been sufficiently integrated in the product development plans of the businesses as major adjustments of the products were often required.

Product Development Model of Case Businesses

The case examinations showed that the businesses were generally adhering to a formal stage-gate model with four stages and three gates. The idea phase for the businesses, however, was less clear and less formalised. Similarly, the idea and concept phases were often less formalised and burdened with a demand for "high speed".

All case businesses confirmed that they had informal product development models and processes running parallel to their formal models and processes. The informal models and processes were particularly intensive in the idea phase and in the middle part of the product development phase. The informal models and processes came into being especially when there was a pressure on time and where the need for creativity was "far above the common measure"; i.e. in cases where product development could not take its optimal course within the formally determined framework.

The persons responsible for the product development were acquainted with the informal product development models and processes and were aware of their influence on the product development success criteria.

It was an outstanding characteristic that the small businesses did not have as many informal models and processes.

Product Development Success Criteria

Generally, the case businesses focused on the short-term success criteria such as time, costs, and performance. The businesses differed widely when prioritising the importance of the success criteria. Similarly, focus on success criteria differed in accordance with the product development stages and gates. Thus, the time and performance factor was generally in focus in the idea phase, the product development phase, and the prototype test phase. The time and costs factor were in focus in the screening phase and the process development phase. Focus on long-term success criteria was very limited.

The managers of the product development of the businesses realised that the informal product development models and processes were important to the business's product development success criteria – especially time and cost but also to a lesser degree CI. The managers of product development were generally not convinced that the performance of the product was influenced by the informal product development models and processes. Furthermore, they were uncertain as to whether CIM and learning were influenced by such models and processes.

Generally, I observed that in situations where competition was fierce and turbulent, pressure on the speed of product development increased considerably. Time is generally prioritised very highly in the implementation phase.

Functions Involved in Product Development Process

Generally, the businesses had a traditional approach to identifying the functions to be included in the product development process. Marketing, sales, management, and product development were the leading functions involved in the initial phases of product development, whereas production and finance seemed to be the leading functions in the middle and – especially production – in the last phases of product development. HRM hardly ever played a part in the product development of the businesses.

It was a distinguished feature that finance only became involved in the product development at a very late point of the concept phase; often not until the concept screening phase. Several businesses only involved finance subsequent to the development of the concept which meant that the finance department could easily turn into a "bottleneck" in terms of time, budget, and acceptance of the development projects.

Management and Choice of Management for PD Projects

It was a telling example that the managers of the businesses' product development were found on the top management level. Furthermore, the choice of management was practically always made by the top management itself. On the part of the management focus was specifically on the short-term success criteria and PDM. In other words, the PDM and PDL level was thoroughly mixed.

The case examination showed that there could be various approaches to the managerial style when the objective was to ensure high speed in product development. The very authoritarian managerial style could be exceptionally advantageous in the middle and the end of the product development phase when there was a need to put extraordinary pressure on time in order to reach the market in time. Similarly both the authoritarian and the flexible and team based managerial style would produce excellent results at the time of concept development or when the product development is "stuck" and there was a need for extraordinary measures to exceed the limits of product innovation.

Generally, in terms of management, the introduction to the idea phase and the actual idea phase seemed to be less organised and less controlled. Though in some cases this may be accidental.

Furthermore, the case examination showed that small businesses tend to manage their product development projects themselves and directly by the top management.

Participation in External Networks

The businesses examined tended to involve their customers in the first phases of product development; specifically the idea phase. Most often the suppliers were involved at a later point of the product development phase. The competitors were hardly ever involved in the product development of the examined businesses. The use of other external product development partners was strongly limited and primarily focused on the idea and concept generation phases. However, there were major variations in the management and managerial style of the businesses when the businesses represented both the hierarchical, traditional managerial style, the flexible managerial style, the very decentralized managerial style, and the team based managerial style.

The case examination showed that businesses placed in markets with fierce competition were less prone to enter into network cooperation.

Use of High Speed Enablers

High speed enablers employed by the businesses were generally the customer enabler, the product development model enabler, and the modularisation enabler. To a lesser extent the network enabler was in focus. The businesses' use of the e-development enabler was limited. However, there were major differences between the businesses' use of the high speed enabler.

9.2.2 Summary of Case Examination

Table 9.113 sums up on the main results of my case examination.

Table 9.113   Empirical results – pilot case study

Dimension	Results
Product Type	Primarily physical products (65%), service (20%), and knowledge and consultancy (5%) Primarily physical processes (70%), digital processes (30%), and virtual processes (0%)
Idea Generation	Before, during, and after the product development phase Primarily by customers, sales, and product development
Consequences to Core of Product	Related product development, seldom alterations of the core of the product
"Goals and Framework"	Mission not always formulated Goals and budget not always formulated Use of personnel/organisational resources formulated and specified Strategy, interface, and contact to network partners normally not formulated. More dynamic pari passu with the arising needs and demands from product development
Strategic Area	Known strategic areas (95%), unknown strategic areas (5%) Low degree of innovation (85%), high degree of innovation (15%) Old markets (80%), new markets (20%) Old customer needs (85%), new customer needs (15%) Old customer groups (90%), new customer groups (10%) Known customer technology (70%), unknown customer technology (30%) Technologically old areas (65%), technologically new areas (35%) Known networks (90%), new networks (10%) Weak competitive environments (10%), medium competitive environment (30%), fierce competitive environment (60%)

(Continued)

Table 9.113 Continued

Dimension	Results
Innovation Degree	
Market	
Customer Needs	
Customer Group	
Customer Technology	
Technology	
Network	
Competition	
Product Life Cycle	Generally diminished
	Often need for product development adjustments
	Often need for product development on products with less than 3 years
	On the market
Product Development Model	Mainly stage-gate model
	Normally 4 stages and 3 gates
	Informal product development models exist in the businesses often as a result of pressure on time and creativity
Management of Product Development	Very focused on short-term success criteria; time, costs, and performance
	Very focused on PDM and hardly ever on PDL
	Top-managed; choice of manager made by top-management itself
	Different managerial styles authoritarian, decentralized, and team based managerial style
Functions Involved in Product Development	Traditional functions involved; marketing, sales, product development, and management primarily in the initial phases. Finance in the concept screening phases. Product typically not involved until the last part of the concept phase and in the product development phase. HRM not involved in product development phase
Prioritizing Success Criteria for Product Development	Performance – high, especially in the idea and prototype phase
	Costs – medium to high especially in the screening phase and the last parts of the product development phase
	Time – High in idea, screening, and product development phase
	CI – hardly ever prioritised
	CIM – hardly ever prioritised
	Learning – hardly ever prioritised

9.3 Summary

Table 9.114 shows the results of verification on the different hypotheses.

Table 9.114 Verification table

Chapter 9 Empirical Results – Pilot Case Studies		Verified/Not Verified
Overall Research Questions to be Verified	Hypotheses to be Verified and Tested	
1. What is network based high speed NPD	HS NPD can be seen from different view (Macro environment, business, product, market, customer, technology, competitive and network view)	Verified
	HS NPD is a matter of right speed and not high speed.	Verified
2. What enablers to NB HS PD can be identified?	Businesses use different HS enablers.	Verified
	HS enablers are identical to the 10 enablers – 1–10	Partly verified
	There can be more than these 10 enablers to HS PD	Verified
	The enablers will play a different role according to the PD situation and project (Secondary focus)	Verified
	The customer enabler, the network enabler and the PD model enabler plays an important role in the upper phase of the HS PD phase.	Partly verified
3. What framework models and processes in the idea and concept stage/gate of high speed product development based on networks can be measured	The HS PD projects can be divided into to radical and incremental PD projects	Verified
	The radical and the incremental PD projects follow different generic HS PD models and processes and can thereby be described by different generic frameworks	Partly verified
4. What success criteria can be used for measuring high speed product development based on networks?	The success criteria for HS PD are dependent on the specific PD project – radical or incremental	Verified

Table 9.114 Continued

Chapter 9 Empirical Results – Pilot Case Studies		Verified/Not Verified
Overall Research Questions to be Verified	Hypotheses to be Verified and Tested	
	HS PD success criteria can be formulated as short term and long term success criteria	Verified
	Time, cost and performance are central success criteria in a short term perspective	Verified
	Continuous improvement (CIM), continuous innovation (CI), and learning are central success criteria in a long term perspective to reach right time, right cost and right performance in NB HS PD.	Not verified

10

Empirical Results – PUIN Focus Group Meetings

This chapter presents the results of the semi-structured focus group meetings held in the PUIN network group focusing on network based high speed product development. The focus group meeting had been refined to comprise 10 focus group meetings held during 2001 and 2002. At these meetings different themes of NB HS NPD have been discussed. The chapter documents the results of these discussions together with the book Network Based Product Development which also documents the empirical results and the industry's own writing on NB HS NPD. The results of the semi-structured focus group meetings showed the businesses' involvement in NB HS NPD and the way in which they worked with NB HS NPD. The chapter gives some in-depth empirical results which the researcher could not have gathered through case research, survey, or other empirical observations.

10.1 Introduction

On the basis of ten exploratory semi-structured focus group studies carried out in SME businesses the empirical focus group results can be presented. All businesses took part in the semi-structured focus group meetings together with 3 researchers from CIP. An example of a meeting agenda and a semi-structured questionnaire can be found in Table 10.1.

In Chapter 2 the methodology of these focus group interviews was described in details. The focus group interviews were supported by questions filled out by the businesses beforehand, news group discussions, and telephone interviews to give additional support and discussions on specific questions.

The businesses' general profiles can be found in appendix together with their website addresses.

Table 10.1   General themes and agendas for focus group meetings

General Theme and Agenda for the 10 Focus Group Meetings
General conditions and Trends in PD
The task of PD – Radical or incremental
Enablers to NB HS product development
The core of a HS PD project
HS PD models and networks
Success criteria of HS PD
Time and HS PD
Speed and HS PD
Cost and HS PD
Performance and HS PD

The individual focus group interview meetings have had 10 different agendas or themes which were:

Each agenda will be described and analysed within the framework model of the network based high speed product development prepared in Chapter 8.

The aim of the chapter is:

- to verify, test and give answers to the research hypotheses and questions set up earlier in Chapter 1.
- to show and verify different NB HS NPD aspects carried out under different characteristics in the field of product development
- to show different SME businesses' solutions to and reflections on NB HS NPD
- to reflect on the different consequences which high speed and right speed would have on different parameters as shown in Table 10.2

Table 10.2   Consequences of high speed/right speed

Consequences	High Speed	Right speed
Time		
Cost/Value		
Performance		
Market fit		
Risk		
Security		

In Table 10.3 the contributions of each focus group meeting to the research questions are shown.

A brief introduction to the focus group meetings and the way in which they were conducted is given below.

Table 10.3 Hypotheses to be verified

Empirical Results – Focus Group Interview

Overall Research Questions to be Verified	Hypothesis to be Verified and Tested	Meeting 1 — General Conditions and Trends in PD	Meeting 2 — The Task of PD – Radical or Incremental	Meeting 3 — Enablers to NB HS PD	Meeting 4 — The Core of a HS PD Project	Meeting 5 — HS PD Models and Networks	Meeting 6 — Success Criteria of HS PD	Meeting 7 — Time and HS PD	Meeting 8 — Speed and HS PD	Meeting 9 — Cost and HS PD	Meeting 10 — Performance and HS PD
What is network based high speed NPD?	HS NPD can be seen from different views (macro environment, business, product, market, customer, technology, competitive, and network view)	X	X								
	HS NPD is a matter of right speed and not high speed.	X	X	X		X	X	X	X	X	X
What enablers to NB HS PD can be identified?	Businesses use different HS enablers.		X	X		X	X	X	X	X	X
	HS enablers are identical to the 10 enablers – 1–10.			X		X	X	X	X	X	X
	There can be more than these 10 enablers to HS PD.			X	X	X	X	X	X	X	X
	The enablers will play a different role according to the PD situation and project (Secondary focus).			X		X	X	X	X		
	The customer enabler, the network enabler, and the PD model enabler plays an important role in the upper phase of the HS PD phase.	X		X	X	X	X	X	X	X	X
What framework models and processes in the idea and concept stage/gate of HS PD based on networks can be measured?	The HS PD projects can be divided into radical and incremental PD projects.	X	X	X	X	X	X	X	X	X	X

(Continued)

Table 10.3   Continued

Empirical Results – Focus Group Interview		Meeting 1 — General Conditions and Trends in PD	Meeting 2 — The Task of PD – Radical or Incremental	Meeting 3 — Enablers to NB HS PD	Meeting 4 — The Core of a HS PD Project	Meeting 5 — HS PD Models and Networks	Meeting 6 — Success Criteria of HS PD	Meeting 7 — Time and HS PD	Meeting 8 — Speed and HS PD	Meeting 9 — Cost and HS PD	Meeting 10 — Performance and HS PD
Overall Research Questions to be Verified	Hypothesis to be Verified and Tested										
	The radical and the incremental PD projects follow different generic HS PD models and processes and can thereby be described by different generic frameworks.		X	X		X					
What success criteria can be used for measuring HS PD based on networks?	The success criteria for HS PD are dependent on the specific PD project – radical or incremental.						X	X	X	X	X
	HS PD success criteria can be formulated as short term and long term success criteria.		X				X	X	X	X	X
	Time, costs, and performance are central success criteria in a short-term perspective.	X						X	X	X	X
	Continuous improvement (CIM), continuous innovation (CI), and learning are central success criteria in a long term perspective so reach right time, right cost and right performance in NB HS PD.	X					X	X	X	X	X

10.2 Focus Group Meetings

The meeting was held either at the CIP Centre in Aalborg or at one of the businesses. When meetings were held at the CIP Centre, one researcher introduced the theme for the focus group meeting by introducing the theoretical work and theoretical view from different researchers in the world on the specific theme. Subsequently, the discussion began in the focus group and the researchers introduced the first question of the focus group meeting. The researchers observed the discussion between the businesses and introduced a new question from the semi-structured questionnaire when appropriate.

When meetings were held at the businesses, the host business showed the visiting businesses around the premises after which they introduced their business. Subsequently, the host business gave a short introduction to the theme on the agenda and explained how they work with this theme in their business. This introduction introduced the focus group interview and discussion which was reflected on and concluded by one of the researchers.

10.3 General Conditions and Trends in PD

The focus group businesses' product portfolio was strongly focused on physical and service products. However, on the new product introduction a stronger focus on knowledge and consultancy products could be seen. The businesses claimed that the products moved from purely focused on physical products to immaterial products.

The businesses saw a trend and reaction from the market on more needs and wants for knowledge and consultancy products as seen in Table 10.4. Many of the businesses were just beginning to penetrate this product potential. The decrease in service product development should mainly be seen as a result of the businesses' former strong development in this area. Additionally, the existing service products could probably already or with incremental development solve the customers' demands for service products.

Table 10.4 Focus on product types

	Physical Products	Service Products	Knowledge and Consultancy
Existing product portfolio	83,5	15,0	1,5
New products	83,0	8,5	8,5

When looking at the businesses' products in another dimension, it appeared that today 86% of the products are physical products, 13% were digital products, and only 1% are virtual products as seen in Table 10.5. This meant that hardly any of the businesses had yet begun to offer virtual products to the market.

Table 10.5   Focus on products and processes

	Physical Products	Digital Products	Virtual Products
Existing product portfolio	86	13	1
	Physical processes	Digital processes	Virtual processes
	83	16	1

Up to 2003, the business had 83% on physical processes and 16% on digital processes.

The focus group interviews showed very clearly that the businesses did not think of the product as a process. The product to process thinking and enabler was not yet introduced in the businesses. The businesses still saw the products as physical "encapsulated" products with a start and an end – products with the classic life-cycle.

The focus group businesses were asked to make general comments on their view of the conditions in "the field of product development". These comments are collected in Table 10.6 which also present a detailed presentation of the comment.

Table 10.6   Comments on conditions in "the field of PD"

The Main Components Context	Characteristics
Market Stable markets Evolving markets Dynamic markets	Most SMEs claimed that they were operating in stable to evolving market with customers who have mostly incremental development in preferences.
Technology Stable technology Evolving technologies Dynamic technologies	The businesses claimed that their markets were under pressure of new evolving and some times unknown technologies. The technology gave the businesses new technological possibilities but the technological possibilities were often ahead of market demand.
Network Stable networks Evolving networks Dynamic network	The businesses general involvement in networks was mainly based on physical and stable, narrow networks often internal and dominated network.
	However, a slightly new evolvement of networks based on a mix of new evolving system of networks – both physical networks and ICT networks were recognized. None of the businesses were joining virtual networks.

Table 10.6   Continued

	Only very few of the businesses joined networks based on a mix of dynamic networks with a high degree of dynamic where network partners constantly come and go. None of the businesses had joined a network with no formal network leader.
Business competence context	The businesses felt that there was a high pressure on support competenccs and that they had to develop on complementary competences either by internal development or by external recruiting in their networks. A high pressure on the businesses' core competences were realised and some of the businesses felt that their competitive advantage on core competences was reduced or diminished by competitors.

This gave the following framework picture as shown in Figure 10.1:

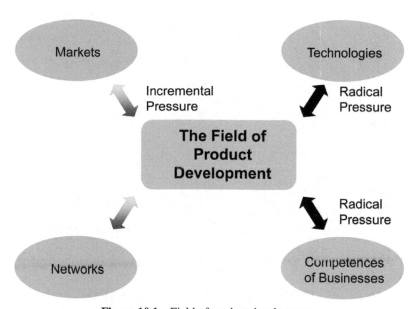

Figure 10.1   Field of product development.

10.3.1 Product Development Tasks

On the basis of the focus group interviews the businesses' task of product development could be verified as seen in Table 10.7.

Table 10.7   Product development task of focus group businesses

	Physical Products	Service Products	Knowledge and Consultancy
Existing product portfolio	83	15	2
New products	81,5	9,5	9
Product development	81	8	11

Of the business's product development tasks 81% could be related to hardware or physical products whereas 19% could be related to service and knowledge products. Obviously, the businesses' product portfolio mainly focused on physical products and service products today as seen in Figure 10.2.

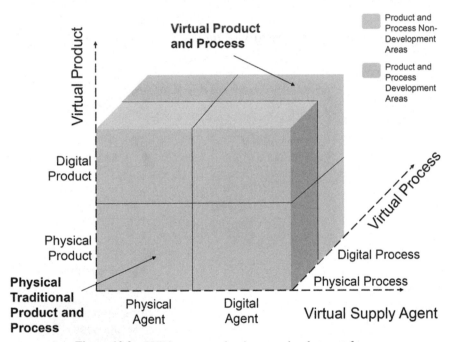

Figure 10.2   PUIN group product/process development focus.

However, the introduction of new products and the product development efforts were now changing and focusing more on knowledge and consultancy products.

The businesses also claimed that product development projects could in general be divided into 80% strategic known and old areas and 20% unknown and new areas as seen in Table 10.8.

Table 10.8 PD projects in relation to strategy

	Strategic Areas		Market	Technology	Network	Business
	Known an Old Areas	Unknown and New Areas				
Lyngsø	40	60	Evolving and dynamic	Dynamic	Evolving and dynamic	Evolving and dynamic
Linco	80	20	Evolving	Evolving	Stable and evolving	Evolving
AKV	95	5	Stable	Stable	Stable	Stable
B&O	80	20	Evolving	Evolving and dynamic	Stable and evolving	Evolving
Danfoss	95	5	Stable and evolving	Evolving and dynamic	Stable and evolving	Evolving
NEG Micon	90	10	Stable and evolving	Evolving and dynamic	Stable and evolving	Evolving
Ansager	90	10	Stable	Stable	Stable	Stable
Scanio	80	20	Stable and evolving	Stable and evolving	Stable and evolving	Stable and evolving
GSI	85	15	Evolving and dynamic	Evolving and dynamic	Evolving and dynamic	Evolving and dynamic
Grundfoss	70	30	Stable and evolving	Evolving and dynamic	Stable and evolving	Stable and evolving
Total Average	80	20				

Nevertheless, there were considerable variations in the businesses' product development activities. It was verified that the focus in PD was very much related to the characteristics of the specific elements on "the field of product development". It was verified that those businesses who face evolving and dynamic characteristics on their elements on the field of product development had more focus on unknown and new strategic areas. They were influenced and pressed by elements from outside to focus their product development on unknown and new strategic areas; in other words, a push towards more radical product development activities. It also seemed as if these businesses felt a higher pressure on speed and time in product development.

When applying the product/market model to the product development projects of the businesses, it appeared that product development projects at the businesses could generally be characterized as incremental product development.

Additionally, the businesses' product development projects were mainly development of old products more than three years old.

The data showed evidence of previous statements about a diminishing product lifecycle. 48% of the businesses' products needed product development after 1 year. In an overall perspective, 49% of the businesses' products needed big redevelopment and 48% of the products needed redevelopment after 1 year's lifetime. This indicates a pressure on product development at the businesses in 2003 as seen in Table 10.9.

Table 10.9   Product development in relation to product

Average need of product development after 1 year	48%
Average need of big redevelopment in products	49%
Average products need for redevelopment after 1 years lifetime	48%

The focus group interviews showed that there were considerable differences in the pressure on product development from one business to another and that the characteristics in the field of PD exceeds influence on this pressure for speed in product development. Dynamic and evolving elements in the field of product development increase the pressure on the product development of the businesses.

The focus group interviews also show that 82% of the product development in the businesses – Table 10.10 – was on known and old customer groups. This verified the incrementalness on account of which the customers were going to buy the product development of the businesses in the future.

Table 10.10 PD in relation to customer groups and needs

	Known and Old Customer Groups	Unknown and New Customer Groups	Known and Old Customer Needs	Unknown and New Customer Needs	Market	Technology	Network	Business Competences
Lyngsø	40	60	40	60				
Llnco	90	10	80	20	Evolving and dynamic	Dynamic	Evolving and Dynamic	Evolving and dynamic
AKV	90	10	95	5	Evolving	Evolving	Stable and evolving	Evolving
B&O	80	20	50	50	Stable	Stable	Stable	Stable
DANFOSS	95	5	95	5	Evolving	Evolving and dynamic	Stable and Evolving	Evolving
NEG MICON	95	5	90	10	Stable and evolving	Evolving and dynamic	Stable and Evolving	Evolving
Ansager	90	10	80	20	Stable and evolving	Evolving and dynamic	Stable and Evolving	Evolving
Scanio	70	30	70	30	Stable	Stable	Stable	Stable
GSI	90	10	80	20	Stable and evolving	Stable and evolving	Stable and evolving	Stable and evolving
Grundfoss	80	20	65	35	Evolving and dynamic	Evolving and dynamic	Evolving and dynamic	Evolving and dynamic
Total Average	82	18	75	26	Stable and evolving	Evolving and dynamic	Stable and Evolving	Evolving

Looking at the product development projects and at the customers' needs, we realized that 75% of the product development projects were related to known and old customers' needs. This also indicated that there was not much radicalness in this variable of product development. However, it must be said that there was variation in the numbers. Again it can be related to the evolving or dynamics of the elements in "the field of product development". The data also gave some indication of which businesses were experiencing major changes of customers needs. Businesses like Lyngsø, B&O, and Grundfos experience considerable changes in customers needs and were also in an evolving and dynamic field of product development.

The above, however, indicated that the businesses deal with rather incremental product development projects and that the businesses know the characteristics of "the field of product development" very well. During the meetings the following comments supported this argument:

> *"Generally speaking, we only perform incremental product development – in 95% of the cases. Advanced models are only used in connection with large, radical product development assignments."*
> (Danfosss)

On the technical level the businesses claimed that 27% of their product development projects involved new technology – radical technology areas and 25% of their projects were carried out in known areas or in development areas in which small, incremental technology adjustments were necessary as seen in Table 10.11.

The businesses said that technology was the main actor increasing pressure on speed in product development. It was also verified that some of the businesses are more pushed by technological evolvement than others, and that some businesses which were not pushed by the market and customer element now are strongly pushed by the technology to speed PD further.

The case research also showed that 72% of the product development projects were related to market areas with fierce and rival competition as seen in Table 10.12

As pointed out by Richard Leifers (2002), it was very interesting to see that businesses developing in rather radical product development areas as e.g., Lyngsø, and Grundfos did not feel so hard and fierce a competition in some market areas. The focus group interviews also verified that businesses dealing with radical product development do not have such a high pressure on speed in product development.

Table 10.11 PD projects in relation to technology

	Known Technology	Known Technology with Small Adjustments (Incremental Technology)	Completely New Technology (Radical Technology)	Market	Technology	Network	Business Competences
Lyngsø	40	30	30	Evolving and dynamic	Dynamic	Evolving and Dynamic	Evolving and dynamic
Linco	15	15	70	Evolving	Evolving	Stable and evolving	Evolving
AKV	80	15	5	Stable	Stable	Stable	Stable
B&O	50	20	30	Evolving	Evolving and dynamic	Stable and Evolving	Evolving
DANFOSS	70	20	10	Stable and evolving	Evolving and dynamic	Stable and Evolving	Evolving
NEG MICON	30	50	20	Stable and evolving	Evolving and dynamic	Stable and Evolving	Evolving
Ansager	80	15	5	Stable	Stable	Stable	Stable
Scanio	70	20	10	Stable and evolving	Stable and evolving	Stable and evolving	Stable and evolving
GSI	10	25	65	Evolving and dynamic	Evolving and dynamic	Evolving and dynamic	Evolving and dynamic
Grundfoss	35	40	25	Stable and evolving	Evolving and dynamic	Stable and Evolving	Stable and Evolving
Total Average	48	25	27	Evolving	dynamic	Evolving	Evolving

Table 10.12 Product development projects in relation to competition

	Markets with Low or No Competition	Markets with Medium or Intensive Competition	Markets with Fierce and Rival Competition	Market	Technology	Network	Business Competences
Lyngsø	30	20	50	Evolving and dynamic	Dynamic	Evolving and Dynamic	Evolving and dynamic
Linco	0	50	50	Evolving	Evolving	Stable and evolving	Evolving
AKV	30	20	50	Stable	Stable	Stable	Stable
B&O	0	0	100	Evolving	Evolving and dynamic	Stable and Evolving	Evolving
DANFOSS	0	0	100	Stable and evolving	Evolving and dynamic	Stable and Evolving	Evolving
NEG MICON	0	0	100	Stable and evolving	Evolving and dynamic	Stable and Evolving	Evolving
Ansager	0	60	40	Stable	Stable	Stable	Stable
Scanio	10	20	70	Stable and evolving	Stable and evolving	Stable and evolving	Stable and evolving
GSI	10	35	55	Evolving and dynamic	Evolving and dynamic	Evolving and dynamic	Evolving and dynamic
Grundfoss	0	0	100	Stable and evolving	Evolving and dynamic	Stable and Evolving	Stable and Evolving
Total Average	8	21	72				

Table 10.13 PD projects in relation to degree of innovation

	No Degree of Innovation – Routine NPD Project	Medium Degree of Innovation – Modified Product Development with Minor Demands on Adjustment (Incremental)	High Degree of Innovation – Innovation with Many Elements of Innovation (Radical Innovation)	Market	Technology	Network	Business Competences
Lyngsø	25	25	50	Evolving and dynamic	Dynamic	Evolving and Dynamic	Evolving and dynamic
Linco	30	35	35	Evolving	Evolving	Stable and evolving	Evolving
AKV	25	25	50	Stable	Stable	Stable	Stable
B&O	30	40	30	Evolving	Evolving and dynamic	Stable and Evolving	Evolving
DANFOSS	50	45	5	Stable and evolving	Evolving and dynamic	Stable and Evolving	Evolving
NEG MICON	30	50	20	Stable and evolving	Evolving and dynamic	Stable and Evolving	Evolving
Ansager	0	80	20	Stable	Stable	Stable	Stable
Scanio	10	20	70	Stable and evolving	Stable and evolving	Stable and evolving	Stable and evolving
GSI	15	45	40	Evolving and dynamic	Evolving and dynamic	Evolving and dynamic	Evolving and dynamic
Grundfoss	5	15	80	Stable and evolving	Evolving and dynamic	Stable and Evolving	Stable and Evolving
Total Average	22	38	40		dynamic	Evolving	Evolving

The businesses considered 40% of their product development projects as having a high element of innovation – radical innovation. This designated therefore a rather radical element of the product development projects at the businesses. When dealing with innovative elements the businesses tended to focus primarily on internal elements – the businesses internal competences – and how the businesses' competences matched the challenge on the tasks of product development put forward by the businesses as seen in Table 10.13.

10.4 The Task of PD – Radical or Incremental

From the comments given above the following picture of the businesses' product development task related to radical and incremental product development could be drawn as shown in Table 10.14.

Table 10.14   PD task in relation to radical and incremental PD

Dimension	Incremental	Radical
Where was the idea discovered	On the marketplace	
Initiator of idea	Customer	
Product type	Hardware 5% Software 95%	
Consequences for product core		Modified core
Placement in product development stage		Concept stage
Innovation degree		Low
Market		Old and mature
Customer needs		Evolving
Customer group	Old	
Customer technology	Old	
Technology		Old and stable
Network		Old
Competence's	Old and known	
Product management	The business	
Competition	High	
Strategic importance	High, important, short term and critical – survival	
Success criteria	Performance – high Cost – high Speed – very high CI – none CIM – some Learning – none	
Product development task	⬅━━━━━➡	
PU model – formal	Stage gate	
Functions involved in initial phase	Sales and production	
Partners involved in initial phase	Business and customer	

AKV and other similar businesses very seldom or only by coincidence worked with radical product development projects as their market, technology, network, and competences were very stable.

Ansager was mainly dealing with incremental physical product development because the market was not prepared to new developments on digital and virtual products nor were they prepared for development on immaterial products. Lyngsø worked with very radical PD projects but mainly on the technological and market specific area. Linco Trading worked mainly with incremental product development but often had to change incrementally products that were established a few years previously mainly because of a fast development on the markets of Linco's customers. NEG Micon, Grundfos and Danfoss perform mainly incremental product development on the technological side. However the demand on all markets was evolving e.g. for the windmills the market was turning to a demand for bigger windmills. NEG Micon were now facing a time for a radical product development on the mills as there was a physical limit to how large a gearbox they could build.

Scanio was mainly dealing with incremental product development on the meat machines whereas B&O was dealing with incremental and sometimes radical product development on the design (performance area). GSI Lumonics was dealing with radical product development on their laser welding machinery. These developments was heavily performed together with their customers.

General Product Development in Focus Group Businesses

The focus group research verified the existence of the following general characteristics of the sources to ideas to product development at the businesses.

The product development ideas came mainly from product development (31%), the sales function (21%), and from the customers (19%). Nevertheless, the variation from business to business was large as can be seen. However, the general picture indicated that the product development of the businesses was generally strongly based on identified needs and wants in the market and the businesses in general were strongly customer/sales oriented. However, there was a trend that an increasing number of ideas in the business were coming from network partners. In general, the businesses had a feeling that sources to new ideas in the future would come more from network partners and that businesses would join more NB PD projects as seen in Table 10.15.

Table 10.15   Sources of product development ideas in general

Sources to Product Development Ideas in General	Lyngsø	Linco	AKV	B&O	DANFOSS	NEG MICON	Ansager	Scanio	GSI	Grundfoss	In Per cent
Customers	36	20	15	0	40	10	20	20	25	5	19
Suppliers	9	5	5	0	0	5	0	10	10	0	4
Marketing	0	5	0	0	10	5	0	0	5	10	3
Finance	0	0	0	0	0	0	0	0	0	0	0
Sales	27	20	50	5	15	15	45	0	20	10	21
Leadership/ Management	9	10	0	15	0	0	0	5	5	5	4
Production	5	10	25	0	5	5	10	10	5	15	9
Product Development	9	20	0	80	20	45	15	50	20	50	31
Human Resources	0	0	0	0	5	0	0	0	0	0	0
Competition	5	20	5	0	5	15	10	10	5	5	8
Others	0	0	0	0	0	0	0	0	0	0	0
Do not know	0	0	0	0	0	0	0	0	0	0	0
Total	100	100	100	100	100	100	100	100	100	100	100

It was verified very clearly that the area of attracting ideas was not particularly in focus of high speed in any of the businesses. The businesses liked to get new ideas but they were not focussing on speeding the processes.

10.5 Core of HS PD Project

The core of the product development projects was generally formulated in the strategic level inside the businesses. Most businesses formulated the core of the PD with an inside-out view. In general, the formulation of the core was made in this way because the businesses' ISO 9000 standard demanded such specification. 7 out of 10 businesses were ISO 9000 certified.

This was further supported by the fact that formal goals and limits (goals, costs, resources etc.) for the product development project were always specified. This is illustrated in Table 10.16.

Table 10.16 Goals and limits to product development

Definition of Goals and Limits to PD Projects	In Percentage			
At the Beginning	Yes	No	Do not Know	Total
Mission	80	20	0	100
Goals	100	0	0	100
Strategy	60	10	30	100
Economic Resources	90	10	0	100
Personnel/Organisational Resources	70	20	10	100
Contact Limits to Network Partners	60	20	20	100

In most businesses the goals and limits for the product development projects in the businesses were defined in details in the following areas as can be seen in Table 10.16.

The businesses maintain that these specifications practically always helped the business to reach the success criteria for the product development project. Generally, the focus group businesses could be characterized as planning oriented businesses with a high focus on ISO 9000 standards. This meant high performance, high quality and stick to the rules of the ISO 9000 standards.

However, there were some differences in the above-mentioned from one business to another as can be seen in detail in Table 10.17.

As indicated in Table 10.17, the product development core was not always initially determined in some of the businesses.

Table 10.17   Goals and limits to product development

Definition of Goals and Limts to Product Development Project	Lyngsø	Linco	AKV	B&O	DANFOSS	NEG MICON	Ansager	Scanio	GSI	Grunfoss
ved begyndelsen	No	Yes	No	Yes	Yes	No	In most cases	Yes	Yes	In most cases
Mission	Yes	Yes	Yes	Yes	Yes	No	Yes	Yes	Yes	No
Goals	Yes	Yes	Yes	Yes	Yes	Yes	Yes	Yes	Yes	Yes
Strategy	Yes	Yes	DNT	Yes	Yes	No	nsir	Yes	Yes	DNT
Economic Resources	Yes	No	Yes	Yes	Yes	Yes	Yes	Yes	Yes	Yes
Personnel/Organisational Resources	Yes	No	Yes	Yes	Yes	No	DNT	Yes	Yes	Yes
Contact Limits to Network Partners	Yes	No	Yes	Yes	DNT	Yes	DNT	Yes	Yes	No

Yes
Yes
No
DNT   Do not know

10.6 HS PD Models

10.6.1 Formal Stages and Gates

In the focus group interview 6 businesses claimed that they had a formal stage gate model. These models were shown and discussed at the focus group interview meetings. The models can be seen in appendix. In addition, the focus group meetings could verify that many of the models were defined absolutely in accordance with the ISO 9001 standard. The stage and gates of the businesses formal product development model are shown in Table 10.18.

The case research showed that in general the businesses' formal product development models had three stages – a concept stage, a product development stage, and a process development stage. 40% of the businesses claimed that in general they had no formal idea stage but the initial phase of the product development process started with the concept stage.

In the screening area, the picture showed that the businesses had three gates – a concept screening gate, a prototype test gate, and a process test gate.

The above focus group interview results proved that the stage gate model existed at the businesses. Nevertheless, the results also proved that the stage gate model was slightly different from our research hypothesis model because the idea stages and gates did not formally exist in many of the businesses. Additionally, in many businesses (40%) there were no concept gates. The businesses go directly to prototyping. During the focus group meetings it was verified that many of the businesses did not use much time on the screening phases and these phases were under high speed pressure.

10.6.2 Informal Stage and Gate

The focus group interview showed that there was an informal model running parallel to the formal model in 60% of the focus group businesses. Some informal PD models were running all the time and in other businesses ad hoc when needed. The businesses confirmed the existence of such an informal product development model. The content existence of the informal product development model in different areas at the businesses are shown in Table 10.19.

It was very interesting to see that the businesses' informal product development model in nearly all businesses contains all stage and gates as defined in the research framework model. The idea stage and gate existed in most businesses only on an informal basis. The businesses claimed that the reason for this was that the idea stage and gate could not "live" in a strict ISO 9000

Table 10.18   Stages and gates of Lyngsø's formal PD model

	Idea	Concept	PD Phase	Process Development Phase	Idea Screening	Concept Screening	Proto Type	Process Testing	
Lyngsø	N	Y	Y	Y	N	Y	Y	Y	ISO 9001
Linco	Y	Y	Y	Y	Y	Y	Y	Y	ISO 9001
AKV	N	Y	Y	Y	N	Y	Y	Y	ISO 9001
B&O	Y	Y	Y	Y	Y	Y	Y	Y	ISO 9001
DANFOSS	Y	Y	Y	Y	Y	Y	Y	Y	ISO 9001
NEG MICON	N	Y	Y	Y	N	N	Y	N	ISO 9001
Ansager	N	N	N	N	N	N	N	N	ISO 9001
Scanio	Y	N	Y	Y	Y	N	Y	Y	
GSI	Y	Y	Y	Y	N	Y	Y	Y	ISO 9001
Grundfoss	Y	Y	Y	Y	N	N	Y	Y	ISO 9001
Total Average Yes	60	80	90	90	40	60	90	80	
Total Average No	40	20	10	10	60	40	10	20	

Table 10.19 Stages and gates of informal PD model of focus group businesses

	Idea	Concept	PU Phase	Process Development	Idea Screening	Concept Screening	Prototype	Process Testing
Lyngsø	Y	Y	Y	Y	Y	Y	Y	Y
Linco	N	Do not know	Y	Y	N	N	Y	Y
AKV	Y	Y	Y	Y	Y	Y	Y	Y
B&O	N	N	N	N	N	N	N	N
DANFOSS	N	Y	Y	N	N	N	Y	N
NEG MICON	N	N	N	N	N	N	N	N
Ansager	Y	Y	Y	Y	N	Y	Y	Do not know
Scanio	N	N	N	N	N	N	N	N
GSI	Y	Y	Y	Y	Y	Y	Y	Y
Grundfoss	N	N	N	N	N	N	N	N
Total Average Yes	40	50	60	50	30	40	60	40
Total Average No	60	40	40	50	70	60	40	50
Total Average Do Not Know	0	10	0	0	0	0	0	10

model with formal procedures etc. Furthermore, the businesses claimed that because of demands of ISO 9000 when they were ready for conceptualising the idea they had to "put" the idea into the formal stage-gate product development model.

Those businesses who had an informal product development model claimed that the informal product development model was important for all listed success criteria in the businesses. This state of affairs was illustrated in Table 10.20 below.

Table 10.20   Importance of informal PD model in relation to success criteria

	Time	Costs	Performance	CIM	CI	Learning
Lyngsø	Y	Y	Y	Y	Y	Y
Linco	Y	N	N	Y	Y	Y
AKV	Y	Y	TSE	TSE	N	TSE
B&O	N	N	N	N	N	N
Danfoss	TSE	TSE	N	TSE	TSE	TSE
NEG Micon	N	N	N	N	N	N
Ansager	Y	Y	TSE	N	TSE	TSE
Scanio	N	N	N	N	N	N
GSI	Y	Y	Y	TSE	Y	Y
Grundfos	N	N	N	N	N	N
Total Average Yes	50	40	20	20	30	30
Total Average TSE	10	10	20	30	20	30
Total Average No	40	50	60	50	50	40

The informal product development models influenced in particular time. Also costs were influenced but not in particular performance. According to the businesses, the long term success criteria were not particularly influenced by the informal PD model.

The Informal Product Development Process

The focus group interviews verified that an informal product development process existed in many businesses and that it carries an impact on all success criteria of the businesses' product development projects. The focus group interview showed more details on the running of informal processes at the idea and concept stage as well as on the influence on time and speed in the product development process.

Internal Functions Involved in Product Development Process

In the focus group interview the following functions showed to be involved at the different stages and gates of the product development process.

The businesses had a rather traditional involvement of functions at the product development stage and gates. Although it was verified that the business was very focused on the involvement of sales, management and the product development department at the initial idea and concept stage and gates.

Sales, marketing, product development, management and production were the main actors at the idea stage of the product development process. In large businesses the marketing functions were more involved at the initial stage and gates in PD. HRM and finance were hardly ever involved in the initial product development phase and in many businesses production only comes in occasionally in the initial phases of the product development stage. The businesses would see many of these functions on this early stage of the PD process. Neither finance or HRM were of relevance here as seen in Tables 10.21 and 10.22.

Table 10.21 Functions participating in PD idea stage

Total	Idea Generation	Marketing	Finance	Sales	Management	Production	Product Development	HRM
	Y	70	0	90	70	60	100	10
	ISC	20	40	10	20	20	0	0
	N	10	60	0	10	20	0	90
	Total	100	100	100	100	100	100	100

Y = Yes
N = No
ISC = In some cases
DN = Do not know

Table 10.22 Functions participating in PD concept stage

Total	Concept Generation	Marketing	Finance	Sales	Management	Production	Product Dev.	HRM
	Y	60	10	70	40	60	80	0
	ISC	10	20	30	50	10	20	0
	N	30	70	0	10	30	0	100
	Total	100	100	100	100	100	100	100

Y = Yes
N = No
ISC = In some cases
DN = Do not know

Product development, sales, marketing, and production were the main actors at the concept stage of the product development process. Product development became more important at this stage and the management function diminishes its participation in the concept stage. HRM and finance were practically not involved at this stage.

Production and product development were the main actors at the product development (prototype stage)of the product development process as seen in Table 10.23. All other participating functions practically play a very small role at this stage. Management was more involved in PD in the small businesses.

Table 10.23   Functions participating in PD stage

Total	Product Development	Marketing	Finance	Sales	Management	Production	Product Dev.	HRM
	Y	20	20	20	30	90	100	10
	ISC	40	0	40	50	10	0	0
	N	40	80	40	20	0	0	90
	Total	100	100	100	100	100	100	100

Y = Yes
N = No
ISC = In some cases
DN = Do not know

In the process development phase production and product development were the main actors. It was very interesting to see the product development functions participating so intensely at this stage. The focus group interviews did not give a clear picture of why the product development function was so involved in this stage as seen in Table 10.24.

Table 10.24   Functions participating in process development stage

Total	Process Development	Marketing	Finance	Sales	Management	Production	Product Dev.	HRM
	Y	10	10	10	20	100	70	10
	ISC	10	20	10	40	0	30	0
	N	80	70	80	40	0	0	90
	Total	100	100	100	100	100	100	100

Y = Yes
N = No
ISC = In some cases
DN = Do not know

When looking at the gates of the product development models in the focus group businesses the hypothesis framework model could not be verified in several areas.

At the idea gate it was primarily the product development functions which were participating. In some cases sales, production, marketing and management were involved but the focus group meeting verified that this was mostly when the product development project had strategic importance or was to some extent radical considering the task of the product development project.

Many of the businesses did not have an idea screening gate or have a very minimised idea gate. As can be seen, the businesses focused more on the concept gate.

As seen in Tables 10.25 and 10.26 the product development function was mostly involved and other function are involved very much when there were some functions or areas that were of importance to the particular internal

Table 10.25 Functions participating in idea gate

Total	Idea Gate	Marketing	Finance	Sales	Management	Production	Product Dev.	HRM
	Y	50	30	60	50	60	90	0
	ISC	30	10	20	30	0	10	0
	N	20	60	20	20	40	0	100
	Total	100	100	100	100	100	100	100

Y = Yes
N = No
ISC = In some cases
DN = Do not know

Table 10.26 Functions participating in concept gate

Total	Concept Gate	Marketing	Finance	Sales	Management	Production	Product Dev.	HRM
	Y	50	20	60	60	70	80	0
	ISC	30	0	20	30	10	20	0
	N	20	80	20	20	20	0	100
	Total	100	100	100	100	100	100	100

Y = Yes
N = No
ISC = In some cases
DN = Do not know

actor. Management also participate more when the strategic importance of the project was high.

The focus group interviews showed, however, that the concept gate was often passed through at high speed and as can be seen in the finance function was still not very involved. The focus group interviews verified that a new product development idea was able to "slip" very far into the product development process before it was met with strict gates to pass. The focus group interviews showed that this was very often due to a high pressure on time. Often, the result could be that businesses get stuck in a product development project and cannot "slip out" of the PD project again or only at great expense.

The prototype gate – Table 10.27 – was verified to be more important to the businesses but it was mainly the production and the product development functions which were involved.

Table 10.27   Functions participating in protype gate

Total	Prototype Test	Marketing	Finance	Sales	Management	Production	Product Dev.	HRM
	Y	20	20	30	20	80	70	0
	ISC	10	20	30	50	20	30	0
	N	70	60	40	30	0	0	100
	Total	100	100	100	100	100	100	100

Y = Yes
N = No
ISC = In some cases
DN = Do not know

At the process gate – Table 10.28 – it was mainly the production and product development function which were involved.

Table 10.28   Functions participating in process gate

Total	Process Test	Marketing	Finance	Sales	Management	Production	Product Dev.	HRM
	Y	30	30	20	20	90	60	0
	ISC	0	0	0	0	0	0	0
	N	70	50	80	30	0	20	100
	Total	100	80	100	50	90	80	100

Y = Yes
N = No
ISC = In some cases
DN = Do not know

Summing up on the stage and gates in the focus group businesses showed that the participation of functions in the product development model was very different from business to business but generally the picture looked as shown in Figure 10.3.

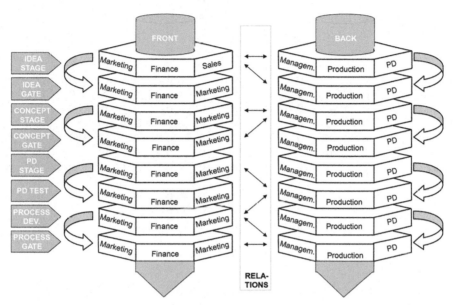

Figure 10.3   Participation of functions.

As can be seen HRM cannot be verified to be involved in the product development process and it could be verified that the financial function plays a very diminished role in the product development process.

PU Management and External Networks Involved in PD

To a large extent the product development projects at the businesses were managed by the customers (49%) with the businesses' product development department in the second place (24%). This gives a strong indication that the product development at the businesses was highly market oriented and based on a network consisting of the customers and business's sales department as indicated in Figure 10.3.

However, there were major differences in this picture depending on the characteristics of the field of product development. As can be seen businesses such as B&O, Danfoss, and Lindholst have a strong internal management of their product development projects as seen in Table 10.29.

Table 10.29   Management of projects at focus group businesses

Management of Project (%)	Lyngsø	Linco	AKV	B&O	Danfoss	Neg Micon	Ansager	Scanio	GSI	Grundfoss	Total
Customer	60	0	85	0	0	100	95	50	10	90	49
Supplier	5	0	0	0	0	0	0	30	10	0	5
Marketing	10	0	0	0	0	0	0	0	5	0	2
Finance	0	0	0	0	0	0	0	0	0	0	0
Sales	25	10	0	0	0	0	0	20	10	0	7
Management	0	0	0	0	65	0	0	0	10	0	8
Production	0	20	0	0	0	0	0	0	10	0	3
Product Development	0	70	0	100	35	0	0		30	0	24
HRM	0	0	0	0	0	0	0	0	0	0	0
Competition	0	0	0	0	0	0	0	0	0	0	0
Common Leadership	0	0	0	0	0	0	0	0	15	0	2
Do not Know	0	0	15	0	0	0	5	0	0	10	3
Total	100	100	100	100	100	100	100	100	100	100	100

The nature of the network partners involved in the product development process can be seen in Table 10.30.

The customers were very involved at the beginning of the product development process in all businesses and at the end of the product development process when prototype tests were made. Surprisingly, the suppliers were not very involved in the product development process. They mainly joined the product development project at the process stage and process gate. The focus group interviews verified that the suppliers join the project at a very late point of time in the product development process.

The competitors were not involved in the business's product development process. The focus group interviews verified this very clearly. There was barely some networking going on around product development e.g. basic product development of new technology.

Other network partners were mainly involved in the upper part of the product development process mainly at the concept development and product development stage.

Table 10.30 Network partners involved in PD process

	Customers					Suppliers					Competition					Other Network				
	Y	N	ISC	DN		Y	N	ISC	DN		Y	N	ISC	DN		Y	N	ISC	DN	
Idea Generation	90	10	0	0	100	50	50	0	0	100	10	90	0	0	100	20	60	0	20	100
Concept Generation	100	0	0	0	100	30	70	0	0	100	10	90	0	0	100	60	30	0	10	100
Product Development	60	40	0	0	100	60	40	0	0	100	10	90	0	0	100	50	40	0	10	100
Process Development Phase	10	80	0	10	100	60	40	0	0	100	10	90	0	0	100	20	60	0	20	100
Idea Screening	50	10	0	40	100	20	60	0	20	100	0	90	0	10	100	10	40	0	50	100
Concept Screening	40	30	0	30	100	20	60	0	20	100	10	90	0	0	100	20	30	0	50	100
Proto Type Test	90	10	0	0	100	40	60	0	0	100	0	90	0	10	100	20	60	0	20	100
Process Test	40	40	0	20	100	70	30	0	0	100	0	90	0	10	100	30	50	0	20	100

Y = Yes
N = No
ISC = In some cases
DN = Do not know

Summing up on the network activities in product development at the focus group businesses:

Table 10.31   HS enablers used in focus group businesses

	HS Enablers	Total				
		Y	ISC	N	DN	Total
1	ICT Communication Enabler	50	40	10	0	100
2	Customer Enabler	80	20	0	0	100
3	PD Model Enabler	30	60	10	0	100
4	Network Enabler	50	40	0	10	100
5	Innovation Enabler	50	40	0	10	100
6	HRM Enabler	30	50	20	0	100
7	Process Enabler	20	70	10	0	100
8	Product to Process Enabler	70	20	0	10	100
9	Modularisation Enabler	70	20	0	10	100
10	Development Enabler	20	40	20	20	100

It was verified that all enablers are in use – Table 10.31 – but the HRM enabler and the e-development enabler were not so much in use. The focus group interview, however, verified that the businesses' use of the enablers was very much on an ad hoc basis and seldom related to the task of product development to be solved or to the characteristics of the "field of product development".

It was also verified that many of the businesses focus on one or two of the enablers, e.g. the modularisation enabler, and use this enabler in all high speed approaches.

Summarising on the enablers to high speed product development it could be verified that there was a use of all HS enablers but the potential of the individual HS enablers seemed not to be use in full scale.

10.7 Success Criteria of HS PD

The focus on success criteria in NB HS NPD were treated by the researchers in a two step way. Firstly, the businesses were introduced to the short term success criteria via Balachandra's theoretical framework. Secondly, the long term success criteria were presented.

The short term success criteria were presented in the course of 3 meetings. Firstly, the time and speed aspect then the cost and finally the performance aspect.

The research verified that the businesses were very focused on short-term success criteria – in particular performance and time as seen in Table 10.32.

Table 10.32 Priorities of success criteria at Lyngsø

Success Criteria	Total					Total
	1	2	3	4	5	
Time	30	70	0	0	0	100
Cost	30	20	50	0	0	100
Performance	90	10	0	0	0	100
CIM	20	50	30	0	0	100
CI	50	20	10	20	0	100
L	20	40	30	10	0	100

The businesses had minor focus on long term success criteria and most focus in a long term perspective was on continuous innovation.

The reason why the businesses did not generally focus on time might be that the businesses are characterised by high quality products. Secondly, as we had stated before businesses operating in radical product development areas did not have a high focus on speed and time. Nevertheless, the focus group interview verified that the priorities of success criteria changed from one business to another as can be seen in Table 10.33.

Table 10.33 General priorities of success criteria at focus group SMEs

Priorities	Lyngsø	Linco	AKV	B&O	Danfoss	Neg Micon	Ansager	Scanio	GSI	Grundfoss
Time	2	2	2	1	2	2	2	1	1	2
Cost	2	2	3	1	3	1	3	1	3	3
Performance	1	1	1	1	1	1	1	2	1	1
CIM	2	2	3	1	2	2	1	2	3	3
CI	1	1	4	1	1	3	2	2	1	4
L	2	2	4	3	1	3	3	2	1	2

Furthermore, the research verified that the prioritising of success criteria changed during the product development stage- and gates.

As can be seen, time was the most considered success criteria to the product development of the businesses. When the product development project reached the concepts screening and proto type test, performance becomes most in focus.

The case businesses were very different in their way of prioritising the success criteria of a product development project. Related to the characteristics on the field of product development and the radicalness of the task of product development, a different prioritising could be verified as seen in Table 10.34.

Table 10.34   Specific priorities of success criterias at focus group SMEs

	Idea	Concept	PD Stage	Process Stage	PD Test	Idea Screening	Concept Screening	Proto Type Test	Process Test
Time	30	30	50	30	30	30	0	10	40
Costs	0	10	20	30	20	30	20	20	20
Performance	20	20	20	20	10	0	60	50	0
CIM	20	10	0	10	0	10	0	0	0
CI	20	20	0	0	20	10	0	0	10
Learning	0	0	0	0	0	0	10	0	0
NA Total	10	10	10	10	20	20	10	10	30
	100	100	100	100	100	100	100	100	100

10.8 Time, Speed and Cost of HS PD

The focus group businesses showed to focus very much on speed and time at the product development stage, at all initial screening gates and finally at the implementation stage. Some of the businesses verified that they used the speed factor as a tool to get more product development work out of the organisation. One business even verified that a diminishing of the pressure on speed in product development would diminish the work significantly in the product development department.

The focus group interview showed that the businesses work very differently with speed in product development. Some businesses worked actively with speed in product development when others only work with speed when they were forced to do so.

The cost of HS PD was not calculated in the businesses involved. Only some businesses had a overall calculation on the cost of HS PH.

10.9 Performance and HS PD

Performance was very much in focus at the last gates of the product development process. The businesses put a high priority on CIM at the idea stage and on CI at the concept stage. This meant that the pressure on time was reduced because the businesses focus on maximal innovation and continuous improvement at the initial stages.

10.10 Reflection on Results of Focus Group Interview

In the following Table 10.35 it was possible to see which hypo these had been verified in the focus group interview.

Table 10.35 Verification table of Chapter 10

Chapter 10 Empirical Results – Focus Group Interviews		
Overall Research Questions to be Verified	Hypotheses to be Verified and Tested	Verified/not verified
1. What is network based high speed NPD	HS NPD can be seen from different views (Macro environment, business, product, market, customer, technology, competitive and network view)	Verified
HS NPD is a matter of right speed and not high speed.	Verified	
2. What enablers to NB HS PD can be identified?	Businesses use different HS enablers.	Verified
	HS enablers are identical to the 10 enablers – 1–10	Partly verified
	There can be more than these 10 enablers to HS PD	Verified
	The enablers will play a different role according to the PD situation and project (Secondary focus)	Verified
	The customer enabler, the network enabler and the PD model enabler plays an important role in the upper phase of the HS PD phase.	Partly verified
3. What framework models and processes in the idea and concept stage/gate of high speed product development based on networks can be measured	The HS PD projects can be divided into to radical and incremental PD projects	Verified
	The radical and the incremental PD projects follow different generic HS PD models and processes and can thereby be described by different generic frameworks	Partly verified
4. What success criteria can be used for measuring high speed product development based on networks?	The success criteria for HS PD are dependent on the specific PD project – radical or incremental	Verified
	HS PD success criteria can be formulated as short term and long term success criteria	Verified
	Time, cost and performance are central success criteria in a short term perspective	Verified
	Continuous improvement (CIM), continuous innovation (CI), and learning are central success criteria in a long term perspective to reach right time, right cost and right performance in NB HS PD.	Partly verified

11

Empirical Results – Survey

11.1 Introduction

The analysis was made on data captured during a study of high speed product development practices in small and medium sized businesses in 2002. The unit of the analysis used in this study was the product development project and the respondent was the person responsible for the SMEs' product development activities.

A questionnaire was developed on behalf of the case and focus group research and collected data through an Internet based survey of 156 Danish, German, Czech, and Italian small and medium sized businesses in 2002. This was done to verify the observations in the case, focus group and other research activities to improve reliability and validity in the research project and to try to reach a triangulation in the research project.

The German, Czech, and British businesses only amounted to 12 businesses.

The survey was presented to 456 businesses via the Internet. Unfortunately, some of the e-mail addresses were not correct or did not exist any longer. This reduced the survey to 379 potential respondents. Approximately 87 businesses did not want to answer due to different reasons, and this left the survey with the following results:

As can be seen in Table 11.1, the response rate (53.5%) was satisfactory compared to the number of businesses who could and would answer the questionnaire sent via the Internet.

The aim of this chapter is:

- to verify, test, and give answers to the research hypotheses and questions set up earlier in Chapter 1.
- to show and verify NB HS NPD models and processes carried out in SMEs.
- to verify different SME businesses' solutions to NB HS NPD.

Table 11.1   Survey response statistics

Survey Response Statistics	No.	Danish	Foreign	Total	%
Original respondent	456	444	12	456	100
E-mail address error	77	77	12	89	19.5
Potential respondent	379	367	12	379	80.5
Did not want to answer out of total	87	87	0	87	19.1
Did not want to answer out of potential	87	87	0	87	23.0
Potential respondents who wanted to answer	292	280	12	292	77.0
Answered out of total	156	144	12	156	34.2
Answered out of potential	156	144	12	156	53.4

- to determine whether there were differences or extras that have not been observed in both case and focus group interviews.
- to reflect on the consequences which high speed and right speed have on different parameters as shown in Table 11.2.

Table 11.2   Consequenses of speed on different parametres

Consequences	High Speed	Right Speed
Time		
Cost/Value		
Performance		
Market fit		
Risk		
Security		

Table 11.3 presents the contribution of the survey to the research questions.

Table 11.3   Hypotheses to be verified in Chapter 11
Empirical Results – Survey

Overall Research Questions to be Verified	Hypotheses to be Verified and Tested	Survey
1. What is network based high speed NPD	HS NPD can be seen from different views (Macro environment, business, product, market, customer, technology, competitive and network view).	X
	HS NPD is a matter of right speed and not high speed.	X
2. What enablers to HS PD can be identified?	Businesses use different HS enablers.	X
	HS enablers are identical to the 10 enablers – 1–10	X
	There can be more than these 10 enablers to HS PD.	X
	The enablers will play a different role according to the PD situation and project (Secondary focus).	X

Table 11.3 Continued

	The customer enabler, the network enabler and the PD model enabler plays an important role in the upper phase of the HS PD phase.	X
3. What framework models and processes in the idea and concept stage/gate of high speed product development based on networks can be measured	The HS PD projects can be divided into to radical and incremental PD projects.	X
	The radical and the incremental PD projects follow different generic HS PD models and processes and can thereby be described by different generic frameworks.	X
4. What success criteria can be used for measuring measuring high speed product development based on networks?	The success criteria for HS PD are dependent on the specific PD project – radical or incremental.	X
	HS PD success criteria can be formulated as short term and long term success criteria.	X
	Time, cost and performance are central success criteria in a short term perspective.	X
	Continuous improvement (CIM), continuous innovation (CI), and learning are central success criteria in a long term perspective to reach right time, right cost and right performance in NB HS PD.	X

11.2 General Conditions and Trends in PD

The survey businesses' product portfolio is strongly focused on physical and service products as seen in Table 11.4. However, on the new product introduction a stronger focus on physical products were seen. The development on service products was diminishing radically.

Table 11.4 Focus on product types

	Physical Products	Service Products	Knowledge and Consultancy
Existing Product Portfolio	81	11	8
New Products	88	4	8

The businesses' products seen in another dimension – Table 11.5 – showed that 93% of the products were physical products, 4% were digital products, and only 3% were virtual products. This meant that hardly any of the businesses had yet gone into offering virtual products to the market.

The business had 92% on physical processes and 5% on digital processes as seen in Table 11.5. The survey showed very clearly that the businesses did not think of the product as a process. The product to process thinking and high speed enabler was not yet introduced in the businesses. The businesses still

Table 11.5   Focus on products and processes

	Physical Products	Digital Products	Virtual Products
Existing Product Portfolio	93	4	3
	Physical Processes	Digital Processes	Virtual Processes
Existing Product Portfolio	92	5	3

think of the products as physically "encapsulated" products with a beginning and an end – products with the classical lifecycles.

The survey businesses were asked to give general comments on their view of the conditions in "the field of product development". These comments are collected in below. Below the Table 11.6 central comments are presented in details.

Table 11.6   General conditions on the field of PD according to the survey

The Main Components Context	Characteristics
Market Stable markets Evolving markets Dynamic markets	Most SMEs claimed that they were operating in stable to evolving market with customers who have mostly incremental development in preferences.
Technology Stable technology Evolving technologies Dynamic technologies	The businesses claimed that their market were under pressure from new, evolving and some times unknown technologies. The technology gave the businesses new technological possibilities but the technological possibilities were often ahead of market demand.
Network Stable networks Evolving networks Dynamic network	The businesses' general involvement in networks were mainly based on physical and stable networks; often internal and dominated network. However, a slightly new evolvement of networks based on a mix of new evolving system of networks – both physical networks and ICT networks were recognized. None of the businesses were joining virtual networks. Only very few of the businesses joined networks based on a mix of dynamic networks with a high degree of dynamic where network partners constantly enter and leave. None of the businesses had joined a network without a formal network leader.
Business competence context	Businesses felt that there was a high pressure on support competences and that they had to develop complementary competences either by internal development or by external recruiting in their networks. A high pressure on businesses core competences were realised and some of the businesses felt their competitive advantage on core competences were reduced or diminished by competitors.

11.2.1 General PD at Survey Businesses

The survey research verified that the sources to ideas to product development at the businesses were as seen in Table 11.7:

Table 11.7 Sources of PD ideas in general

Sources to Product Development Ideas in General	Total
Customers	22
Suppliers	3
Marketing	4
Finance	0
Sales	16
Leadership/Management	8
Production	9
Product Development	26
Human Resources	0
Competition	7
Others	4
Do not know	3
Total	100

The product development ideas came mainly from product development (26%), the customers (22%), and from the sales function (16%). Nevertheless, the variation from business to business was huge as can be seen. However, the general picture indicates that the product development of the businesses was generally strongly based on identified needs and wants in the market. The businesses in the survey seem in general to be strongly customer/sales oriented.

It was verified very clearly that the area of attracting ideas was not in particular focus of high speed in the businesses. This will be explained at a later point in this book. The businesses were not focussing on speeding processes or of attracting new ideas.

11.3 Product Development Task

On the basis of the survey, the task of product development of the businesses could be verified as follows:

The businesses' new product generation – as seen in Table 11.8 – could be related to an increase in hardware or physical products from 81% to 88%, whereas service and knowledge products had decreased from 11% to 4%. The businesses' product portfolios mainly consisted of physical products (81%)

Table 11.8   PD Tasks of survey businesses

	Physical Products	Service Products	Knowledge and Consultancy
Existing Product Portfolio	81	11	8
New Products	88	4	8
Product Development			

and service products today (11%). This was expected also to be the case in the near future. However, the introduction of new products and the product development efforts were now changing and the businesses were focusing more on knowledge and consultancy products as indicated in Figure 11.1.

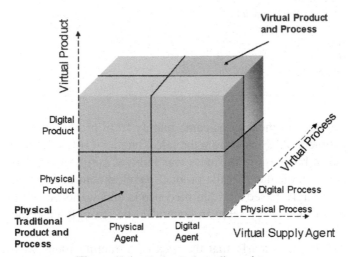

Figure 11.1   Turban's three dimensions.

The businesses also claimed as seen in Table 11.9 that in general product development projects could be divided into 81% strategic known and old areas and 19% unknown and new areas.

Table 11.9   PD projects in relation to strategic areas

	Strategic Areas		Total
	Known and Old Areas	Unknown and New Areas	
Total	81	19	100
Std. Div.	14.5	14.4	
Min	40	0	
Max	100	60	

Still there were major variations in the businesses' product development activities as can be seen in the standard deviation. It could be verified that the focus in PD was very much related to the characteristics of the specific elements in "the field of product development". When looking at the businesses' industry data and background it was verified that those businesses who face evolving and dynamic characteristics on their elements on the field of product development had more focus on unknown and new strategic areas. They were influenced and pressed by elements from outside to centre their product development more on unknown and new strategic areas and a to push towards more radical product development activities. It also seemed as if these businesses felt a higher pressure on speed and time in product development.

When applying the product/market model to the product development projects of the businesses, it appeared that product development projects at the businesses could generally be characterized as incremental product development.

The businesses' product development projects mainly concerned development of new products (51.7%). It was interesting to note that 24.4% of the businesses' new products need major adjustments after 1 year as seen in Table 11.10.

Table 11.10   PD in relation to product

	Old Products More than 3 Years with a Need for Small Adjustments	Old Products More than 3 Years with a Need for Major Adjustments	New Products Older than 1 Years with a Need for Small Adjustments	New Products Older than 1 Years with a Need for Major Adjustments	
Total in %	21.1	27.1	27.3	24.4	100
S Dev.	17.4	24.2	25.5	23.6	
Min	0	0	0	0	
Max	60	100	100	80	

The survey also showed that 51.1% of the businesses' existing product portfolios needed huge adjustments. This indicated a high pressure on product development and a diminishing product life cycle.

The survey research showed that there were many differences in the pressure on product development from one business to another and that the characteristics in the field of PD influenced this pressure on product development to a very large extent.

The survey also showed that 82% of the product development in the business was on known and old customer groups as seen in Table 11.11.

Table 11.11   PD in relation to customer groups and needs

	Known and Old Customer Groups	Unknown and New Customer Groups	Known and Old Customer Needs	Unknown and New Customer Needs
Total	82	18	74	26
S Dev.	14.7	14.6	19.6	19.3
Min	30	0	30	0
Max	100	70	100	70

This verifies incrementalness for which customers were going to buy the businesses product development in the future.

Looking at the product development projects and at the customers' needs, we realized that 74% of the product development projects were related to known and old customers' needs. This also indicates that there was not much radicalness in this variable of product development. However, it must be said that there was variation in the numbers. Yet again it could be related to how evolving or dynamic the elements on "the field of product development" were.

The above verifies that the businesses dealt with incremental product development projects on this dimension and that the businesses knew very well which characteristics "the field of product development" had.

On the technical level – Table 11.12 – the businesses claimed that 14% of their product development projects involved new technology – radical technology areas and 86% of their PD projects were carried out in known areas or in development areas in which small, incremental technology adjustments were necessary.

Table 11.12   PD projects in relation to technology

	Known Technology	Known Technology with Small Adjustments (Incremental Technology)	Completely New Technology (Radical Technology)	Total
Total	61	25	14	100
S Dev.	22.1	17.0	14.6	
Min	15	0	0	
Max	100	95	70	

The survey businesses seemed not to be pushed so much by the technology as earlier maintained in the case and focus group interviews. As can be seen by the statistical data it was verified that there were major variations in the data. Furthermore, some of the businesses were more pushed by technological evolvement. Some businesses which were not pushed by the market and customer elements were now strongly pushed by technology.

The survey research also showed that 54% of the product development projects were related to market areas with fierce and rival competition as seen in Table 11.13.

Table 11.13 PD projects in relation to competition

	Markets with Low or No Competition	Markets with Medium or Intensive Competition	Markets with Fierce and Rival Competition	Total
Total	9	37	54	100
S Dev.	18.2	31.4	35.6	
Min	0	0	0	
Max	70	100	100	

The businesses considered 28% of their product development projects as having a high element of innovation as seen in Table 11.14. This designated a not so high radical element of the product development projects as previously seen in the case and focus group interview.

Table 11.14 PD projects in relation to degree of innovation

	No Degree of Innovation – Routine NPD Project	Medium Degree of Innovation – Modified PD with Minor Demandson Adjustments (Incremental)	High Degree of Innovation – with Many Elements of Innovation (Radical Innovation)	Total
Total	36	36	28	100
S Dev.	25.7	18.4	26.6	
Min	0	0	0	
Max	100	80	100	

However, the innovative element focused on the businesses' internal competences – and how the businesses' competences matched the challenge on the tasks of product development – gave a picture of a future high challenge and pressure on this dimension.

11.4 Task of PD – Radical or Incremental

From the above-given comments the following picture of the businesses' product development task related to radical and incremental product development could be drawn as seen in Table 11.15.

Table 11.15   Incremental and radical product development in SMEs

Dimension	Incremental	Radical
Where was the idea discovered	Known – On the market place	
Initiator of idea	Known – By customers, sales and product development department.	
Product type	Known – Hardware 83% Service 8.5%. Knowledge and consultancy 8.5%	
Process type	Known – Physical processes 83% Digital processes 16% virtual processes 1%.	
Strategic areas	Known and old areas (81%) unknown and new areas (19%).	
Innovation degree	Low	
Market	Old and well known markets	
Customer group	Old and well known (82%) unknown (18%).	
Customer needs	Known and Old Customer Needs Slightly (74%) Unknown and new customers needs (26%) slightly evolving.	
Technology		Known Technology (61%) – Known Technology with Small Adjustments (Incremental Technology (25%) – Completely New Technology (Radical Technology) (14%) evolving.
Competition		Markets with Low or No Competition (9%) – Markets with Medium or Intensive Competition (37%) – Markets with Fierce and Rival Competition (54%).
Network	Old and narrow	
Innovation-challenge and press on businesses competence		No Degree of Innovation – Routine NPD Project (36%) – Medium Degree of Innovation – Modified Product Development with Minor Demands on Adjustment (Incremental) (36%) – High Degree of Innovation – with Many Elements of Innovation (Radical Innovation) (28%).
Product development task	⬅➡	

As can be verified, the product development task was rather well defined, and the pressure on product development comes mainly from the technology and the challenge to the competences of the businesses as indicated in Figure 11.2. The high pressure from competition could not be verified to be related to radical development on customers' needs and wants.

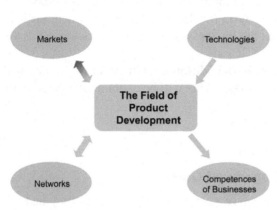

Figure 11.2   Field of product development.

Consequently, the picture of the field of product development could be illustrated according to the results of the survey.

11.5 Core of HS PD Project

The core of the product development projects was generally formulated at the strategic level inside the businesses. Most businesses formulated the core of the PD with an inside-out view.

This was further supported by the fact that formal goals and limits (goals, cost, resources etc.) for the product development project were always specified. This is illustrated in Table 11.16.

Table 11.16   Goals and limits to product development

Definition of Goals and Limits to Product Development Project	Yes	No	Do Not Know	Total
Mission	63	31	6	100
Goals	94	6	0	100
Strategy	50	38	13	100
Economic Resources	94	6	0	100
Personnel/Organisational Resources	75	25	0	100
Contact Limits to Network Partners	59	38	3	100

The goals and limits for the product development projects in the businesses were in most businesses defined in details in the areas as shown in Table 11.16.

The mission and contact limits to network partners were seldom formulated. The strategy was often formulated later in the product development project.

The businesses maintained that the specifications helped the businesses to reach the success criteria for the product development project. Generally, by the empirical data the businesses could be characterized as very planning oriented businesses.

However, there were some differences in the above -mentioned statements from one business to another. This will be explained in detail in the following paragraph.

11.6 HS PD Models

11.6.1 Formal Stages and Gates

The businesses in the survey claimed that in general they had a formal stage gate model. The stage and gates of the businesses' formal product development model are shown in Table 11.17.

Table 11.17   Stages and gates of the survey businesses' formal PD model

	Idea	Concept	PD Phase	Process Development Phase	Idea Screening	Concept Screening	Prototype Test	Process Testing
Yes	88	92	96	88	57	57	87	77
No	12	8	4	8	35	30	4	23
Under constructions	0	0	0	4	9	13	9	0
Total	100	100	100	100	100	100	100	100

The survey showed that generally the businesses' formal product development model had four stages – an idea stage, a concept stage, a product development stage, and a process development stage. A quite different result from that produced by the case and focus group interviews.

In the screening area, the picture showed that the businesses had four gates – an idea gate, a concept screening gate, a prototype test gate, and a process test gate.

However, there were uncertainties when it came to the idea and concept screening gates as quite many businesses claimed that they did not have such gates. The existence of prototype and process gate was extensively confirmed by the businesses.

The above survey results proved that the stage gate model existed at the businesses. Nevertheless, the results also proved that the stage gate model was slightly different from our research hypothesis model because the idea and concept gates did not formally exist in many of the businesses. It seemed as if many businesses had put tremendous efforts into and importance on the prototype stage and gate.

Thus, the formal product development model could be verified as existent in the SMEs.

11.6.2 Informal Stages and Gates

The survey research also showed that there was an informal model running parallel to the formal model in 60% of the focus group businesses. Some informal PD models were running all the time and in other business ad hoc when needed. The businesses confirmed the existence of such an informal product development model. The content existence of the informal product development model in different areas at the businesses are seen in Table 11.18.

Table 11.18 Stages and gates of focus group businesses informal PD model

Informal Models	Idea	Concept	PD Phase	Process Development Phase	Idea Screening	Concept Screening	Prototype Test	Process Testing
Yes	71	62	86	52	35	30	80	50
No	29	29	14	38	55	60	20	45
Under Constructions	0	10	0	10	10	10	0	5
Total	100	100	100	100	100	100	100	100

It is very interesting to see that the businesses' informal product development model in nearly all businesses contained all stage and gates as defined in the research framework model. The existence of informal idea stage and PD stage was very significant in most businesses. Informal gates were not so much in use in the businesses except in the prototype gate. The survey could not explain the reason for this.

The businesses who had an informal product development model claimed that the informal product development model was important for all listed success criteria in the businesses but particularly the time criteria. This state of affairs is illustrated in Table 11.19.

Table 11.19   Importance of informal PD model in relation to success criteria

	Time	Costs	Performance	CIM	CI	Learning
Yes	48	29	24	33	24	30
No	10	14	19	19	14	25
ISC	43	48	43	38	43	30
Do not know	0	10	14	10	19	15
	100	100	100	100	100	100

Also cost was influenced but not in particular performance. The long-term success criteria were not particularly influenced by the informal PD model except CIM.

11.6.3 The Informal PD Process

The survey verified that an informal product development process existed in many businesses and that it carried an impact on all success criteria of the businesses' product development projects.

11.6.4 Internal Functions Involved in PD Process

In the survey the following functions showed to be involved at the different stages and gates of the product development process. The businesses had a rather traditional involvement of functions at the product development stage and gates. Nevertheless, it seemed as if the business was very focused on the involvement of sales, management and the product development department at the initial idea and concept stage and gates as seen in Table 11.20.

Product development, sales, management, production and marketing were the main actors at the idea stage of the product development process. HRM and finance were practically not involved in the initial product development phase.

Product development, marketing, management, sales, and production were the main actors at the concept stage of the product development process as seen in Table 11.21. Product development are still most important at this stage and the management and sales function diminish their participation in the concept stage. The marketing function increased its involvement. HRM and finance were still hardly involved in this stage.

Table 11.20   Functions participating in PD idea stage

Total	Idea Generation	Marketing	Finance	Sales	Management	Production	Product Dev.	HRM
	Y	38	8	65	54	43	95	0
	ISC	27	73	0	8	11	3	92
	N	32	19	35	35	43	3	3
	DN	3	0	0	3	3	0	5
	Total	100	100	100	100	100	100	100

Y = Yes
N = No
ISC = In some cases
DN = Do not know

Table 11.21   Functions participating in PD concept stage

Total	Concept Generation	Marketing	Finance	Sales	Management	Production	Product Dev.	HRM
	Y	49	9	43	44	43	83	0
	ISC	23	77	6	15	23	3	91
	N	26	11	51	41	34	14	3
	DN	3	3	0	0	0	0	6
	Total	100	100	100	100	100	100	100

Y = Yes
N = No
ISC = In some cases
DN = Do not know

Product development and production were the main actors at the product development (prototype stage) of the product development process as seen in Table 11.22. All other participating functions were practically not involved at this stage.

Management was more involved in the PD stage than other functions such as marketing and sales. Finance and HRM were still not very involved.

In the process development phase production and product development were the main actors as seen in Table 11.23. It was very interesting to see the product development function's considerable participation at this stage. The survey did not give a clear answer to the question why the product development function was so involved at this stage.

Table 11.22   Functions participating in PD product development stage

Total	Product Development	Marketing	Finance	Sales	Management	Production	Product Dev.	HRM
	Y	22	14	30	35	68	95	3
	ISC	30	5	32	43	24	3	0
	N	46	78	32	19	5	3	92
	DN	3	3	5	3	3	0	5
	Total	100	100	100	100	100	100	100

Y = Yes
N = No
ISC = In some cases
DN = Do not know

Table 11.23   Functions participating in process development stage

Total	Process Development	Marketing	Finance	Sales	Management	Production	Product Dev.	HRM
	Y	11	11	9	31	75	69	3
	ISC	17	19	31	36	22	22	0
	N	69	64	57	31	3	6	94
	DN	3	6	3	3	0	3	3
	Total	100	100	100	100	100	100	100

Y = Yes
N = No
ISC = In some cases
DN = Do not know

When looking into the gates of the product development models in the focus group businesses, several areas did not succeed in verifying the hypothesis framework model.

At the idea gate as seen in Table 11.24 it was primarily the product development function which was participating. In some cases management, sales, production, and marketing were involved but the survey verified that this was mostly when the product development project had strategic importance or was rather radical considering the task of the product development project.

Many of the businesses did not have an idea screening gate or had a very minimised idea gate. As can be seen the businesses focused more on the concept gate.

Table 11.24 Functions participating in idea gate

Total Idea Gate	Marketing	Finance	Sales	Management	Production	Product Dev.	HRM
Y	25	9	41	45	38	75	0
ISC	47	6	41	32	25	16	0
N	25	81	16	16	34	9	97
DN	3	3	3	6	3	0	3
Total	100	100	100	100	100	100	100

Y = Yes
N = No
ISC = In some cases
DN = Do not know

The product development function was the function most involved. Other function were very involved when there were functions or areas that were of importance to the particular internal actor. Management also participate more when the strategic importance of the project was great.

The survey showed, however, that the concept gate was often passed through at high speed, and as can be seen in Table 11.25 the finance function was still not involved. The survey also verified that a new product development idea was in danger of "slipping" very far into the product development process before it encounters strict gates to pass.

Table 11.25 Functions participating in concept gate

Total Concept Gate	Marketing	Finance	Sales	Management	Production	Product Dev.	HRM
Y	45	13	33	40	38	73	0
ISC	31	0	50	43	21	17	0
N	21	83	13	10	41	10	97
DN	3	3	3	7	0	0	5
Total	100	100	100	100	100	100	100

Y = Yes
N = No
ISC = In some cases
DN = Do not know

The prototype gate was verified to be very important to the businesses but it was mainly the product development and the production functions which were involved.

In the process gate it was still mainly the production and product development functions which were involved.

The rest of the functions were not particularly involved.

Summing up on the stages and gates in the survey we understand that the participation of functions in the product development model was very different from one business to another.

As can be seen, the HRM function cannot be verified to be involved in the product development process as seen in Tables 11.26 and 11.27. It can also be

Table 11.26   Functions participating in protype gate

Total	Prototype Test	Marketing	Finance	Sales	Management	Production	Product Dev.	HRM
	Y	14	5	65	27	70	86	0
	ISC	17	5	35	43	22	11	3
	N	61	81	35	30	8	3	89
	DN	8	8	3	0	0	0	8
	Total	100	100	100	100	100	100	100

Y = Yes
N = No
ISC = In some cases
DN = Do not know

Table 11.27   Functions participating in process gate

Total	Process Test	Marketing	Finance	Sales	Management	Production	Product Dev.	HRM
	Y	45	13	33	40	38	73	0
	ISC	31	0	50	43	21	17	0
	N	21	83	13	10	41	10	97
	DN	3	3	3	7	0	0	3
	Total	100	100	100	100	100	100	100

Y = Yes
N = No
ISC = In some cases
DN = Do not know

verified that the financial function played a very diminished role in the product development process. This was quite surprising.

PU Management and External Networks Involved in PD

To a large extent the product development projects at the businesses were managed by the customers (56%) with the management of the businesses in second place (12%) as seen in Table 11.28. This gave a strong indication that the product development at the businesses was highly market-oriented at the time of the research was carried out.

Table 11.28 Management of projects in survey businesses

	Management of Project (%)			
	Total	Std. Div.	Min	Max
Customer	56	44	0	100
Supplier	12	28	0	75
Marketing	7	29	0	80
Finance	1	14	0	45
Sales	0	3	0	10
Management	12	41	0	100
Production	1	6	0	15
Product Development	9	36	0	100
HRM	0	0	0	0
Competition	0	0	0	0
Common Leadership	2	19	0	60
Do no know	0	0	0	0
Total	100			

However, there were major differences in this scenario. This is mainly related to the characteristics of the field of product development.

The nature of the network partners involved in the product development process can be seen from.

In all businesses the customers are very involved at the beginning of the product development process and at the end of the product development process when prototype tests are made. The supplier comes in at a later point of time in the product development process. Usually, they come in for the product development, the process stage, and the process gate. The focus group interviews verified that the suppliers come in very late in the product development process.

The competitors are not involved in the businesses' product development process although the survey verified that the competitors are more involved than verified by the case and focus group interview.

Other network partners as seen in Table 11.29 showed to be much more involved in the survey research than in the other research. However, they are mainly involved in the upper part of the product development process mainly the idea, concept, and product development stage.

Table 11.29   Network partners involved in PD process

	Customers					Suppliers					Competition					Other Network				
	Y	N	ISC	DN		Y	N	ISC	DN		Y	N	ISC	DN		Y	N	ISC	DN	
Idea Generation	75	25	0	0	100	28	72	0	0	100	8	81	0	11	100	39	47	0	14	100
Concept Generation	67	31	0	3	100	24	73	0	3	100	6	83	0	11	100	39	50	0	11	100
Product Development	56	44	0	0	100	73	27	0	0	100	6	83	0	11	100	43	43	0	14	100
Process Development Phase	19	75	0	6	100	53	47	0	0	100	6	83	0	11	100	25	61	0	14	100
Idea Screening	72	28	0	0	100	44	56	0	0	100	6	83	0	11	100	22	58	0	19	100
Concept Screening	49	40	0	11	100	17	80	0	3	100	3	81	0	17	100	20	60	0	20	100
Proto Type Test	31	60	0	9	100	14	83	0	3	100	3	83	0	14	100	20	60	0	20	100
Process Test	67	33	0	0	100	59	38	0	3	100	0	83	0	17	100	28	58	0	14	100

Y = Yes
N = No
ISC = In some cases
DN = Do not know

HS Enablers

The use of high speed enablers showed that all enablers were considered at the businesses but especially the customer enabler was in focus along with HS enablers Nos. 8 (product to process) and 9 (modularisation).

It was verified that all enablers were in use as seen in Table 11.30; however, the HRM enabler and the e-development enabler were not so much in use. The focus group interview still verified that the businesses' use of the enablers was very much on an ad hoc basis and not very much related to the task of product development which had to be solved or to the characteristics of the "field of product development".

It was also verified that many of the businesses focus on one or maybe two of the enablers e.g. mainly the customer and the modularisation enabler. They used these enablers in all high speed approaches and all through the product development process.

Table 11.30	HS enablers used in the survey businesses				
	Very Much	In Some Cases	No	Do Not Know	Total
ICT Communication Enabler	15	45	33	6	100
Customer Enabler	71	26	3	0	100
PD Model Enabler	21	42	24	12	100
Network Enabler	32	53	9	6	100
Innovation Enabler	18	56	18	9	100
HRM Enabler	9	29	56	6	100
Process Enabler	15	62	18	6	100
Product to Process Enabler	27	45	21	6	100
Modularisation Enabler	29	34	17	20	100
E-Development Enabler	9	37	37	17	100

Summarising on the enablers to high speed product development it could be verified that there was a use of all HS enablers but the potential of the individual HS enablers seemed not to be use in full scale.

11.7 Success Criteria of HS PD

The research verified that the businesses were very focused on short-term success criteria – particularly performance and time. The businesses had minor focus on long-term success criteria and in a long term perspective focus was primarily on continuous innovation as seen in Table 11.31.

Table 11.31	Priorities of success criteria at survey SMEs				
Priorities	Time	Cost	Performance	CIM	CI	L
1	41	22	46	27	27	19
2	32	32	38	32	30	38
3	14	30	3	19	24	22
4	11	11	5	16	14	8
5	3	5	5	0	5	11
NA	0	0	3	5	0	3
	100	100	100	100	100	100

Nevertheless, the survey verified that the priorities of success criteria change from one business to another as can be seen in Table 11.32 during the product development stage and gates.

As can be seen, time and performance was the central success criteria during the product development process as seen in Table 11.32. When the product development project reached the screening gates time – and indeed

Table 11.32   Specific priorities of success criteria at the surevey SMEs

Priorities		Idea	Concept	PD Stage	Process Stage	Idea Screening	Concept Screening	Proto Type Test	Process Test	Implementation
Time		21	21	38	21	27	21	24	18	52
Cost		0	12	18	41	3	6	12	27	9
Performance		24	24	35	15	18	18	53	38	24
CIM		18	15	0	12	15	18	0	0	3
CI		32	18	3	0	9	6	0	0	3
L		3	3	3	3	6	9	9	12	3
Na		3	9	3	9	21	24	3	6	6
Total		100	100	100	100	100	100	100	100	100

NA = Not answer

Question wording Hvilke succeskriterier er mest afgørende for Deres produktudviklingsprojekter i følgende faser?

at the time of prototype test – performance came most in focus. Time was the most important success criteria in the implementation phase.

The survey businesses differed very much when it came to putting priorities on the success criteria in a product development project. The survey verified that there was hardly any focus on long-term success criteria.

11.8 Time, Speed and PD

The focusing on time dominated the product development stage particularly in all initial screening gates, and the implementation stage.

Speed and HS PD was a major issue and challenge to the SMEs involved, and most businesses involved focused on this challenge in their product development. The businesses defined time and speed within physical time and they were very much focus on time and speed from idea to market introduction – "time to market".

Time and speed in most businesses were transformed to cost and direct cost and cost had a central placement in the formulation of success criteria to a product development project.

Performance was very much in focus at the last gates of the product development process.

On the long time success criteria area the businesses had a somewhat more poor focus. The businesses put a although high priority on CIM at the idea stage and on CI at the concept stage. This meant that the pressure on time was reduced because the businesses focus on maximal innovation and continuous improvement at the initial stages.

11.9 Reflection

On behalf of the survey the following hypotheses could be verified and not verified as seen in Table 11.33.

Table 11.33 Verification table of Chapter 11
Empirical Results – Survey

Overall Research Questions to be Verified	Hypotheses to be Verified and Tested	Verified/Not Verified
1. What is network based high speed NPD	HS NPD can be seen from different views (Macro environment, business, product, market, customer, technology, competitive and network view)	Verified
	HS NPD is a matter of right speed and not high speed.	Not Verified
2. What enablers to NB HS PD can be identified?	Businesses use different HS enablers.	Verified
	HS enablers are identical to the 10 enablers – 1–10	Partly verified
	There can be more than these 10 enablers to HS PD	Verified
	The enablers will play a different role according to the PD situation and project (Secondary focus)	Not Verified
	The customer enabler, the network enabler and the PD model enabler plays an important role in the upper phase of the HS PD phase.	Partly verified
3. What framework models and processes in the idea and concept stage/gate of high speed product development based on networks can be measured	The HS PD projects can be divided into to radical and incremental PD projects	Verified
	The radical and the incremental PD projects follow different generic HS PD models and processes and can thereby be described by different generic frameworks	Not verified
4. What success criteria can be used for measuring high speed product development based on networks?	The success criteria for HS PD are dependent on the specific PD project – radical or incremental.	Partly Verified
	HS PD success criteria can be formulated as short term and long term success criteria.	Partly Verified
	Time, cost and performance are central success criteria in a short term perspective.	Verified
	Continuous improvement (CIM), continuous innovation (CI), and learning are central success criteria in a long term perspective to reach right time, right cost and right performance in NB HS PD.	Not verified

12

Empirical Results – Others

This chapter presents other empirical results which had been collected during the research. These other empirical results had been collected supplementary to the formal research plan and could be seen as support and increased empirical verification and knowledge to the formal PhD project. As explained in this research project, the SMEs involved had informal and parallel product development processes and activities. Similarly, this research offered the possibility of carrying out informal research activities outside the formal research – program. Such activities allowed me to observe and participate in other product and project development activities. In this last empirical chapter I have chosen to demonstrate some of these activities together with my visits and discussions with SME businesses and research colleges. The reason for this is mainly based on the fact that the empirical results support and increase the knowledge about NB HS NPD.

12.1 Introduction

While working on my research project I had the opportunity to participate in activities which empirically support and increase the knowledge about NB HS NPD already verified in Chapters 9–11. The activities to be described in this chapter were divided into three main groups:

1. Business observations and empirical data

> Visit to Tele Danmark Internet
> Visit to Licentia Group
> Visit to Sideros

2. Research projects on NPD HS NPD

> The Dolle case
> The PITNIT research group
> Stay at Polytecnico di Milano

Participating in the TOM research project
The DISPU research project
The SMER research project

3. Action research learning projects

The TIP project
The BESTCOM – EU Project

These activities provided important empirical data to NB HS NPD on e.g.

- other HS PD models than the ones I had previously observed.
- observations and knowledge in an international perspective demonstrating how PD models and processes were carried out at high speed in four European countries.
- observations on product development models and process on products which were not physical products – mainly software and service products.
- knowledge of how HS product development were carried out in research.
- further information of the consequences of NB HS NPD.

I therefore choose to include these observation and empirical data in my research.

The aim of the chapter was:

- to verify, test and give answers to different part of the research hypotheses and questions set up earlier in Chapter 1. Please see Table 12.1.
- to show and verify NB HS NPD models and processes carried out in other product development projects under pressure of speed.
- to verify other different product development groups, e.g. SMEs' solutions to NB HS NPD.
- to verify if there were differences or extras that had not been observed in either case, focus group interviews, or survey.
- to reflect on which consequences high speed and right speed had on different parameters in other environments. The parameters in focus were shown in Table 12.2.

In each case the contribution to the research questions were shown in the Table 12.1.

12.2 Visiting Tele Danmark Internet

During the research project I had the opportunity to visit Tele Danmark Internet in Spring 2001. Tele Danmark Internet or TDC (www.tdc.dk) had been the market leader on the Danish telemarket for several years and had just some

Table 12.1   Hypotheses to be verified in Chapter 12

Empirical Results – Other Empirical Results		Visit to Tele Danmark Internet	The PITNIT Research Group and the Dolle Case	My stay at Politecnico di Milano	Participating in the TOM Research Project Milano	Visiting the Sideros Business	The TIP Proejct	The BESTCOM Project	The DISPU Research Project	The SMER Research Project
Overall Research Questions to be Verified	Hypothesis to be Verified and Tested									
What is network based high speed NPD?	HS NPD can be seen from different views (macro environment, business, product, market, customer, technology, competitive, and network view).	X	X	X	X	X	X	X	X	
	HS NPD is a matter of right speed and not high speed.	X		X			X		X	
What enablers to NB HS PD can be identified?	Businesses use different HS enablers.	X	X	X		X	X	X	X	X
	HS enablers are identical to the 10 cnablers – 1–10	X				X		X	X	X
	There can be more than these 10 enablers to HS PD.	X		X					X	
	The enablers will play a different role according to the PD situation and project (Secondary focus).	X		X			X	X	X	
	The customer enabler, the network enabler, and the PD model enabler plays an important role in the upper phase of the HS PD phase.	X	X	X		X	X	X	X	

(Continued)

Table 12.1   Continued

Empirical Results – Other
Empirical Results

Overall Research Questions to be Verified	Hypothesis to be Verified and Tested	Visit to Tele Danmark Internet	The PITNIT Research Group and the Dolle Case	My stay at Politecnico di Milano	Participating in the TOM Research Project Milano	Visiting the Sideros Business	The TIP Proejct	The BESTCOM Project	The DISPU Research Project	The SMER Research Project
What framework models and processes in the idea and concept stage/gate of HS PD based on networks can be measured?	The HS PD projects can be divided into radical and incremental PD projects.	X		X		X	X	X	X	
	The radical and the incremental PD projects follow different generic HS PD models and processes and can thereby be described by different generic frameworks.			X		X	X	X	X	
What success criteria can be used for measuring HS PD based on networks?	The success criteria for HS PD are dependent on the specific PD project – radical or incremental.	X		X			X	X	X	
	HS PD success criteria can be formulated as short term and long term success criteria			X			X	X	X	
	Time, costs, and performance are central success criteria in a short-term perspective	X		X		X	X	X	X	
	Continuous improvement (CIM), continuous innovation (CI), and learning are central success criteria in a long term	X		X	X		X	X	X	

Table 12.1 Continued

perspective so reach
right time, right cost
and right performance
in NB HS PD.

Table 12.2 Impact on different prametres by speed

Consequences	High Speed	Right Speed
Time		
Cost/Value		
Performance		
Market fit		
Risk		
Uncertainty		
Continuous improvement		
Continuous innovation		
Learning		

few years ago decided to increase with a new SBU and move into the internet market. TDC had a turnover of many billions DKR and was represented on the Danish stock exchange. The American business SBC was one of the main owners of the TDC business.

TDC were represented by the personnel director Lars Hansen and manager of a special task force which should help Tele Danmark Internet's current develop managers and teams to speed their product development – which meant a very high focus from TDC on the HRM enabler as a main enabler to keep high speed in product development and to support his speed.

During the interview I had the opportunity to discuss network based high speed product development with four persons within Tele Danmark Internet. These were product development manager Preben Meyer, Human Resource Manager Lars Hansen, Product development Manager Stig Bøgh Carlsen and Manager of task force Lars Bundgard.

The research fiamework of the research project was used for the interviews with Tele Danmark Internet.

12.2.1 PD Task – PD at HS in Radical New Market

The product development tasks were many in the TDC and were always related to the introduction of new products to the market. In this case the product development challenge to the management at TDC was to focus on keeping the process running – and at high speed. This meant that TDC was always into new rather radical product development task.

12.2.2 Field of Product Development

Tele Danmark Internet was a very interesting business to visit because they were at that time – 2001 and 2002 – under an immense pressure from nearly all components on the field of product development especially the market, technology, and network to carry out high speed product development. At the same time a new foreign investor in TDC increased the pressure on Tele Danmark Internet to speed product development further and to introduce more new products on the market at a higher speed.

The market for Internet products was extremely dynamic and turbulent in Spring 2001 because of a very evolving market with several businesses offering different products to the market. As a service provider Tele Danmark Internet faced an enormous increase in turnover, new products, new employees and development into new market. It was very important for the business to continuously innovate new products to the market and to gain first mover advantage. At the same time the Internet **technology** along with the tele technology were evolving at a speed that had hardly been seen before. Furthermore, **the network** which Tele Danmark was involved in and was dealing with was also increasing, and they were consequently forced into more unknown networks at an exceptionally high speed. Finally, **the competences** of Tele Danmark Internet were under high pressure. At the same time, this forced the manager into some areas to join more with the sub-suppliers.

Taking all these aspects into account the product development managers and the product development leaders were under extremely high pressure because of dynamic and fast moving markets, technology, networks, and competences. In the competence area Human Resource Manager Mr Lars Hansen was in a situation where it was a question of the rate and speed of how many new employees he could adapt to the organisation to develop to support the development and support of new products to the market.

For a long time the managers of Tele Danmark Internet had been able to develop new products at the right time gaining the first mover advantage but suddenly, they faced a situation of stagnation within the business's competence to develop new products.

"We have seen the creativity and the development of new products have stagnated. We feel it has something to do with the fact that

there are too many employees and too many rules to work on."
(Lars Hansen TDC)

To overcome this, Tele Danmark Internet established a special task force managed by Lars Bundgaard where four people were full time applied to service and nurse different product development managers and teams. This task force had one main goal – to keep the product development managers on track and help them not to fall into stagnation or slow speed.

Looking into the model of product development of TDC it could be verified that it was mainly dealing with **a stage-gate approach**. Another important observation was that many product development projects were managed by young managers and the teams included many new, young and inexperienced product development participants.

Lars Hansen also noticed that the organisation faced a challenge of moving from a rather entrepreneurial young organisation where every body new everybody and were highly motivated the entrepreneurial, organisational environment to a more stabilised organisation. The change could be characterised as a change in focus on success criteria from short term speed and long time continuous innovation to short term cost and performance together with long term continuous improvement and learning.

Summary on TDC

The task of TDC's product development was now changing into a more stabilised situation where TDC would have the opportunity to change focus from short term success criteria to long-term success criteria. An increased focus on continuous improvement and learning together with a focus on right speed was theoretically appropriate. The first two areas were presently being cultivated at TDC.

12.3 Visiting Licentia Group

Licentia (www.licentia.dk) was one of Europe's largest producers of knock-down furniture with factories in Denmark, Sweden and England. Licentia's turnover was more than 1 billion DKR. The Licentia group was owned by the holding business Bækgaard holding in Ikast, Danmark.

At a seminar at the University in Aalborg I had the opportunity to talk to the product development manager at Licentia, Eva Paarup. Furthermore, during three visits to Licentia I had the opportunity to discuss product development with the managing Director Lars Thorrild. At the same time Licentia joined the Bestcom project which gave me a close insight into the product development activity of the Licentia group.

12.3.1 Field of Product Development

The market for knock-down furniture had been under an extremely high price pressure for a long time. This had caused a further pressure on high speed product development. Many international producers had turned their focus to more and faster product development.

The technology and increased used of new materials gave major new opportunities and challenges to Licentia's product development. **The network** component was also evolving as new networks was attended to support the increased demands for new competences to support the product development process. Licentia's **competence** especially in the product development department was under a high pressure because they were in lack of knowledge about what was the new trend in the market and further in lack of a faster access to knowledge about new trends in the market.

Product Development Task – Multiplying Incremental PD with HS

The product development task was as in a general speaking very incremental but a high pressure from customers and sales department along with a pressure on including new materials and new technology in the products gave some major challenge to the product development department to speed product development more. Licentia had also in the last two and three years merged with a furniture business in England and Sweden which challenged the product development department to do product development in network with the product development department in the merged businesses.

Licentia had chosen to solve the above-mentioned challenge by focusing on the product development model enabler – especially on introducing the stage-gate model in Licentia. Licentia had also chosen to focus on the customer enabler which meant that the marketing manager would be more integrated to the product development process and be responsible for "the customers voice" into the product development process.

Licentia had further decided to focus on the cost of product development and develop a cost model which would be able to show how much direct cost had been used on each product development project. The direct cost were collected from idea stage to the product development project was introduced to the market.

Licentia focus very much on the **short term success criteria** time and cost. Licentia wants to focus on long term success criteria but were very much stuck in their new decided product development model and further the high pressure on time in product development.

12.4 Visit to Sideros

The Sideros business (www.Sidaros.it) was the second largest Italian producer of woodburning stoves and exports more than 85% of its turnover to most of Europe. The Sideros business had a strong cooperation with a Spanish producer of stoves. Sideros had for a long time had a strong cooperation with businesses in Modena which produce hand painted tiles which were used to decorate the stoves.

Field of Product Development

The Sideros product development department had for a long time been into a rather stable **market** all though with high pressure on price competition but a rather stable and slightly falling market. The **technology** had also been very stable along with the **network cooperation** in the industry. Sideros was cooperating with some long time network partners very much placed in the local area where the Sideros business was placed.

The competence of Sidaros was until 2003 very much focused on a stable efficient industry business. Product development had until now been very much incremental with some very stable introduction 1 time a year of new products.

Product Development Task – New Stove, New to Market, Developed at HS

The challenge to Sideros on the product development area was on more areas. Firstly to integrate new technology into the stoves because of new possibilities to produce stoves which were more efficient than the old existing ones. This challenge was a movement into some new rather radical product development areas. Furthermore, the market trend was mowing towards some increased pressure from the market on a higher speed on introduction of new products. The market was therefore changing from a stabilised market to an evolving and to some extent dynamic market.

Some new to the market products had entered the market and put further an increase to pressure on high speed in product development. Product life cycle was shrinking.

Sideros Solution

The Sideros business choose to solve the challenge by increasing their network activities – focus on the network enabler. An increased use of network and unknown network to gain access to competences that Sideros did not already have inside the business was in focus. Furthermore, the Sideros business choose to force the product development department to use a new product development model – rapid prototyping. This was not a normal procedure in the business but as the market was moving very fast the Sideros management felt they had to take new product development methods in use. Therefore the product development enabler was also used.

However, the Sideros business had still some consideration to the previous product development because the products was developed at a very high speed which could include that some faults e.g. not yet had been discovered.

The pressure on speed had at the same time resulted in some increased costs which Sideros hoped they could win again by coming so soon to the market.

The high speed in Sideros product development activity and the risk of still having faults in the newly developed products prevented Sideros to penetrate new markets because they were frightened to be dragged into a "first mover bad advantage" situation.

12.5 Dolle Case

The Dolle case was elaborated during the research project as a part of the PITNIT research project. The PITNIT project will be comment later in this chapter. The case of DOLLE can be seen in the case book – Organizing for Network Information Technologies – Cases in Process integration and Transformation (Hørlück et al., 2001).

The Dolle case shows how e-development could be developed and I had through my cooperation with the DOLLE business (www.DOLLE.dk) the opportunity to follow at close sight how the architecture both the network and the e-development software were developed in the Dolle business.

The main results of this work is shown in Table 12.3.

12.6 PITNIT Research Group

The PITNIT research project was a typical development project with similar terms and conditions as product development projects in the industry.

Table 12.3   Results of development at Dolle

Main Results of the DOLLE Case	Main Results of DOLLE Case in Details
Development of e-development in networks demands a strong trustful network	The case showed that the Dolle e-development network system demanded a strong and trustful network because the projects touch all parts of the network partners internal systems and procedures.
The software part of an e-development project can be overcome but "the soft part of the network cooperation" can be difficult to overcome if the network partners do not have trust in each other.	The Dolle case showed where the barriers was to more NB HS NPD. It showed very clearly that it was not a matter of software integration and development but instead "the soft part" of the network cooperation
When developing a e-development system where customers also are a part o the development team, then businesses should focus on how the customers perceive the product.	The cooperation with Dolle showed very clearly that Dolle had a major challenge to elaborate a e-development system that matched the customers terminology and there way of seeing and developing the product. This was the difficult part of the development of a e-development software.

12.6.1 Field of Product Development

The field of product development was in this case rather stable. The researchers had no pressure from the **market** except a deadline for some delivery which was fairly reasonable. The customers to the results of the research was initially the ministry of science but later the industry. Both the **technology** and the **network** was fairly stable however the **competence** were evolving at a high speed because each individual researcher were developing their competences both individually and together with other network partners.

Product Development Task – a Knowledge and Research Product

The goal of the PITNIT – project was to describe, analyze and offer practical guidelines for the integration and transformation of industrial processes that were enabled by new network information technology. The project integrated researchers from engineering, social sciences, and information systems.

The key research challenges were

1. the merger between a number of process innovation concepts and associated IT.
2. the extended enterprise that emerges from a multitude of different co-operating organizations and associated IT.

3. network-based interaction with the environment using new IT for marketing purposes.

To address these three challenges the research group alternated between a practical level and a theoretical level. First, the research group was engaged in a joint case description of businesses using new network IT – The research group decided to focus on the network enabler. The cases were analyzed from the three theoretical perspectives. On this basis the research group formulated a new interdisciplinary framework. Finally, the resulting normative guidelines were evaluated. The research methodology is depicted as an illustration in the Figure 12.1.

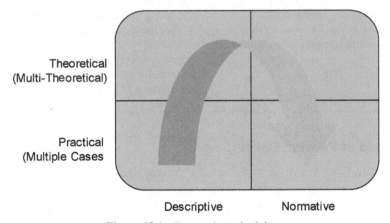

Figure 12.1   Research methodology.

The project results in the following "products":

1. Cases
2. An interdisciplinary vocabulary
3. Multi-theoretical frameworks
4. New inter-disciplinary normative guidelines (can be seen in the case book "Cases in Process Integration and Transformation", Hørlück et al., 2001).

The "product development process" used a typical stage-gate model and reached all deadline in time. However there were was crisis to reach the time schedule but buy some extra efforts and some informal product development activities the products was delivered within time. The product was judged as a success as it matched the performance demands stated earlier in the process.

12.7 Stay at Polytecnico di Milano

During my research project I had the opportunity to visit Polytecnico di Milano to do some further research together with Italian researchers. During my stay I discovered major findings which helped me very much in my research to find new way of understanding NB HS NPD.

The first discovery was the theory of **"on the market product development activity"**. Professor Mariano Corso showed me through his case on an Italian Scooter business how this business left a major part of its product development process stay "on the market". This business let their customers "play" with their products on the market and observed carefully their movements and "play" with the product. The business were well aware of the customers need and want to change the product into some personal need satisfaction. The customers change The Scooter businesses products into something that would match their individual needs.

All these valuable findings and observations were collected at the Scooter business and the best of these were transformed into further improvement of the old products or even new products.

The time aspects was further discussed and elaborated on at my stay in Italy. Before my discussion with my Italian colleagues I was convinced that physical time was possible to measure to product development projects. Additionally, I was convinced that all product development projects had a start and an end.

However I must say that I change my opinion on this when I saw how the many Italian businesses work with the product development time not in a physical perspective but more as in a process perspective. I had often wondered about since then if this way of thinking about time is related to south European style of living but in my opinion it convinced me and gave me some new ideas how to work with time and speed in product development. These has been comment before in this research project and will further in details be comment in the next chapters to come.

12.8 Participating in TOM Research Project

The Tom project was another typical research development project with the same terms and conditions as a product development project in the industry would have at that time. However, to a large extent the PD project was based on network development – the network enabler and was under a much higher pressure of time than the PITNIT – research project. I often found that this pressure was very much coming out of the structure of the project and further the manager and team of the research who were very much focused on developing several articles e.g. at a high speed.

12.8.1 Product Development Task

The product development task was stated as follows:

Starting from the identification of the emergent approaches and from the analysis of their effects within specific industrial and organisational contexts, the project will develop interpretative models and management tools intended to support managerial actions. Project objectives may therefore be articulated at the interpretative and the supportive levels.

At an interpretative level, the objective is to identify and describe the emergent configurations of technological, organisational, and management tools, to identify the determinants of such configuration adoption, as well as analyse their impact on performances. In particular, the objectives are as follows:

1. *to describe processes through which knowledge, in its different forms, is assimilated, created, transferred, stored and retrieved;*
2. *to identify the organisational mechanisms, the Information and Communication Technologies (ICT) and the Management Systems through which firms can influence such processes;*
3. *analyse internal coherence between technological, organisational and managerial choices;*
4. *analyse relationships between technologies, organisational mechanisms and Management tools and, then, identify the coming out of exhaustive configurations for Knowledge Management;*
5. *identify relationships between different configurations and contingent characteristics at industrial, geographical and organisational level;*
6. *identify each configuration effect in terms of innovative capabilities and quality of working life;*

At a supportive level, on the other hand, the objective is to identify implications for Knowledge Management analysis and re-design in terms of:

1. *support in ICT choice and coherent adoption in order to foster a more effective Knowledge Management, given the firm contingent situation and the improvement objectives/priorities;*
2. *guide to analysis and improvement of Organisational Mechanisms – in terms, for example, of structures, network roles and mechanisms – in order to foster a more effective Knowledge Management, given the firm contingent situation and the improvement objectives/priorities;*
3. *guide to the analysis and improvement of Management Systems – in terms, for example, of procedures, performance measurement systems, wage and reward systems – in order to foster a more effective Knowledge Management, given the firm contingent situation and the improvement objectives/priorities;*
4. *analysis and improvement of the whole configuration for Knowledge Management.* (Source: The Tom project description)

As can be seen the product development task was very ambiguous and must be said to be of a rather radical character. Further as stated before the research was under a high pressure on time. This time pressure was mainly coming from those who were funding the project and internal the researcher group particularly "the manager" of the research group.

The research was not yet finish when I left Italy in 2002 and some few articles and conference papers on the very first research results had just been published when I finished this research project in 2003. However it seemed as if the research group meet the timetable. As far as I could observe by joining the research group I found the reason to why the "research team" could perform at such a high speed due to two main enablers. Firstly a strong and often hierarchical management of the research project, which was necessary to steer so many researcher who were all placed in different universities all over the north and mid Italy – the management enabler. Secondly and strong focus on the network enabler where different competences and resources were used at exactly the right "spots" in the research project. Finally I observed a increased use of the ICT enabler were all participant participating in the research could connect, find all necessary materials and communicate with each other electronically. Further the manager and the individual members could watch each others activity to the electronic research place together with a continuously development of the results of the research project which was shown and visual to all research partners at "the electronic research place". I claim this last part as similar to an advanced used of electronic development or the use of the e-development enabler.

12.9 DISPU Research Project

During the research project time I joined another research project called the DISPU research project. The DISPU project were focusing on distributed product development with a support of E-development software. The research was carried out in 3 different stages.

Firstly a focus group discussion with representative from industry, university and national consultancies. Secondly the research project established a conference for SMEs and e-development software supplier to do a discussion in large and small groups about the use or non use of e-development software to support product development and high speed product development. Finally a survey was developed and send out to more than 1000 SMEs asking them several questions on product development.

The main result of this research can be seen in Table 12.4.

Table 12.4　Results of DISPU survey

Main Results of DISPU	Main Results of DISPU in Detail
The use of e-development software is very poor in SMEs	Some SME use some CAD, CAM software in their product development but the use of these systems are very poor and it is not integrated to the businesses other system.
The SME's do generally not develop new products with e-development software tools together with network partners	The DISPU research showed very clearly that the development on new products in network supported by e-development software was very poor.
Development of new products in networks supported by e-development software demands that network partners trust each other.	The DISPU research project showed that many SMEs did not develop in network based on e-development software because they did not trust or saw themselves in a competitive dangerous situation if joining and implementing such a cooperation. Obviously the price of e-development software was a barrier to further implementation of network based e-development but the major barrier was trust to network partners. The DISPU project showed some few examples where networks had overcome this barrier by signing a strategic alliance and in these networks e-development was a major tool in the network based product development and further help the businesses to develop new products faster.

12.10 SMER Research Project

The purpose of the SMER survey was to give a descriptive status of E-business in SMEs in Northern Denmark. One area of this survey focused on product development and in particular product development in networks and with the use of e-development software tools. The survey was done on more than 500 SME businesses with a answer rate of 19%.

The SMER research showed different results and the main results related to NB HS NPD are shown in Table 12.5.

Table 12.5 Results of SMER research project

Main Results of SMER	Main Results of SMER in Details
The use of e-development software in networks is very poor	The SMER research showed very significant that the use of e-development software in networks was very poor.
The product development model most used in SMEs is the stage-gate model	The SMER research verified that thee stage-gate model is the most used product development model
The SME businesses focus mostly on narrow networks	The SMEs involved focus on narrow network partners mostly the geographic nearby. The research showed that the SMEs had more trust to the narrow network partners as they were known and easy to access.
The product development projects are mostly on incremental product development projects	The SMER research showed very significantly that most product development projects was on incremental product development.

12.11 TIP Project

The TIP Project was an inter-organisational product development project carried out as a cooperation between four institutions.

12.11.1 The Aarhus School of Architecture

One of the four institutions was the School of Architecture in Aarhus. From this institution industrial design students toke part in the project. They used the project as their final assignment of their study.

12.11.2 The Aarhus School of Business

BA students from the Aarhus School of Business also participated. The TIP project was often used as the students' final bachelor report in the 6th semester.

12.11.3 The School of Engineers in Horsens and Aarhus

From HHH a number of machine engineer students participate.

In the TIP project new products were developed at high speed in a network consisting of the above-mentioned students and often in cooperation with businesses from the Danish industry. All three groups of students were "forced" to cooperate on all aspects of a product development process – analysis of customer needs, customer use of the product, technical and economical restrictions and possibilities, production technologies, project management etc. within a time span of only four months.

The project began with a "kick start" where all students meet for the first time at the beginning of September. The aim of the first meeting was to encourage the students to learn from each other and to form new product development teams. Additionally, the "kick start" meeting addressed the issue of conflict solving, product-/business development and other relevant topics.

Through September the students worked with idea development, idea screening, and analysis of market possibilities.

From October to December several courses on management of product development and product development were held e.g.:

- Idea generation
- Product development leadership and management
- High speed in network based product development
- Product patenting

Several visits to product development departments of Danish businesses were made to discuss and learned about other product development projects.

12.11.4 Experience and Observations from TIP Project

The TIP project gave me the opportunity to monitor 18 product development teams over three years. I observed their performance when doing NB HS NPD and observed their reaction and solutions to how to carry out NB HS NPD.

It must be said that all groups were formed by students and that some would say that this was not a realistic picture of what was going on in industry. However, it must be said that the amount of work the students had to do in these product development groups was no less than what was done in the "real" industry. Furthermore, the students were also "pressed" by other study activities. Thus, their situation very much resembles the situation found in industry. No one joining a product development team was relieved of other activities and responsibilities.

During my observations I made several findings which I will comment upon in the following Table 12.6.

Table 12.6   Observations from TIP project

Observations	Details
Use of high speed enablers is low.	The use of high speed enablers seemed to be low. Although the students had access to and knowledge of several high speed enablers, they hardly ever used them
The planning of the use of high speed enabler is poor, coincidental or non-existent.	The students mostly did not plan to use highs speed enablers. If anyone used the high speed enablers, they only did it by coincidence and often at a late point of time in the product development process.
The students often felt stuck in the product development process and wanted to go back and restart.	Many product development projects seemed to be stuck in the product development process and seemed to have gone the wrong way. The students even wished that they had not moved so far and could not see any way back. The pressure on time prevented them from going back and redo their work.
Speed and time pressed the students into a line of product development which they did not want.	Some of the students claimed that they were forced further into the product development "tunnel" because of the pressure on speed and time. They knew it was not an optimal way but they felt that it was not possible to go back.
The students always adhered to a stage-gate model.	None of the product development projects used another model than the stage-gate model, even though they had been introduced to other PD models at the beginning of the TIP project.
Those PD projects which seemed have difficulties at the beginning of the project turned out to be those with the best results.	As an observer of the product development process it was very peculiar to watch product development projects with serious problems in the first phase of the project turn out later to be the ones to come up with the best results.
The students seemed to loose time in the initial phase and in the middle of the product development phase.	It was significant that the students' product development projects lost time in the initial phase. The students forgot to stay and finalise a good product development architecture and plan. Therefore they encountered serious time problems later in the product development process. Also in the middle of the product development process it seemed as if the students lost time and motivation or access to solutions.
The students do not use all their competences from the beginning of the product development project.	In all projects we could observe that the students did not use all their competences at the beginning. Especially the students of business economics waited for the designers and engineers to come up with "an answer". Later it was the opposite when the designers waited for the students of business economics to give answers from the market.

12.12 BESTCOM – EU Project

12.12.1 Objectives

The Bestcom project was an EU 5th framework program with an activity corresponding to more than 3 million Euro.

The overall objective of the BESTCOM project was to implement best practice on E-business strategies and solutions in 11 European businesses in order to improve their competitive advantages. Furthermore, the project should develop best practice guidelines for use in regional business centres in three European regions, with special emphasis on the enlargement of country participants to further their interaction and networking with EU partners. Working with the various technological models and solutions and organisational alterations, the businesses would be enabled:

a. to define E-business strategy and choose the best business and technical solutions,
b. to handle the implementation of the chosen solution in best practice way,
c. to train employees at all levels in the organisational changes and the new technology.

Specific objectives were to achieve competitive advantages, notably to secure success in making new sales channels, more cost efficient production, and new service opportunities.

12.12.2 Description of Work

The project had four phases as described in the contract signed with EU:

- The analytical phase which identifies specific user requirements reflecting the businesses' future business processes enabling them to operate in the new economy. Evaluation of adequate solutions to match the needs will be carried out. It will include available technology as well as EU RTD results.
- The planning phase will elaborate the implementation plans for each business to assure that their business and technical needs are met by the selected systems. The training plans will be prepared for all levels in the businesses.
- Implementation plan will cover all the actual deployment, monitoring and evaluation and refinement of the solutions and practices. It includes changes in management and redesigned business processes.
- The evaluation and dissemination phase will ensure increased awareness of the project's best practice experience with E-business implementation.

It will be ensured through presentations, seminars and workshops both internally and externally to the consortium. Special attention will be given to the Czech participants to reach a larger business community in the candidate country. Dissemination will be ensured through various networks, e.g. business advisory centres, chambers of commerce, labour offices as well as through publications on both web-site and press. Meetings and targeted workshops will be organised with the above.

The consortium consisted of 11 businesses, 3 catalysts, and 1 project coordinator. The businesses represent various sectors and sizes (with a majority of SMEs) and had different profiles in terms of technological status and business practices. The group represented a sufficiently critical mass to obtain measurable results. The three catalysts represented experienced consulting businesses and a well established educational institute to ensure the implementation of technical and business solutions according to best practice, and to assist the businesses' transition into innovative digital businesses.

Milestones and Expected Results

- M1: SME user requirements completed and evaluated
- M2: SME contracts with suppliers and sub-suppliers
- M3: Implementation plans reviewed and approved for implementation
- M4: Implementation review and E-business review
- M5: Final report dissemination and validation

The E-business solutions enable the businesses to reduce entry costs, establish new sales channels, new service opportunities and gain first mover advantages to stay competitive. Three regional clusters for innovation had been established.

During the Bestcom project I discovered the following findings related to NB HS NPD as seen in Table 12.7:

Table 12.7 Findings during BESTCOM project

Observations	Observations in Details
The network enabler is difficult to use trans-nationally.	The Bestcom project showed that it is very difficult to use the network enabler trans-nationally if the network partners do not see any need for or benefit of working together. The Bestcom project showed very clearly that businesses in the three countries focused on the narrow network although they could gain major benefits from working together in trans-national networks.

(Continued)

Table 12.7   Continued

Observations	Observations in Details
High speed in product development becomes an issue when businesses are pressed on finance.	The Bestcom project showed that when businesses become pressed on finance, they begin to develop new products and to evolve existing products and projects at high speed. However, this will often be too late and additionally, it may turn out not to be not business economically optimal.
Businesses that perform right speed have a good architecture behind their project development together with a manager who focuses on long-term success criteria.	The Bestcom project showed that the businesses who had developed a good architecture behind their project also were able to develop new projects and products at an optimal speed – right speed. Also, these businesses often had a manager who focused on long-term success criteria.
SMEs develop new products and projects with a stage-gate model.	All SMEs in the Bestcom project turned out to develop their projects with a stage-gate model. This was interesting as this model was chosen exactly because they wanted to achieve high speed.

12.13 Reflection

Table 12.8 shows the results of other research activities related to the PhD hypothesis.

Table 12.8   Verification table of Chapter 12

Empirical Results – Other Empirical Results		
Overall Research Questions to be Verified	Hypotheses to be Verified and Tested	Verified/Not Verified
1. What is network based high speed NPD	HS NPD can be seen from different views (Macro environment, business, product, market, customer, technology, competitive and network view).	Verified
	HS NPD is a matter of right speed and not high speed.	Verified
2. What enablers to NB HS PD can be identified?	Businesses use different HS enablers.	Verified
	HS enablers are identical to the 10 enablers – 1–10.	Partly verified
	There can be more than these 10 enablers to HS PD.	Verified
	The enablers will play a different role according to the PD situation and project (Secondary focus).	Verified

Table 12.8 Continued

	The customer enabler, the network enabler and the PD model enabler plays an important role in the upper phase of the HS PD phase.	Partly verified
3. What framework models and processes in the idea and concept stage/gate of high speed product development based on networks can be measured	The HS PD projects can be divided into to radical and incremental PD projects.	Verified
	The radical and the incremental PD projects follow different generic HS PD models and processes and can thereby be described by different generic frameworks.	Partly verified
4. What success criteria can be used for measuring high speed product development based on networks?	The success criteria for HS PD are dependent on the specific PD project – radical or incremental.	Partly Verified
	HS PD success criteria can be formulated as short term and long term success criteria,	Partly Verified
	Time, cost and performance are central success criteria in a short term perspective.	Verified
	Continuous improvement (CIM), continuous innovation (CI), and learning are central suc cess criteria in a long term perspective to reach right time, right cost and right performance in NB HS PD.	Not verified

PART V

Comparing Theoretical Framework and Empirical Results

This part presents the comparison of the theoretical framework and the empirical results of the research project. The part also presents the learning aspects of the research project. Chapter 13 will present the comparison of the theoretical framework and the empirical results. The chapter will give answers to the hypotheses initially set up in the research project. A reflection and analysis of this will be presented continuously during the chapter. Chapter 14 presents the learning perspectives of the research project and offers suggestions for further research referring to the learning which can be drawn from the project.

13

Comparing Theoretical Framework Model, Hypothesis, and Empirical Results

This chapter compares the theoretical framework model, hypotheses, and the empirical results of the research project to give answers to and verify the research questions previously outlined. Furthermore, the chapter seeks to reflect on the findings of this comparison. The chapter will be divided into five parts. Firstly, an answer to the main question – *What is high speed NPD?* – will be given. Secondly, I will answer the question – *What enablers to HS PD can be identified?*. Thirdly, I will answer the question – *What framework in the idea and concept stage/gate area of high speed product development based on networks can be measured?* Fourthly, I will answer the question – *What success criteria can be used for measuring high speed product development based on networks?*. An answer will be given to each main question. Finally, the impacts of NB HS product development will be commented on, analysed and put into perspective.

13.1 Introduction

On the basis of the theoretical framework and the empirical findings analysed and described in the previous chapters, this chapter seeks to answer the research questions initially raised in the research project. In addition, this chapter tries to compare and verify the theoretical hypothesis model with the empirical findings. Finally, the chapter tries to analyse and reflect on the findings to clarify the impacts on businesses and product development in the context of NB HS NPD.

The main research questions of the research project were shown in Table 13.1.

The chapter will be structured according to the above overall research questions and according to the overall research framework model shown in Figure 13.1.

Table 13.1   Overall research questions of PhD project

Overall Research Questions	Hypotheses to be Tested
1. What is high speed NPD?	What is time and speed in NB HS NPD? HS NPD can be seen from different points of view (Macro environment, business, product, market, customer, competitive and network view) HS is central in the second phase of the PD process – the PD phase. HS NPD is a matter of right speed and not high speed.
2. What framework in the idea and concept stage/gate of high speed product development based on networks can be measured?	1. The HS PD projects can be divided into radical and incremental PD projects/tasks. 2. The radical and the incremental PD projects follow different generic HS PD models and processes and can thus be described by different generic frameworks. 3. HS PD projects always have a formulated PD core 4. A HS PD model follows a different PD model than the ordinary PD model of the business.
3. What enablers to HS PD can be identified?	1. HS enablers are identical to the 10 enablers – 1–10. 2. There can be more than these 10 enablers to enable HS PD. 3. Businesses use different HS enablers 4. The enablers will play different roles according to the PD situation and project (Secondary Focus) 5. The customer enabler, the network enabler and the PD model enabler play an important role in the upper phase of the HS PD phase.
4. What success criteria can be used for measuring high speed product development based on networks?	1. The success criteria for HS PD are dependent on the specific PD project – radical or incremental. 2. HS PD success criteria can be formulated as short-term and long-term success criteria 3. Time, cost, and performance are central success criteria in a short-term perspective 4. Together with time, cost and performance continuous improvement, continuous innovation, and learning are central success criteria in a long-term perspective. 5. NB HS NPD demands businesses to distinguish between PD leadership and PD management.

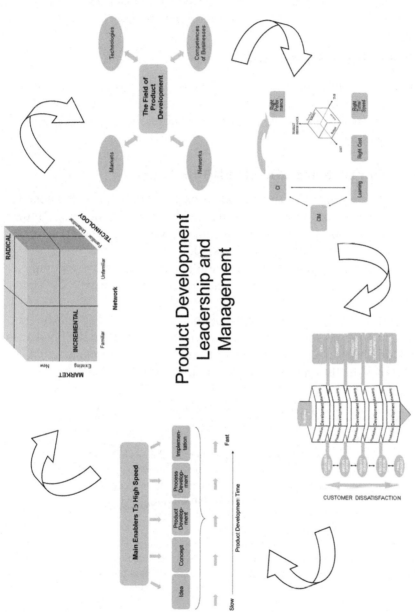

Figure 13.1 The overall research framework model.

The empirical data and the findings give the possibility to answer each question and verify the hypotheses and the framework model. This supports, increases, and puts into perspective the result of the research.

13.2 What Is High Speed NPD?

In the course of the theoretical literature study the PhD project showed that there was a necessity to distinguish between time and speed. The theoretical study and the practical study showed and verified differences in the definition of both time and speed.

13.2.1 Defining Time and Speed in NB HS NPD

A definition of speed in NPD turned out to be more difficult to formulate than originally expected. Initially, it seemed as if it was very obvious. Both researchers and industry seemed to know what they were talking about and how they defined speed in relation to NB NPD.

Subsequent to a longitudinal, theoretical literature study and to an empirical pilot research further clearness came to the definition of NB high speed NPD.

The theoretical study showed that the definition of time and speed should be made in a relative perspective where time and speed should be defined on the basis of the observer's point of view of the product development activity. Table 13.2 shows the differences in views on time NB HS NPD.

Table 13.2　Definition of time

Views on Time from Different Actors in NB HS NPD	Theoretical Definition of Time	Practical Definition of Time
The macro view – the society view either national or community view	From the point in time when the society, the nation, the community e.g. recognize the need or demand for a new product to the point in time when the product is introduced to the society.	Inside out but not with a SME business optimal view – physical time
Market view	From the point in time when potential customer needs and demands are recognized (often before customers have realised the demand) to the point in time when the total product is delivered.	Inside out but not with a SME optimal business view – physical time

Table 13.2 Continued

Technical view	From the point in time when a new product is technically possible to develop to the point in time when it is delivered to sales and production.	Inside out but not with a SME business optimal view
Network view	From the point in time when the network agrees on developing a new product to the point in time when the network decides to introduce the product to the market – the time when it is optimal to the network partners	A multi inside out or a mix of outside in and inside out view. Still not with an optimal business view
Business view	The point in time when the business decides to develop a new product to the point in time when the business decides that it is optimal to introduce the product to the market.	Inside out view but not with an optimal business view
The customer view	The point in time when needs or wants are identified by the customer to the point in time when the product is ready and available on the market for consuming.	Outside in view but not with an optimal business view – right time
The competitor view	The point in time when the competitor recognises the want or need for a new product to the point in time when the product is introduced to the market by the competitor	Outside in but not with an optimal business view – right time
The new marketing view	The point in time when the business possibility is recognised to the point in time when it is business optimal to the business to introduce the product to the market.	Outside in but with an optimal business view

As can be seen there can be several different views on time; the business view, the customer view etc. Additionally, there can be several different ways in which we choose to work with time; low speed, high speed, right speed etc. Businesses in the research mainly think about time as physical time but theoretically time is more complex to define. The businesses' use of time is mostly not based on a business economically optimal definition of time. Firstly, it is not optimal to the market entry point, secondly it does not take into consideration alternative time – time of informal processes, time of coming too early or too late to a market. Furthermore, the businesses define time for product development from an inside out view and perspective, and they define time as physical time.

In this definition, time is verified in the research to be transferred by businesses to cost – and direct cost. The businesses place the product in the interval between too early and too late. They were therefore "lucky" if they "hit" the right time to introduce a new product. This meant that they often lost anyhow because either they introduce the product to the market too late or too early.

The research verified that these costs were not calculated in the businesses but it was verified that these costs or loss of values could have a large influence on businesses and especially on ROI of PD projects.

As can be seen the prioritising of the view on time can be different according to e.g. market, business, competitor, and competition. However, the research showed that:

- Some businesses saw business possibilities before customers – they saw the idea before it was brought into the formal idea stage.
- Many businesses often developed products before their customers demanded the products, which meant – "at an in-optimal time" – seen from a business economic point of view.
- Businesses who saw a competitor implementing a new product first, recognized and harvested a business possibility first. In other words, they realise the difficulties it created to speed the product development. Firstly, they were often "stuck" in a situation where the competitors had gained first mover advantage and had harvested the major part of the market beforehand. Secondly, they faced the risk of increasing the cost of product development because of increasing informal product development processes and thereby high alternative cost.
- In most cases businesses used a definition of time in NPD corresponding to the verified definition on time called the business definition. This was shown in Figure 13.2.

The choice of time perspective gave the involved business new limits and potentials to time and speed in product development. As can be seen, the time limit is much tighter when the business view the potential at a later point of time than their customer and their competitors. This stresses the importance of over-viewing and analysing continuously the field of product development. It also stress the importance of choosing the right speed in the point of entry of ideas to the business and right speed in the product

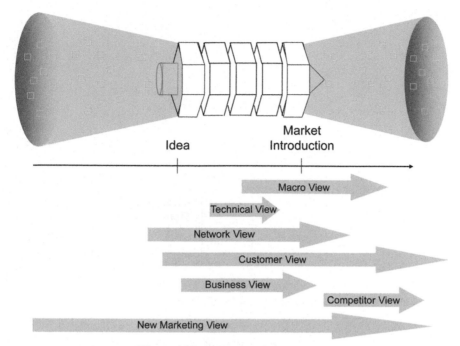

Figure 13.2 Different views on PD.

developments upper part of the product development process – idea and concept development.

The research showed significant consequences on cost – both direct and alternative cost when businesses did not choose the right speed. Businesses which use the new marketing view where they focus on perceived value and alternative cost as well as value direct cost, had significantly less cost and better performance in product development than SMEs using the business view, the customer view or the competitor view. Furthermore, businesses who focused on right speed had significantly better results than other businesses. The curve illustrated in Figure 13.3 shows the result of the above-mentioned.

Definition

HS MI = High speed Market introduction
RS MI = Right speed market introduction
NS MI = Normal speed market introduction

The Figure 13.3 showed how the cost curve develops with different views on speed.

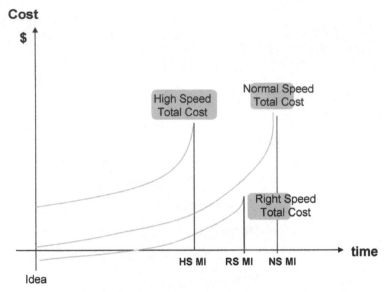

Figure 13.3   Cost curve of different PD views and speed strategies.

SMEs with a business view – an inside out view – experienced increased costs, both direct and alternative, when speeding time, because product development demands too many resources to speed time. Such resources were required either to reach the market before it "slipped" away, or to e.g. fight the competitors because the market was already covered by competitors. The business view was very much identical to the result in the curve shown as the high speed cost curve.

The research showed very significantly that the business view created more failures to product development and increased costs to repair and solve product failures after the product had been introduced to the market. The same can be verified by the technological view. If the technological view was used, the product would often be developed too complicated and often not in relation to the optimal point of entry.

The research verified that the new marketing view could offer major competitive advantage if used appropriately. If businesses strictly used the new marketing view in their product development, a competitor would "perceive" the time and speed in product development of this particular business as very high and would face a major challenge to compete. The competitors would often use "the HS model" to compete. This, however, would increase their cost of NPD and would thus result in increased competitive advantage to the business focusing on RS. I called this the "dead lock of HS NPD".

The empirical research showed that very few businesses used the new marketing view on product development. Most businesses used the business or the technical view and only a few used the market view. The market view was mostly focused on the competitors' and customers' value and not on perceived value. Some of the secondary cases showed the practice of the new marketing view (Case No. 1 Zara, Case No. 62 Ryanair, Case No. 14 Nike).

The empirical research only showed few examples of businesses which used the network view. I claimed that the network business view was not optimal to product development because the network view used an inside out or a multi-inside out/outside in perspective. This view had a potential "born in" conflict as the network partners would usually not have the same business goals or views on the optimal speed of product development. Businesses were seldom focused on the business optimal time to introduce new products. This was mainly due to a different calculation on alternative time and cost.

As a consequence, I claimed that my research had verified that in this context businesses should view time and speed in network based product development as:

> *"The time transformed into value and perceived value minus cost both direct and alternative which a product development project takes from identification of the potential product idea to the time the product is taken out of the market.*

This is shown in Figure 13.4.

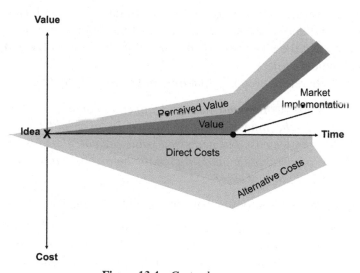

Figure 13.4   Cost value curve.

"The business economic optimal time is when the product is delivered to the market and when this introduction maximize

Net profit = (Direct and perceived alternative value) – (Direct and alternative cost)

This is shown in Figure 13.5.

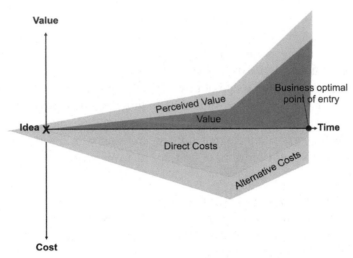

Figure 13.5   Business optimal time of market introduction.

The research verified that speed could only be defined when businesses know an exact start and an exact end of a product development project. The research verified that the businesses could not clarify the start or the end of a product development project.

The research therefore proposes that time in PD could be defined as:

"The net profit per working hour from identification of the potential product idea to the time, when the product is taken out of the market". Alternative cost and perceived value are calculated from the business optimal point and from both before and after the business optimal point of entry.

Furthermore, the research proposes that Time in PD is defined as:

"The relative time according to which view the viewer has of the NPD project – either the Macro view, the business view, the market view, the technical view, the network view or the new marketing view"

The last definition will be further commented on later in this chapter.

The empirical study showed, however, that not many SMEs calculated the time for a product development project. They just did it. Those SMEs who define time did it within the concept of physical time used for PD projects. As an example, time could be defined in relation to the amount of time it takes to develop a new windmill or the amount of time it takes to develop a new chicken slaughtering machine.

SMEs defined one product development project to run at a higher speed than another when it takes e.g. 1 working day or 1 working month less to develop a new product than last time a similar PD project was carried out.

The research showed that SMEs did not focus on the difference on the status, position, or characteristics of "the main components in the field of product development".

Time in NPD was therefore verified in SMEs:

> *"as the total physical time – man-hour, working days etc. that it takes to develop a new product from the idea enters the businesses' product development system – idea stage – and to the product is introduced and implemented to the market".*

The time was often transformed into cost; but direct cost. The time when this was calculated was often after the PD project had been fulfilled. Businesses seldom used estimations or ROI analyses on this area beforehand.

Speed in NPD was empirically verified to be defined as:

> *"the time (working days or man-hour) one product development process takes from an idea comes up in the business to the product is implemented to the marked" compared to the time (working days or man-hour) another product development process takes from an idea comes up in the business to the product is implemented to the marked" The empirical research verified that a product development project is said to run with a higher speed 1) if it reach the market before the other product development project 2) one product development project measured in time (working days) are developed faster than another product development project."*

The first empirical argument shows that businesses do not calculate the cost of reaching the market. The second argument shows that businesses do not calculate the alternative cost or the perceived value before or after the market introduction.

Therefore, the theoretical definition is quite different to this calculation as it was the time when it was business economically optimal and efficient to implement the product. Consequently, I claimed that it was theoretically impossible to define speed of a product development project within physical time.

However, the SMEs made the comparisons on speed as shown in Figure 13.6.

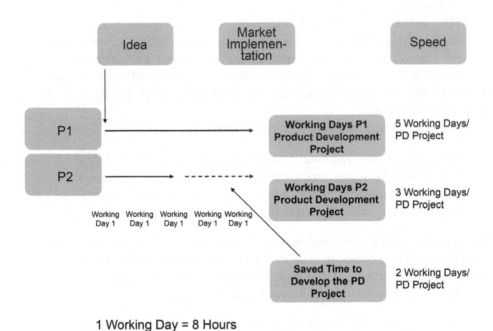

Figure 13.6 Emperical definition of speed in PD.

As an example, product development project P1 runs at a speed of 5 working days and product development P2 at a speed of 3 working days. P2 saves 2 working days in product development compared to P1. P2 runs product development faster or at a higher speed than P1.

The involved SMEs transferred physical time into cost and direct cost (wages, materials, etc.) which the product development consumed to develop a new product to the market.

Speed in NPD was defined as saved direct cost (wages, materials, etc.) which one product development project spends on developing a new product to the market compared to a similar product development project.

One product development project (P1) was measured as running faster than another (P2) when comparing the costs of the two NPD projects as seen in Table 13.3.

Table 13.3 Definition on speed verified in the emperical research

	Measurement	Working Days	Cost (Direct Cost)
NPD Time for project 1	Cost of NPD project	5 working days	€ 200,000 EURO
NPD time for project 2	Cost of NPD project 2	3 working days	€ 100,000 EURO
Speed faster in project 2	Saved cost and speed	2 working days	€ 100,000 EURO

The empirical data very clearly verified that this was how the SMEs of my research define time, speed, and speed in NPD. The SMEs' definition was – as can be seen – quite different from the theoretical definitions listed in Table 13.4.

Table 13.4 Practical and theoretical definitions of speed and time in PD

	SME Definition and Empirical Research	Theoretical Definition
Time	Physical time from idea start to market implementation	Relative time dependent on which view is taken and the business optimal point of entry
	Direct costs	Cost (both direct and alternative cost) – value and perceived value form before idea comes up to the product is taken out of the market.
Speed	Physical time saved Direct costs saved from one PD project to another	Relative time saved. Net profit gained from one product development project to another.
Cost	Direct costs	Direct costs and alternative costs
Performance	A product that match the businesses' view of good performance	Right performance

Moreover, it could be verified that businesses used the SME measurement to compare one product development project to another, the performance of one product development team on time and speed to another NPD team, or their business's ability to develop new products faster than their competitors.

However, as stressed and verified earlier very seldom did the basis or the characteristics of the field of product development for one product development project equal those of another. The research showed that the market, the technology, the network, and the competences of the businesses were always different and changed from one PD situation to another.

Therefore, both theoretically and practically one development project of e.g. a windmill could not be compared in terms of time or speed to another development project of a windmill when "the main components of PD":

- were different from one NPD project to another
- changed continuously and individually over time

The empirical results verified that it was not possible to compare PD projects in terms of physical time and speed when the conditions of one PD project were never the same as the condition of another. Therefore, I proposed that time and speed used to measure product development activities were deliberated in another way.

The research verified that SMEs made a mistake when they measured high speed in NPD within physical time and direct costs. The research also very well verified that SMEs made a mistake when they measured high speed with a focus only on lowest possible costs as seen in Figure 13.7.

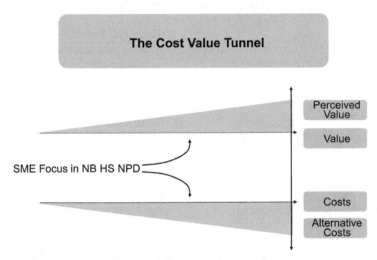

Figure 13.7   Cost value tunnel.

Source: Bohn and Lindgren, 2002.

Instead, my research project verified that it was more preferable to focus also on value and perceived value which was verified to result in a major decrease in costs, and an increase in net profit and competitive advantage as shown in Figure 13.8.

Therefore it could be verified that NB HS NPD defined within physical time and speed gave no meaning to business economics optimisation because it focused only on costs and only on direct costs. The research verified that businesses which forget to calculate both direct costs and alternative costs, value and alternative value of product development projects came too late or too soon to the business optimal time for a market introduction. These businesses could be define as "slow speed" businesses.

The measurement of speed had therefore to be seen primarily in relation to:

1. the task of the PD project as illustrated in incremental or radicalness of the task
2. the complexity of the product development projects or to "the field of product development"
3. the value and cost of speeding the PD project
4. the perceived value and alternative cost of speeding the product development project.

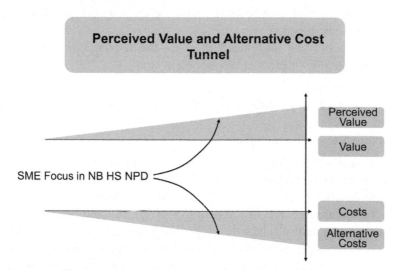

Figure 13.8 Perceived value and alternative cost tunnel.

When talking about comparing speed in two product development projects, the task of defining speed and time becomes very complex. When businesses want to do this in any case I proposed that they define the individual product development task along with the characteristics of the market, technology, network and the competence of the business in both PD projects. The measurement on the businesses' ability to develop new products faster and at higher speed in one PD project than in another depends the above mentioned generic elements.

However, the empirical research showed very clearly that any product development project faces a new "product development field". To compare one product development project or task with another demands that "the two fields of product development" are equal. The empirical study verified that this is never the case, and I therefore question the relevance of businesses comparing product development projects in terms of physical time and speed.

How can product development projects then be compared in relevant business economic terms – and in time and speed?

Can NB HS NPD not be defined? and were businesses' measurement of time and speed of product development projects worthless?

The answer to the first question is "Yes"; indeed it is relevant to measure NB HS NPD in time and speed. Yes, indeed it is worth measuring NPD in time and speed. However, time and speed must be defined in another way and must be seen in relation to the businesses involved, the product or the product development task to be developed, the market and the technology with which the product development has to be conducted, and the network on which the product has to be developed.

The definition of time and speed must be taken from the "the main components on the field of product development" and

- their initial position
- the way in which they change during the time of a product development project
- their interaction with each other

Secondly, the research project had to be defined within the framework of radical and incremental product development. As can be seen in the empirical research most businesses defined their product development projects in the range of incremental product development (85–90%). It could be relevant to compare one incremental product development project with another incremental product development project within time and speed, whereas it was more complex to compare incremental PD projects to radical projects and radical PD projects to radical projects as shown in Figure 13.9.

The empirical data in my research were mainly focused on incremental product development projects and situations in which the main components were fairly stable and where it was preferable to measure PD within time and right speed. I argued that time and speed in this area of product development should have a business economic relevance to the business. Time and speed must be defined and viewed in relation to right time introduced to the field of product development. However, the view and definition of right time in NB HS NPD is always different seen from the different views of product development. This is illustrated in Figure 13.10.

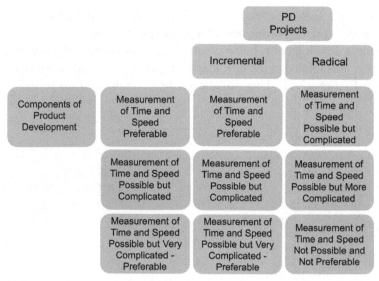

Figure 13.9   Measurement of time and speed in PD projects compared to the PD task and "the components of PD."

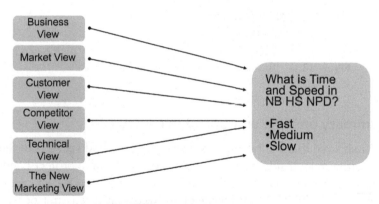

Figure 13.10   Different views on time and speed in PD.

However, the business optimal point of time and speed – right time and right speed – will always be the same seen from the new marketing point of view. For each business there was only one optimal point to enter the market with the new product. Such an optimal point was defined as:

> *"when the market is business economically optimal to the individual business to introduce the new product"*

Thus, I claim that such time and speed were relative as argued before and depended on the task of product development and the characteristics of the field of product development. I also claimed that this point in relation to my scientific view was related to the new marketing view of product development as shown in Figure 13.11. This will be further commented on in the next paragraph.

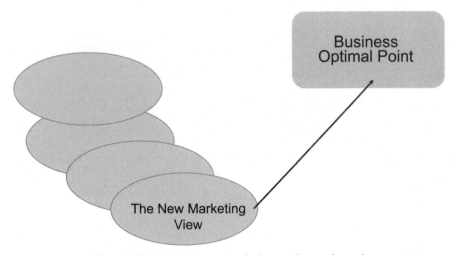

Figure 13.11   Business optimal view on time and speed.

13.2.2 Summary on Time and Speed in NB HS NPD

The research showed that the businesses' views on time and speed take their point of departure in an inside out perspective. In my research project I chose the new marketing view on the definition of time and speed – an outside in view. An analysis of the empirical data showed that it was not possible to verify that SME businesses follow the proposed view on time and speed. Furthermore, the research showed that businesses were narrowly focusing on costs and direct costs instead of on both costs and value including direct and alternative costs and perceived value when measuring time and speed. The result of the PhD project gave the following picture seen in Figure 13.12 of the difference between the practical and the theoretical measurement of speed and time.

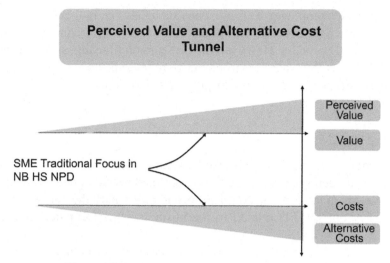

Figure 13.12   SMEs focus on cost and value in NPD.

13.2.3 HS NPD Seen from Different Points of View

Theoretically, HS NPD could be seen from different points of view, macro environment, business, product, market, customer, competitor, and network view as seen in Figure 13.10.

The Business View

My research verified that businesses mainly focus on HS NPD from a **business view**. The business focus was concentrated on the business's ability to diminish physical time and increase speed in NPD seen from an internal business view – an inside out view focusing on primarily cost and direct cost.

The Macro View

The research verified that high speed in product development is often increased by businesses without a view to the macro environment. The businesses hardly ever had a view to the macro environment. The lack of verification on this area of the research was due to a lack of direct focus in the research on this view, but also to the fact that none of the case businesses used that view on NB HS NPD. Literature and articles showed a strong pressure from the macro environment to speed product development. Both from national, international, political, economic, and research environments there was a high pressure on businesses to speed product development. This must be said to have an indirect influence on the way in which businesses perceived and made their effort and initiation on speeding the PD process. However, the pressure on time and speed of the businesses' PD from the macro environment were seldom based on optimised business economic calculations or on pure calculation and analysis of the customers' demand.

The Market View

The market view could be divided into to two parts – one view which looks at time and speed in relation to the customers' wants and needs for new products, and one view which looks at the competitors and benchmark the competitors' ability to carry out HS PD related to the businesses' performance on this topic.

The customer view showed a radical increase in the demands, wants and needs verified in literature and secondary cases. However, the empirical research showed only incremental demands for new products.

The competitor view showed high pressure on product development but still related to the incremental customer demands. This seemed to be "wamped" or incorrect information on real customer demands. It can therefore be verified that there was seldom a reason to put pressure on speed in PD.

The Customer View

The customer view was characterized by a customer want perspective and not by a customer need or demand perspective. My research verified that such a view increases the pressure on speed in PD and diminishes the lifetime of products. Even products that had not fully lived out their life were removed from the market. However, in many cases this could be a business optimal decision to many businesses as shown in Figure 13.13.

The Competitor View

When a business toke the competitor view they tended to believe that the competitive situation outside the business called for and demanded a strong pressure on speed in their product development. However, as the research verified the pressure on speed in product development was seldom related to actual radical needs in the market place. Instead it was founded in incremental needs. This often caused product development which was too fast and therefore resulted in first mover bad advantage.

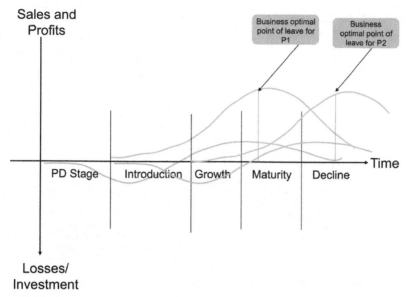

Figure 13.13 Business optimal point of leave related to lifecycle from inception to demise.
Source: Lindgren P. 2003.

The Technological View

The pressure on speed in PD seen in a technological perspective turned out to be more complicated. The research showed that businesses were under high pressure from technological innovation and possibilities. However, businesses often failed to analyse the technological challenge of the product development project. The consequence was that they speed the PD too much because of perceived low technological innovation degree. An increasing demand for the mixing of different technologies together with the businesses' lack of ability to analyse the need and use of technology to solve the product development task, pressed the involved businesses into rather NB IIS radical product development projects. Such projects were decisive when businesses also experience the pressure of time. The research showed that risk and uncertainty increase when technological radical innovation was combined with a high pressure on speed and unknown networks as shown in Figure 13.14.

The Network View

The research showed very clearly that the networks employed were mainly narrow physical networks, mostly internal, and customer supplier network relations. The network view was seldom analysed and used when deciding

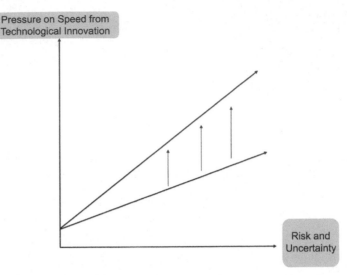

Figure 13.14   Risk and uncertainty related to pressure on speed.

on speed. The network view was mainly used when considering the following questions:

- Do the existing internal and external network partners have the competences to develop the new product?
- Does the business need to attract and integrate new network partners?
- Does the business need to repel networks partners both initially and during the course of the product development process?

The above-mentioned questions demand a strong product development leadership. At a later point in this book, I will deal exhaustively with product development leadership. Furthermore, developments on the global market showed and stressed the importance of beginning the implementation of a increased network view to gain high speed product development projects. This involved an answer to the following questions:

- Is the business ready to do network?
- Does the PD project need a broad or a narrow network?
- Is the network ready to develop the new product?
- Can the current network perform the product development task at the right speed?

The overall answer to these questions was that the businesses were not using the network enabler enough. There was still much to learn here for SMEs.

The lack of use of the network enabler was a major barrier to speed product development further.

The Business's Competence View

The research verified that from the business's perspective – inside out – businesses focused primarily on time and speed in product development. The outside in view was only found in a few businesses.

However, looking at the business's competence to carry out product development and their ability to speed product development further it was verified in the research that most businesses said that they could not increase speed in PD any further. This was claimed to be mainly due to:

- barriers in the stage-gate models
- physical barriers in process
- risk aversion on faults due to more high speed.

The research showed that most businesses in this case did not pay attention to **the product view**. Most businesses did not make an in-depth analysis of the product and the consequences to the core of the products or the core benefits of the products when putting pressure on time and speed in PD. Many businesses did not analyse the perceived needs of the customers. Many businesses did not use the possibility of digitalising the products and of using the e-development enabler together with changing the product to a process – process enabler. The product development therefore turned out to be perceived as more radical than intended and expected.

The research verified that some product development projects seemed at first sight incremental in nature but turned out to be radical because the pressure on speed in product development changed the core of the product and the core benefits of the product. In some cases this provoked:

- faults in NPD
- unexpected reactions from the market and the customers
- slow speed seen form the customer, market and the new marketing view
- the real needs and demands of the new product

The above-mentioned stressed the importance of carefully analysing beforehand:

- the competencies of the businesses to match the individual tasks of product development projects
- the consequences of PD on the core of the product.

The New Marketing View

In the secondary cases, the primary case, the focus group interviews, the survey, and in other research activities carried out in this research project I verified very few examples of businesses with a view of NB HS NPD that equalled the new marketing view. The new marketing view considers:

> *the optimal time and the optimal speed of product development as correlated to performing product development with a speed and time that perform a product development process that match the business optimum time of entering the market.*

The new marketing view considers time and speed not only with a focus on value and cost but also with a focus on perceived value and alternative cost.

When businesses focus simultaneously on value, perceived value, cost and alternative cost in PD, the research showed very clearly that businesses attain better performance, higher speed, diminished cost, increasing profit, faster ROI, and major competitive advantage both on a short-term and a long-term basis. Both the secondary data and the empirical data (Case No. 1 Zara, Case No. 62 Ryanair, Case No. 38 Lyngsø the focus group PUIN) showed that these businesses achieve major business results on the global market.

Summary

There are more theoretical views to HS NPD but very few of these were used empirically up to 2003. Table 13.5 gives an overview of the empirical and theoretical views which were verified in the research.

Table 13.5   Different views on high speed PD

Different Views on HS PD	Verified/Not Verified
The business view	Verified
The Macro view	Not verified
The Market view	Verified
The customer view	
The competitor view	
The Technological view	Verified
The Network view	Not verified
The Product view	Not verified
The New marketing view	Verified in very few secondary businesses or product development projects and in one primary case. Not verified empirically.

13.2.4 HS Central to PD Phase

At an earlier point in this book I stated that my hypothesis was that high speed was central to the second phase of the PD process – the PD phase. The research painted a picture that differed slightly from my hypothesis.

The research focused on time and speed in the three general phases of the product development phase:

1. Time and speed in the innovation phase
2. Time and speed in the product development phase
3. Time and speed in the process phase

However, it must be pointed out that the main focus had been on the two first general phases of the product development. The research could verify that in some businesses time and speed were in focus in all phases of the product development process.

The hypothesis of my research project was that time was not in focus in the innovation phase, speed and time were very much in focus in the second phase, and costs were very much in focus in the third phase. As seen in Table 13.6 the research verified a much more complex picture and result in this area.

Table 13.6   Success criteria and genreal PD phases

Focus	Innovation Phase	Product Development Phase	Process Development Phase
Hypotheses	Innovation	Time	Cost
Case	Innovation and performance	Time	Cost and time
PUIN focus group	Innovation and performance	Time and performance	Time and cost
Survey	Innovation and time	Time and performance	Time and cost
Hypothesis verified/not verified	Verified but with and ad to performance and slightly to time	Verified with and ad to performance	Verified with and ad to time

Time and Speed in Innovation Phase

To some extent, time and speed were in focus in the innovation according to the survey. This hypothesis could only be partly verified. The case research showed that most businesses focused on innovation in the innovation phase but some businesses focused on speed and time in this phase. The research showed that focus on time and speed was related to the business in question, the PD task, but also to the position of the main components in "the field of product development". The analyses and the businesses' internal analyses of the perceived position of the field of product development influence the pressure on time and speed in PD. The task of product development project also influenced the focus on time in a product development project. If the product development task was characterised as very radical, there was less focus on time and speed, and focus on performance and innovation was more highlighted. If the product development project task was a more incremental character, there was more focus on time and speed. If businesses perceive the characteristics of the field of product development as stable and if the task of product development was rather incremental, then there was also a pressure on time and speed in this phase. This is shown in Table 13.7.

Table 13.7   Pressure on time related to field of product development and PD task

	Stable	Evolving	Dynamic
Market	Yes	Yes	No
Technology	Yes	Yes	No/yes
Network	Yes	No/yes	No
The business competence	Yes	No/yes	No
Task of product development			
Radical	Yes	No/yes	No
Incremental	Yes	Yes	Yes

Generally, businesses did not speed PD in the grey areas because:

1. they perceive risk as higher in the grey area
2. businesses did not focus on speed in fuzzy areas

However, the research verified that businesses with high speed competences and with network partners who had high speed competences could suddenly add speed to NPD to gain competitive advantage. Speed in NPD therefore also depended on the ability and motivation of the businesses and the network partners for speeding NPD. The technological evolution could also be a driver to high speed.

Another area which proved to be important when observing time and speed in product development was the strategic importance of the product development project. If the strategic importance was high to the business, high pressure was generally exercised on time and speed also in the innovation phase.

Summing up on the pressure on time and speed in the innovation phase showed that time and speed could be in focus in this phase, if:

- the position in the field of product development was stable
- the task of product development was incremental
- the product development task had strategic importance to the business
- the competences and motivation of the businesses were focused on high speed to gain a competitive advantage.
- the competences of the network partners were focused on high speed
- the "field of product development" drive the NPD process to high speed

In other situations, innovation and performance of a new idea would be in focus in the initial phases. Therefore, my hypotheses was not verified completely because time can be in focus in the initial phase.

13.2.5 Time and Speed in Development Phase

The focus on time in the development phase was also more complex than presented in the hypothesis. The research verified that time was not always in focus in the development phase. In the concept generation stage many businesses still had major focus on innovation and performance. In the concept gate all businesses were very much focused on time. The product development phase showed a fuzzy picture of some businesses focusing on time and other on performance depending on:

- whether the position and characteristics on the field of product development were stable
- whether the task of product development was incremental
- whether the businesses run the screenings phase
- whether the product development task had strategic importance to the business
- whether the business wanted to gain radical competitive advantage

13.2.6 Time and Speed in Process Phase

The research verified that time and speed were not particularly in focus as the primary success criteria in SME businesses in the process phase up to 2003. Cost and performance were in most cases in focus in the product development process phase. Many businesses focused on "encapsulating" the product with the best performance to the market.

13.2.7 Time and Speed Related to PD – Stages and Gates

The above-mentioned focus areas on time and speed concerned the general phases of the product development project.

The research could, however, also show how the businesses focused on time and speed in the individual stages and gates of the product development process.

The hypothesis of the research project was that time was important especially in the middle of the product development phase and thus in the middle of the product development stage and gate. The result of the research showed the following characteristics as seen in Table 13.8:

Table 13.8 Time and speed related to stage and gate

Product Development Stage and Gates	Idea Stage	Idea Gate	Concept Stage	Concept Gate	Prototype Stage	Prototype Gate	Process Development Gate	Process Development Gate
Hypotheses	Innovation	Innovation	Innovation	Innovation	Time	Time	Cost	Cost
Case research	Innovation and time	Time	Innovation (performance) and time	Time	Performance and time	Performance and time	Cost	Cost, performance and time
PUIN focus group	Innovation and time	Time	Innovation (performance and time)	Time	Performance and time	Performance and time	Cost & Performance	Cost, performance and time
Survey	Innovation and time	Time		Time	Performance and time	Performance and time	Cost & Performance	Cost, performance and time
Verified/not verified	Partly verified with an add to time	Not verified	Partly verified	Not verified	Partly verified with an add to time	Partly verified with an add to performance	Partly verified with an add to performance	Partly verified with an add to performance and time

As can be seen, the research showed a more complex and fuzzy picture of the focus on time and speed than expected in the hypothesis. There were no considerable differences in the answers in the different research types, but the research showed that the hypothesis could not be verified. Time and speed can be in focus in all stage and gates dependent on the characteristics mentioned above.

Nevertheless, speed and time are significantly in focus in the idea screenings and concept screenings phase according to the empirical research. Often the screening phases are very much focused on time and speed. This again confirms that time and speed can be in focus in the innovation phase.

In the prototype gate it was significant that focus was very much on cost and performance.

13.2.8 HS NPD – A Matter of Right Speed, Not High Speed

Earlier in this chapter I argued that measurement of time an speed in product development had to be seen in relation to the components of the field of product development (the market, the technology, the network, and the competences of the businesses) and the task of the PD project (radical or incremental). The PhD project asserts that the definition of speed in relation to product development must at least up to 2003 be seen in relation to the market and with the new marketing view. PU projects should be measured against business economic terms. In each specific product development case the management must ask the questions.

When and What is the right time and right speed of a product development project?

Speed and Time Measurement in SMEs

The research showed very clearly that businesses measure the time and speed of a PD project on the basis of physical time. This physical time used to develop a product was transformed into direct cost. My research showed, however, that the transformation of time and speed into cost was seldom made in small businesses. My research showed that this was due particularly to two problems:

1. SMEs had difficulties in defining a beginning and an end to a PD project especially when talking about incremental PD projects.

2. SMEs did not directly calculate time in terms of costs on product development projects. If they did so, time was mainly calculated in terms of direct costs.

At a previous point in this research project I commented on the difficulties of defining a start and an end to a product development project. My research very clearly showed the difficulties for the businesses to define such beginnings and ends.

The discussion of a beginning and an end of a PD project was related to the discussion of defining product development instead as a process without a beginning and an end or with many beginnings and many ends. The research showed that businesses mainly define their product development projects as projects with beginnings and ends and not as product development processes. The research showed that this was a result of strong focus of the businesses on the stage-gate terminology in product development which demanded a beginning and an end.

However, if businesses cannot define the beginning and the end of their product development projects – especially in incremental product development projects – then time and speed cannot be defined within physical time and the following questions can be put:

- Is time and speed then a relevant measurement in NPD and is it relevant to measure time and speed?

Additionally,

- Can businesses manage their product development process on the basis of physical time and speed in physical time?

Furthermore,

- Can businesses manage their product development process with the use of their existing product development models?

The answers to these questions must be "No; physical time and speed is not a relevant measurement when used "in the ways" it was used by SMEs and many researchers up to 2003". Businesses and researchers needed to develop a new definition and measure of time and speed and some new models to handle product development. Businesses must try to focus more on reaching right speed and right time to market because physical time focus on an inside out perspective which will often not give a competitive advantage – at least not up to 2003. I also argue that businesses need new decision models and normative guidelines to choose product development model targets at right time and right speed.

Right Time and Right Speed

I claimed that right time and right speed:

- is very much related to the ability of the product development managements to define the product development task "read the field of product
  development" both initially and as the product development progresses
- is related to the finding of the optimum time – right time – to introduce
  the product to the market
- is related to the matching of NPD process with right speed and time

I claimed that this ability was strongly related to learning.

Managers of product development needed to develop a strategic design of
the way in which NB HS NPD should be used. They needed to understand
why NB HS NPD should be used, and to realise what NH HS NPD models
and processes should be used.

The strategic design of NB HS NPD had to be strongly related to product
development knowledge and product development knowledge creation at the
management level within the business. The managers of businesses had to learn
about NB HS NPD and to develop knowledge of HS in product development
leadership. Yet, learning in NB HS product development management was by
far not easy as the case research showed.

In the case and survey research I observed different types of speed in NB
HS NPD as seen in Table 13.9.

Table 13.9 Types of speed in NB HS NPD

Types of Speed	Characteristics	Cases
Idea to market introduction speed – "time to market speed"	The ability to speed the NPD project from idea to market introduction	Cases Nos. 39, 41, 1 and 63
Stage and gates speed	The ability to speed the single stage and gates within the product development project	Cases No. 39 and 1
Transfer speed	Speed from one stage to another gate	Case No. 1
Complex speed	The ability to speed complex NPD projects	Case No. 64
Concurrent speed	The ability to speed several NPD projects at the same time	Cases Nos. 49 and 64.
Market speed	The ability to speed incremental NPD on the market.	Cases Nos. 55 and 65
Radical Speed	The ability to speed radical PD projects	Case No. 66

Up to 2003 we had only fragmented knowledge of and research on the types of speed and speed tools available and appropriate in different situations of product development. Learning had to be established in all areas of high speed product development to find models of speed in NB NPD and to find when it was advantageous to "hurry slowly". The research showed that when characteristics in market, technology, network, and the competences of the businesses were in a certain position, a slow speed could be advantageous as learning of market, technology, network, and competences develop, proceed, and get ready for the new product.

The empirical data showed how businesses choose to hurry slowly because of e.g. market competition, lack of competences or a strategic decision to wait until technology and market are stable.

I assert therefore that the speed of PD had to match the special conditions of the PD task, otherwise the speed of PD would be too slow or too fast which will give businesses increased cost and a competitive disadvantage.

Even so, the question of speed was more complicated than outlined above. During the product development process the speed sometimes has to be increased, and sometimes to be slowed down. Some of the main components in the field of product development can turn out to influence and make radical changes to "the game of product development". Therefore, businesses often have to change speed during the product development process. The manager of PD had to decide continuously on "right speed" during the PD process.

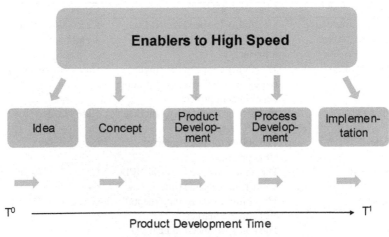

Figure 13.15   Speed in NB PD.

Right speed in product development had to be learned. The critical issue before talking about speed in product development was the ability of the management to analyse "the game of product development " and to learn from one product development project to another which speed was advantageous to this specific product development situation. Even more critical was the ability of the product development managers to learn throughout the product development process. The last learning area concerned the development process from idea to market introduction as well as the span of time after market introduction.

The question of how to establish learning of speed in product development across networks in the product development process was important and was verified in the case research as a major problem to SME. This is illustrated in Figure 13.16.

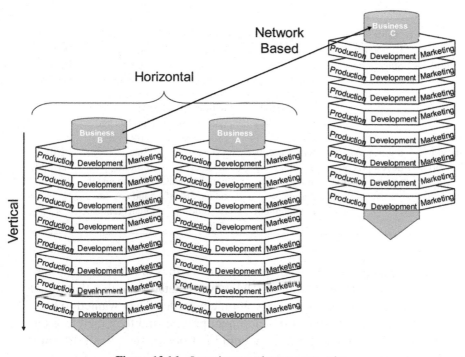

Figure 13.16 Learning speed across networks.

1. Right speed initially in the NPD project
2. Right speed along the NPD project
3. Right speed related to other PD projects
4. Right speed related to projects of network partners

This became the focus of research projects carried out after 2003. However, my research before 2003 did verify that businesses did not define right speed in general and certainly not in relation to points 2–4.

Theoretical Product Development Model and Empirical Model

Whether businesses had a formal stage-gate product development model was of interest for the research to verify the PD model and conditions together with characteristics in SMEs. My hypothesis was that the stage-gate model existed in the businesses and that businesses used this model. This hypothesis could only be partly verified.

In the research the major part of the businesses answered that they had a formal product development model. The research showed that the formal stage-gate product development model existed. This is illustrated in Table 13.10.

Table 13.10   Confirmation of existence of formal PD model

	Hypothesis	Case Research	Focus Group	Survey	Verified
PU model – formal	Stage-gate	Overall stage-gate – but combined with an informal model and processes	Overall Stage-gate combined with informal models and processes	Overall Stage-gate combined with informal models and processes	Partly verified
Functions involved in initial phase	Sales, product development and production	Sales, product development and to some degree management	Sales, product development, management to some degree	Sales, product development and management	Partly verified

However, a large number of businesses said that they did not have a formal, written product development model. The reasons stated were that they often felt restricted by a formal product development model. They wanted to have more flexibility in their product development process and they therefore did not write it down. They felt that they could not obtain such flexibility in a formal product development model.

However, the research showed evidence of a formal "not written down" product development model in the businesses who answered *No* to having a formal product development model. Such models had four stages – an idea, a concept, a product development, and a process development stage. This confirmed the hypothesis of the research that a major group of businesses follow a stage model also when they believe themselves to be without a written, formal product development model.

On the other hand, the picture became less clear where the gates were concerned. The survey showed that the businesses did not give a very high priority to the idea screening or the concept screening gate. Instead, they gave a high priority to the prototype and process screening gate. This indicates that there was not much focus on "reading and screening the product development game and task" within the initial product development process. It also indicated that many ideas and concepts "slip" easily through to the prototype development stage. The research verified that this could be fatal to the business in terms of time, cost, and performance.

The research also showed that businesses often get "stuck" in their formal product development model as e.g. the stage-gate model. This caused problems with flexibility and change of speed later in the product development process. The businesses tried to solve this problem by creating informal product development models and processes. The informal product development model and process could be described as running along or beside the formal product development model or process generating both direct, indirect, and alternative cost and value. However, the businesses did not know how much cost and value were generated in their informal models both management, employees and other invitation the informal models and processes.

The businesses confessed that the importance of the informal product development model in different areas was significant when discussing time, speed, cost, and performance but also CIM, and CI on learning.

Furthermore, the research verified that the pressure on high speed forced the development of informal product development models and processes in businesses. 52,9% of the businesses in the survey said that they applied informal product development models concurrently with the formal product development processes. The managers in the case businesses and focus group responsible for product development revealed to me the contents of the informal product development model.

My research showed that the informal processes and models were very much existent in the product development, the prototype, and the proto test stage and gate at which point in time, time and speed were very important.

The hypothesis was that the informal models and processes were often out of strategic product development control. Therefore, in most cases the models and processes were not providing the businesses with right performance, right cost, or right speed because they were focused on the process inside the product development model without continuous interaction with "the field of product development". Furthermore, the learning gained by the businesses from such informal models and processes was not transferred to continuous improvement (CIM) or to continuous innovation (CI) of the formal product development models and processes of the businesses. This was because the informal models and processes were "one of a kind", and because the knowledge transfer to the formal product development model and process was not existing or formalised.

Businesses had realized that stage-gate models were effective for some product development tasks and situations. However, dynamic and flexible product development models seemed to be more effective. Furthermore, the stage-gate models were not related to "the field of product development", as seen in Figure 13.17.

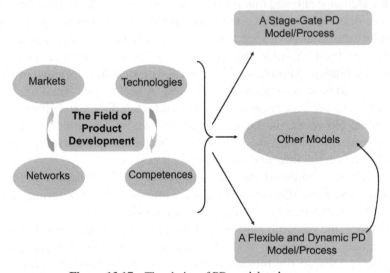

Figure 13.17   The choice of PD model and process.

Source: Lindgren & Bohn, 2002.

My observations show that product development managers up to 2003 faced the difficult task of deciding which product development model and process were most suited for the specific product development task.

The choice of model and process turned out to have a major influence on the possibility of speed in product development in at least three ways. First, how much speed can be achieved during the product development project. Secondly, how can speed be changed and how much change of speed does the business need in the product development process. Thirdly, how much speed can be gained in variation on behalf of the original product development concept.

Businesses can gain speed both in a stage-gate model and in a flexible product development model. However, the costs of changing speed and performance of the product differ and are influenced by the choice of the product development model as can be seen in Figure 13.18.

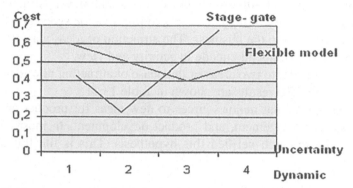

Figure 13.18 Costs of flexible and stage-gate product development models related to uncertainty and dynamics.
Source: Lindgren inspired by R. Verganti, 2002.

Firstly, the flexibility of speed possibilities is not the same in the two models because the cost of changing during the process was higher with a stage-gate model. Secondly, the possibilities of performance of the final product differ from one model to another in accordance with the point in time at which the change of speed in product development is required.

Managers of product development who had developed a strategic design of how to use NB HS NPD model, who understood why NB HS NPD should be used, and who had realised what NH HS NPD models and processes should be used were better at performing right speed and right time in NB HS NPD. Nevertheless, the strategic design of the NB HS NPD model had to be strongly related to product development knowledge and product development knowledge creation at the management level within the business and the business's network. The managers of businesses had to learn NB HS NPD and to develop leadership of NB HS NPD. Yet, learning in product development management was far from easy as my case research showed. Additionally, learning could only be reached by the SMEs who focus and involve themselves in the learning process of NB HS NPD. It follows that researchers and SMEs had to find new PD models to gain right speed in NB NPD.

13.2.9 Point of Entry and Sources to NB HS PD Ideas

The research showed that most businesses focused on attracting new ideas but not specifically at high speed.

The area of ideas coming into the business was not subject to high speed or increased proactiveness. Furthermore, the area was seldom very structured. This reduced the possibility of the businesses to improve the speed at which new ideas were attracted to the business. The attraction of ideas was often a process of inside out, and not an outside in activity and view.

The research looked into two areas of the phase of attracting new ideas to the business with HS. The results are shown in Table 13.11.

The hypothesis was that major sources to new ideas for product development were sales, management, and product development. In most cases the findings of the research verified this hypothesis. This is illustrated in Table 13.11.

Table 13.11   Where was the idea discovered?

Dimension	Hypothesis	Case	PUIN Focus Group	Survey	Hypothesis Verified/Not Verified
Where were the idea discovered	On the market place	on the market place, inside the business and by network partners	on the market place, inside the business and by network partners	on the market place, inside the business and by network partners	Partly verified
Initiator of idea	Internal	All network partners and internal	All network partners and internal	All network partners and internal	Partly verified

As can be seen, the answer to both questions present a more fuzzy picture of the discovery of ideas. My hypothesis could only be partly verified.

13.2.10 After the PD Process Time

The research project did not show significant evidence of time and speed being a success criteria to businesses after the product development process had been finished.

This was maybe due to the fact that the PhD project did not have a particular focus on this area. This was focus for research projects after 2003.

However, the research offered the following results outlined in Table 13.12.

Table 13.12 Businesses' focus on success criterias on the market

	Literature Search	Case Research	Focus Group Interviews	Survey	Verified
Time	(+/−)	No	No specific focus	No specific focus	No
Cost	Yes	Yes	No specific focus	No specific focus	(Yes)
Performance	Yes	Yes	Yes	Not in focus	(Yes)
CI	Yes	Not in focus	Not in focus	Not in focus	Not in focus
CIM	Yes	Not in focus	Not in focus	Not in focus	Not in focus
Learning	Yes	Not in focus	Not in focus	Not in focus	Not in focus

As can be seen, the time criterion was not very much in focus when the product was on the market. The explanation was that businesses had often structured their product development in a way that left the developed product to the sales department and the customers after the process gate. Therefore, there was no pressure to speed product development "on the market". Moreover, the product development was seen as a stage-gate and not a process. As a consequence, businesses had not focused on product development, the existence of product development "on the market" and high speed product development "on the market".

Analysing the use and handling of radical and incremental product development projects and looking into the ratio of incremental and radical product development projects in SMEs gave a clear impression that there was a major potential in 2003 for businesses to implement new models and processes of handling product development.

Firstly, the research verifies that businesses would profit from focusing more on pressing more product development down "on the market" and doing more direct prototyping especially on incremental product development.

Furthermore, the research verified that many product development projects should be "kept" on the market and should never enter the formal product development model as shown in Figure 13.19.

Secondly, businesses would profit from using more flexible and agile product development models. Many of these models we did not know about in 2003 or were just testing as prototypes in specific industries. Businesses should be able to choose product development models in accordance with the position of "the field of product development" as seen in Figure 13.19.

Several researchers had put forward models for speeding up NPD but few of them had in 2003 suggested which models to choose in different product

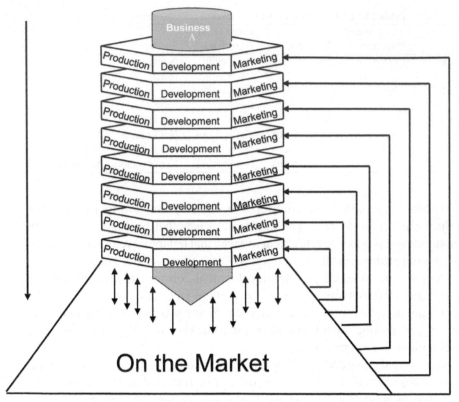

Figure 13.19   PD on the market.

development situations. Two mainstream NPD models had been proposed. "The Stage-Gate" model and "The Water-fall" model had proved to be extremely efficient when market, technology, network, and the competences of the businesses were stable or to some extent evolving. Especially the car industry (Case No. 72 Ford, Case No. 36 VW), the furniture industry (Case No. 67 Licentia, Case No. 58 Tvilum) the pump industry (Case No. 54 Grundfos), the windmill industry (Case No. 59 NEG Micon), and several industries producing mainly hardware (Case No. 6 TC, Case No. 69 DANCASE, Case No. 70 BM) had shown until 2003 to have profited from using stage-gate product development models.

The "flexible" product development models – especially the software industry and industries facing dynamic markets, technologies, networks, and competences as e.g. Case No. 57 TDC, Case No. 38 Lyngsø, Case No. 68 Metza, Case No. 50 Microsoft Internet (Macormarc, Verganti, Iansiti, 2001) had turned out to profit from using more "flexible" and dynamic product development models.

The question for the PD manager was to determine which PD model was best and appropriate to the product development model and process to gain right speed. Summing up on my research until 203 showed characteristics that must exist in order for the business to determine which NPD models to choose and the appropriateness to high speed as seen in Table 13.13.

Table 13.13 PD models and appropriateness to high speed PD

	Stage Gate Model	Flexible Model
Characteristics		
Markets	Familiar markets	Unfamiliar markets
Technology	Familiar Technology	Unfamiliar Technology
Network	Physical networks and stabilised ICT networks	Dynamic networks, ICT – networks, Virtual and dynamic networks
Competences	Stable and physical competences	Dynamic and virtual competences
Product	Products are mainly hardware	Products are mainly processes Software, services,
Strength	When main components can be characterised as stable and in some case evolving on the product development field.	Flexible to sudden change in the main components on the product development field.
Weakness	Inflexible to sudden change on the product development field	When product development turns out to be stable for a long period.
Opportunities	When market, technology, network and competence turn to stabilise	When market, technology, network and competence turn to be dynamic and virtual
Threats	"Trapped in a dynamic process" either in market, technology, network or competence – performance does not match demand of market.	"Trapped in a stable process" either in market, technology, network or competence – too much cost.
Time for change of NPD – model and speed	Going from stabilised to dynamic PD – characteristics When products turn to processes	Going from dynamic to stabilised PD – characteristics When processes turn into products – standard modules

The proposal of tools in the two types of models should only be considered as guidelines up to 2003, as further research had to be done.

Long-Term Versus Short-Term Success Criteria

According to research, businesses focused on short-term success criteria such as time, cost, and performance up to 2003. Businesses hardly ever focused on CIM, CI or learning. Therefore, it was difficult to gain long-term success criteria such as right cost, right performance, and right time. It was difficult to achieve CIM because the product development process had to be improved continuously both initially within the product development process, across product development projects, and on the market place. It was difficult to achieve CI because businesses had to innovate new products and seek innovation possibilities both at the start of a product development project, along the product development process, and when the product had been introduced to the market.

The research verified that major focus in businesses was on short-term success criteria along the product development process from idea to market introduction as seen in Table 13.14.

Table 13.14   Short-term and long-term success criteria

NB HS NPD Short-Term and Long Term Success Criteria	Hypothesis	Case Research	PUIN Focus Group	Survey	Verified/Not Verified
High Speed – Time	Yes	Yes	Yes	Yes	Yes
Cost	Yes	Yes	Yes	Yes	Yes
Performance	Yes	Yes	Yes	Yes	Yes
Continuous improvement	Yes	Not verified	Not verified	Not verified	Not verified
Continuous Innovation	Yes	Not verified	Not verified	Not verified	Not verified
Learning	Yes	Not verified	Not verified	Not verified	Not verified
Right Time – Right Speed	Yes	Not verified	Not verified	Not verified	Not verified
Right Cost	Yes	Not verified	Not verified	Not verified	Not verified
Right Performance	Yes	Not verified	Not verified	Not verified	Not verified

The research verified that a focus on long-term success criteria proved to be difficult for SME to practice because they did not link it to a strategic focus on product development learning and knowledge creation.

The research showed that many businesses were "stuck" in a predefined product development model which was often the stage-gate model. The research showed that the existing product development model did not "fit" the relevant product development situation for the future, and the businesses were not able to carry out or chose a flexible product development design or a flexible choice of product development models and processes because of e.g. ISO standards, business policy, competences lack of PUL. The research verified that many businesses (employees and managers) tried to break the formal product development models. They participate in informal PD models and processes to reach "the want of speed" either from customers, network partners, employees, or management. The cost and value of these informal models and processes were not calculated.

The following statements verified the above:

"If we were to develop new products according to our ISO 9000 standard, products would never reach the market." (Lyngsø)

"I went down into production and forced my employees to break the rules of our product development model. I forced them to make faults to speed the development process. The employees were not happy due to their high quality feelings. They reached the high speed. It cost a lot – but we achieved a first mover advantage on the market" (AKV Langholt)

It seemed as if there was a strong demand for a more flexible design in the product development of the businesses. It also appeared that there was important knowledge in the informal product development models and processes of the businesses which were not used and could be valuable to progress the speed and diminished the costs of NPD and reach long term success criteria.

The knowledge in the informal PD models and processes was worth nothing if the knowledge was not made available in an open form. Furthermore, the knowledge must be transferred to other NPD projects and NPD networks before it could be used to reach right speed in the businesses' product development activities. When PUL was implemented in businesses, and when learning interacted with CIM and CI, my hypothesis was that long-term success criteria such as right performance, right cost, and right speed could be reached.

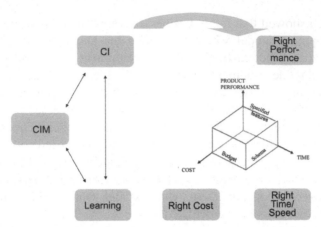

Figure 13.20   Relationship between long term success criteria in network based product development.

Source: Inspired by Lindgren & Bohn.

13.3 Measuring Framework in Idea and Concept Stage-Gate of HS NB PD

The following hypotheses were to be tested:

1. The NB HS PD projects can be divided into radical and incremental PD projects
2. The radical and the incremental PD projects follow different generic HS PD models and processes and can therefore be described by different generic frameworks
3. The HS NPD project always have a formulated core
4. A HS PD model follows another PD model than the business's normal PD model

13.3.1 Division of NB HS NPD

The research verified that product development projects/tasks could be divided into radical and incremental product development projects by different criterions.

The research also very clearly verified that most product development projects (85–90%) were incremental. This was a higher share than expected. However, there were major differences between the radicalness and incrementalness in regard to different criteria.

The most radical component was technology. All others were very much incremental as seen in Table 13.15.

Table 13.15   Incremental and radical PD in SMEs

	Hypothesis	Case	PUIN	Survey	Verified/ Not Verified
Market	Incremental	Mostly Incremental	Mostly Incremental	Mostly Incremental	Verified
Technology	Radical	Mostly Incremental	Mostly Incremental	Medium	Partly verified
Network	Radical	Mostly Incremental	Mostly Incremental	Medium	Not verified
Innovativeness	Radical	Mostly Incremental	Mostly Incremental	Medium	Not verified

13.3.2 Strategic Areas and NB HS PD Tasks

One of the objects of my research was to verify empirically the general task of the PD tasks in SMEs. The research showed that the picture of the PD tasks is far less complicated than the hypothesis had indicated. NB HS NPD was carried out in a "field of product development" which the involved SMEs judge as less complicated in relation to PD than the hypothesis had proposed (Please see Chapters 10–13).

Firstly, the research showed that most product development projects (approximately 85–90%) are generally incremental and very few are in the areas of radical product development. More specifically the research showed that most product development projects were focused on known customer needs, known customer groups, and well known and familiar markets. On these dimensions product development in the SMEs involved in the research was therefore very much incremental and could not support the original hypothesis of the PhD project that PD projects were generally radical.

On the technical dimension, however, the product development task tended to be found in the unfamiliar and radical area. This was also the case with customer technology or the choice of production technology to produce the new products. This meant that in general businesses face a more radical challenge on the technical side of the product development projects than on the market side. Still, my research showed differences in this radical element from one product development project to another. Furthermore, there was no indication that the perceived radical product development task was related to market wants or market needs.

Additionally, neither the perceived innovation degree was very often found in the radical innovative area. Only very few of the product development projects were perceived as being in the radical innovation area as seen in Table 13.16.

Table 13.16 Strategic areas and PD task

Dimension	Hypothesis	Case	PUIN Focus Group	Survey	Hypothesis
Innovation degree	High	High/Medium	Low/Medium	Medium/low	Not Verified
Market	New	Old	Old	Old	Not verified
Customer needs	New	Old	Old	Old	Not verified
Customer group	Old	Old	Old	Old	Verified
Customer technology	Old	Old	Old	Old	Verified
Technology	New	Old/new	Old/new	Old/new	To some extent verified
Network	Old and geographical narrow networks	Mainly old networks. When radical new networks	Mainly old networks. When radical new networks	Mainly old networks. When radical new networks	To some extent verified
Competence's	Middle	Middle	Middle	Middle	Verified
Product management	mostly the business	mostly the business	mostly the business	mostly the business	verified
Competition	Middle	Middle/high	Middle/high	Middle/high	Verified
Product development task	Radical	Mostly incremental	Mostly incremental	Mostly incremental	Not verified

The network dimension was also mainly concentrated on incremental newness of networks. Moreover, the pressure by new PD projects on the competences of the businesses was also seen as incremental and very familiar.

The competition on new product was perceived as being in the middle to high competitive areas. There was a strong indication in the empirical data that the competitive situation was a major driver to high speed product development.

The pressure on the competences of the businesses was perceived in the businesses as middle to low. This meant that the product development projects were not perceived as pushing the competences of the businesses out in areas of radical product development.

All in all, this gave an overview of the SMEs' product development projects as mostly focused on incremental product development projects in the area of related product development and product development to known markets and market segments.

It could therefore be concluded from the research that a very small amount of the product development projects were into the area of radical product development or/and in the area of diversification (Kotler, 2000). The hypothesis of the research on the very radical task and characteristics of product development projects in SMEs were therefore not verified.

Summing Up on Radical and Incremental PD Projects and Tasks

An analysis of the results of the empirical data showed very clearly that PD tasks in businesses up to 2003 were incremental. A reflection on the high degree of incremental product development projects indicated that the product development models in businesses today were too complexly organised and too time-consuming. Furthermore, it indicated that there was a potential to speed product development projects needs further than was the case up to 2003. It also indicated that there was a huge need for a change in leadership in product development. I will comment on this at a later point in this book. Finally, a strong indication of a need to implement more flexibility in the choice of NPD models and processes seemed to be another result of my empirical research.

It must also be stressed that most of the businesses involved were developing on hardware products and hardware parts of the products. A major potential on the immaterial, digital, and virtual side of the product seemed to be of importance to the businesses in their future development activities.

13.3.3 Generic Framework of PD Projects

The research verified that the radical and the incremental PD projects did not follow different generic HS PD models and processes in SMEs. The SMEs followed the same generic formal product development models and could therefore be described by different generic frameworks.

However, my research also showed that radical and incremental product development projects in businesses were not organised and did not follow different product development models. Generally, incremental and radical product development projects were treated equally in the businesses.

Still, my research verified that when businesses formulate product development projects to run at high speed, then the product development projects follow different generic HS PD models and processes. Such models and processes were supplemented by informal PD models and processes and could therefore be described by different generic frameworks either stage-gate model, supplemented with an informal model or rapid proto-typing models. The flexible model could not be verified in the empirical research.

The research showed very clearly that the HS PD model in the businesses followed another model than the normal and often certified PD model in the businesses involved.

13.3.4 HS Product Development Models

Businesses which put pressure on speed in their PD project used:

- Informal PD models and processes
- Rapid prototyping
- More involvement by top managers
- More involvement from network partners
- More development activities taking place outside the core of the product task
- A core of the product that is more dynamic
- The pressing of more product development activities down to "on the market activities"
- The building of a "service or nursing organisation" outside the formal PD organisation to maintain and increase HS in PD

The research verified very clearly that HS PD models and processes had different forms from those of the businesses' formal product development models and processes.

Table 13.17 Core of NB HS NPD

Core	Hypothesis	Literature Search	Case Research	Focus Group Interviews	Survey	Verified
Mission	Yes	Not verified	Partly verified	Partly verified	Partly verified	Partly verified
Goals	Yes	Yes	Not verified	Partly verified	Partly verified	Partly verified
Strategy	Yes	Yes	Verified but later in the process	Verified but later in the process	Partly verified	Partly verified but later in the process
Economic Resources	Yes	Not verified	Not verified	Not verified	Not verified	Not verified
Personal/ Organisational Resources	Yes	Not verified	Partly verified	Partly verified	Not verified	Not verified
Contact Limits	Yes	Not verified	Partly verified	Partly verified	Not verified	Not Verified

13.3.5 HS PD Projects Always Have a Formulated PD Core

The core of the businesses' product development was hypothetically said to consist of five parts mission: goals, strategy, economic resources, personal/organisational resources and contact limits to network partners.

However, this hypothesis was formulated as the optimal and idealistic foundation of a business developing new products. The research showed quite another picture when businesses were carrying out NB HS NPD as shown in Table 13.17.

Mission of Product Development Project

The hypothesis was that businesses always formulated a mission for their specific product development projects. This hypothesis was verified significantly by all empirical studies. This is shown in Table 13.18.

Table 13.18   The core of SMEs' PD model

Does the Business Always Formulate a Mission for the Product Development Project?	Yes	Verified
Hypotheses	All	
Case research	All	Yes
PUIN focus group	All	Yes
Survey	All	Yes

None of the empirical studies gave any indication of what the mission was about or of the contents of the mission. The case research and the focus group interviews verified that the mission was seldom written down and was just orally communicated to all participants in the product development project. The research could not give an answer to the way in which the mission was used to guide the product development project but businesses were convinced that if the mission was formulated, it had an influence on the speed of the PD project.

Nevertheless, the case research and the PUIN focus group interview gave a strong indication that businesses with ISO certification had their mission written down and known by participants of the product development project later on in the PD project. The formulation of the mission did not take place until the product development idea had entered the formal product development model. This was another indication of an informal PD model and process in the HS PD model and process of the businesses.

The research showed that the core of product development projects was under extremely high pressure when businesses wanted to speed the product development process. Sometimes businesses even forgot to formulate the core or they formulated it later in the product development process. Therefore the borders of the HS product development project were often fuzzy and in these projects often unclear. Often the core was so inexact that it resulted in bad performance on the success criteria; especiallysideways on time.

Goals of Product Development Project

Hypothetically, the goals of the businesses' product development projects were always formulated. The empirical study verified that this hypothesis could only be partly verified. This is illustrated in Table 13.19.

Table 13.19   Goal formulation in PD

Does the Business Always Formulate Goals for the Product Development Project?	Yes	Verified
Hypotheses	All	
Case research	All	Yes
PUIN focus group	All	Yes
Survey	%	Yes

In general, when businesses carry out PD the goals were always formulated. The research verified, however, that when businesses wanted to develop new products at high speed, the goals were most often not formulated.

Strategy of Product Development Project

Hypothetically, the strategy of the businesses' product development projects was always formulated before or at the very beginning of the product development project. As illustrated in Table 13.20, the empirical study could not verify this assertion.

Table 13.20   Goal formulation in PD

Does the Business Always Formulate the Strategy for the Product Development Project?	Yes	Verified
Hypotheses	Yes	
Case research	No	No
PUIN focus group	No	No
Survey	No	No

The empirical case study showed that in general businesses formulate the strategy late in the product development project. This was also the case in high speed product development projects.

Personnel and Organisation

When considering personnel and organisational aspects, my hypothesis was that the resources of personnel/organisation were always defined prior to the initiation of a product development project. This hypothesis could only be partly verified by the empirical study. The results are listed in Table 13.21.

Table 13.21   Organisational resources and PD

Does the Business Always Formulate the Resources of Personnel and Organisation?	Yes	Verified
Hypotheses	Yes	
Case research	Yes but firstly at the concept stage	(yes)
PUIN focus group	Yes but firstly at the concept stage	(yes)
Survey	%	(yes)

All empirical studies showed a very limited interest in involving competitors in the product development projects. The empiricism also showed strong restrictions and traditions determining where and when in the product development phase customers, suppliers, and other network partners should be included in the development.

The research showed that the network partners involved were mainly customers and suppliers and definitely not competitors.

Generally, there seemed to be much more potential in the network component for the SME. However, a more intense use of the network component will demand additional development of the businesses' ability to work in networks and to use the network component (Kræmmergaard et al., 2002).

13.4 Identifying Enablers to HS PD

The following hypotheses were elaborated initially in the research project:

1. HS enablers are identical to the 10 enablers – 1–10
2. There can be more than these 10 enablers to HS PD
3. Businesses use different HS enablers
4. The enablers will play a different role according to the PD situation and project (secondary focus)
5. The customer enabler, the network enabler, and the PD model enabler play an important role in the upper phase of the HS PD phase

13.4.1 HS Enablers Identical to the 10 Enablers

One of the hypotheses of this research was that the HS enablers were identical to the 10 enablers (1–10) identified in the literature study.

The research was able to verify 8 of the 10 HS enablers. The results are listed in Table 13.22.

Table 13.22   Use of HS enablers in SMEs

High Speed Enablers	Literature Search	Case Research	Focus Group Interviews	Survey	Verified
1. The ICT Enabler	Yes, in a few businesses	Yes, in a few businesses and very little	Yes, but in few businesses	Yes, but in few businesses	(+)
2. The customer Enabler	Yes	Yes	Yes	Yes	Yes
3. PU model enabler	Yes	Yes – but mainly stage-gate	Yes – mainly stage-gate	Yes – mainly stage-gate	(+)
4. The Network Enabler	Yes	Yes – but mainly customer and to some extent suppliers Limit networks	Yes – but mainly customers and to some extent suppliers Limit networks	Yes – but mainly customers and to some extent suppliers Limit networks	(+)
5. The Innovation Enabler	Yes	No but in few businesses	No but in few businesses	No but in few businesses	–
6. The HRM Enabler	Yes	Yes – but few businesses	Yes – but few businesses and limit efforts	No	(–)
7. The Process Enabler	Yes	No	No	No	No
8. The Product to Process Enabler	Yes	To some extent – few businesses and limit efforts	No	To some extent	(+)
9. The Modularisation Enabler	Yes	Yes, but not fulfilled	Yes, but not fulfilled	Yes, but not fulfilled	(+)
10. The E-development Enabler	Yes, in a few businesses	Yes, in very few businesses	Yes, but in very few businesses	Yes, but in very few businesses	(–)

As can be seen in the comments on each HS enabler there were differences in the use of the enablers.

The ICT Enabler was quite surprisingly not used very intensely or effectively in most businesses. This stressed a large potential within businesses to use the ICT enabler more and better in network based product development. Therefore, the research could not verify extensive use of the ICT enabler in NB HS NPD. The result was very surprising and is deemed critical related to the future trends of PD which demands intensive use of the ICT enabler.

The Customer Enabler could be verified as a major enabler for businesses in order to speed product development. Businesses generally invited the customers in at a very early point of time in the product development process to speed product development, to narrow the focus on real needs of the customer, and to higher the performance of the product in order to meet customer demands. However, the customer was invited to join the PD at the upper part of the PD process, not so much at the lower part of the PD process, or "on the market".

The PD Model Enabler was used in the businesses but mainly in a very restricted way. Most businesses used the PD model enabler with a view to optimising the stage-gate model. Some businesses used the "rapid prototyping model" where businesses miss out on some stage and gates into rapid prototyping. Others performed simultaneous or parallel product development where some product development processes were developed parallel.

No business tried to use other product development models e.g. flexible models, either because they did not know about them, or because they did not know how to shift among product development models.

The research was able to verify the following product development models as seen in Table 13.23 used to speed PD:

Table 13.23   High speed PD models verified in the research

Hypothetical HS PD Models	Case	Focus Group	Survey	Other	Verified
Stage-gate model (Parallel or simultaneous PD)	Yes	Yes	Yes	Yes	Yes
Flexible models (Verganti model)	No	Partly	No	Partly	Partly verified
"Task force model" (TDC task force model)	Partly	Partly	No	Partly	Partly verified
"On the Market model" (Corso on the market model)	No	No	No	Partly	Not verified
Informal HS PD Models	Yes	Yes	Yes	Yes	Verified
"Lindholst model" (Linco HS model)	Yes	Yes	No	No	Partly verified

Table 13.23 Continued

"10% overflow model" (Grundfos model)	No	Partly	No	No	Partly verified
The network HS model(Italy model)	No	No	No	Partly	Partly verified

The above shown in Table 13.23 verifies that businesses did use other PD models to HS the product development but its very fragmented.

The Network Enabler was also very much in use in the businesses but mainly focused on customer and to some extent on the supplier network. The competitors were hardly ever integrated in the NPD process, and other organisations were mainly integrated in the NPD process at the beginning of the process.

The Innovation Enabler could not be verified in the research.

The HRM Enabler could only be verified in very few businesses, and the use of this enabler was very limited. Most businesses do not see the HRM enabler as an enabler to speed product development.

The Process Enabler could not be verified in any of the businesses.

The Product to Process Enabler was to some extent used in the businesses but still in very few businesses and only to a very limited degree.

The Modularization Enabler was used in many businesses but it was not fulfilled in its use. Furthermore, many businesses reported that they had major problems implementing the modularization enabler. The problems mostly concerned the definition and implementation of an optimal product architecture that could support the use of this enabler.

Modularisation was an important enabler to HS. However, in many businesses this enabler had not yet been of great success. I claimed that this was due to an inside out view. Modularisation should be done in accordance with the market. The research verified that the businesses which used the modularisation enabler with an outside in perspective were more successful in the use of this enabler.

The E-Development Enabler was also used very little in the businesses up to 2003, and it seemed as if there was a large potential in this area for speeding up the product development in SMEs by using this enabler.

13.4.2 More Than 10 Enablers to Enable HS PD

Through the research the PhD project verified two extra enablers to HS product development that could not be found in the literature study as seen in Table 13.24.

Table 13.24   Extra enablers in NB HS NPD

Extra Enablers	Literature Search	Case Research	Focus Group Interviews	Survey	Verified
Informal product development models and processes	No	Yes	Yes	Yes	Yes
Management	(–)	Yes	Yes	Yes	Yes

The Informal Model and Process Enabler was verified in almost all businesses. This enabler was verified by businesses as having major importance to speed product development. Furthermore, this enabler was used officially or unofficially by the management level to speed the product development process.

The Management Enabler was used to speed the product development process in most businesses. This enabler was used in different ways e.g. by managers involving themselves directly in the product development project to make the organisation and the surrounding environment see that the product development project had a major management focus and was of high strategic importance.

The management enabler was also used to "kick start" the speed of a product development project when starting or when a critical situation occurred.

Different Enabler Roles According to PD Situation and Project

The research showed that the different HS enablers play different roles in the process of speeding the product development process.

Firstly, it was verified that HS enablers are used at different times. The customer enabler is e.g. mainly used at the beginning of the product development process where the modularization enabler was used at the end of the product development process. The Figure 13.21 shows where in the product development process the HS enablers were identified.

Secondly, some of the HS enablers were identified as being overall enablers that were used more or less at any point in the product development process. Others were running as supporting the HS enablers used.

The research also verified that HS enablers used in a wrong way can result in bad performance, increasing cost, and even slower speed in product development. This could be HS enablers used at the wrong place or HS enablers used in a wrong mix of HS enablers. The research verified that it is important

Use of High Speed Enabler in Networks

HS enablers used in the survey businesses	Idea - phase	Concept - phase	Prototyping - phase	Implementation - phase
ICT Communication Enabler	Yes – but very limited	Yes – but very limited	Yes – but very limited	Yes – but very limited
Customer Enabler	yes	yes	yes	Yes – but very seldom
PD Model Enabler	Yes – but very limited	Yes – but very limited	Yes – but very limited	Yes – but very limited
Network Enabler	Yes – but limited	Yes – but limited	-	-
Innovation Enabler	No	No	No	No
HRM Enabler	Yes – but limited to very few businesses	Yes – but limited to very few businesses	Yes – but limited to very few businesses	Yes – but limited to very few businesses
Process Enabler	No – not verified	No – not verified	No – not verified	No – not verified
Product to Process Enabler	No	Very limited	Very limited	Very limited
Modularization Enabler	-	yes	yes	yes
E-Development Enabler	No	Very limited	Very limited	Very limited
Informal Product Development Model	yes	yes	yes	yes
Management	yes	yes	yes	yes

Figure 13.21   HS enablers in network.

for managers responsible for the businesses product development activities to choose the right HS enablers.

The research also showed that HS enablers had a different speed value to product development depending on:

- the characteristics and development of the main components on the field of product development
- the time at which a HS enabler was used to speed the product development process

The research verified that HS enablers used in a combination could give synergy effects on speed in product development but also "synergy bad effects" when used wrongly.

13.4.3 Customer, Network and PD Model Enabler

Through the research it was verified that in most cases the customer enabler was used primarily to speed the product development process. By using the customer enabler optimally a more precise and accurate need structure could be identified. Such a structure could effectively help the businesses from developing wrong products to speeding the PD process. The customer enabler was seldom used in the final phase of the product development process.

The network enabler was often used in the idea and concept stages and gates but mainly in a narrow perspective. Mostly narrow and close network partners were used, and only very seldom were unknown physical networks used. Neither were the SMEs involved in the research using electronic and virtual networks very much.

The product development model enabler was used in the businesses but in a very narrow and limited way. The businesses were very focused on using their traditional stage-gate model but trying to use it with parallel processing. However, it was verified that these stage and gate models had reached the limit to further speeding of the PD process. Businesses cannot speed the process further using the stage-gate philosophy of PD. The SMEs could only speed the PD process by allowing informal PD models and PD processes running parallel to the formal stage-gate models.

New PD models needed to be developed and implemented in the businesses.

13.4.4 Summary on Enablers to NB HS NPD

Summing up on the enablers to high speed product development the research verified eight originally identified HS enablers and two extra enablers. The empirical research could not verify two of the enablers identified in the secondary case research. The use of the HS enablers differed from one business to another but some of the enablers – especially the customer enabler – are used more often than others.

The research verified that businesses use the HS enablers very differently but that the customer enabler, the product development model enabler, and to some extent the network enabler were the most important enablers in use. The research verified that it was very critical to PD management to choose from and mix HS enablers. It was also verified that there is a large potential in an additional and more intensive use of HS enablers.

13.5 Success Criteria for Measuring NB HS PD

The research verified that success criteria in NB HS PD were more complex to define than the hypotheses had identified. Businesses define their success criteria on product development much more individually and much more in relation to the characteristics of the PD project. Businesses up to 2003 often prioritise the success criteria differently and even change the prioritising as the product development process evolves.

13.5.1 Success Criteria Dependent on Specific PD Project

At a previous point of this thesis I stated that product development projects could be divided into either incremental or radical product development projects and that the research showed that most PD projects could be characterised as incremental.

Dividing product development projects into incremental and radical development projects was nevertheless seldom done in businesses up to 2003 at least not in a focused or analytical way. The research presented the following picture seen in Figure 13.22 of the way in which businesses handle this process.

Ability to Analyse

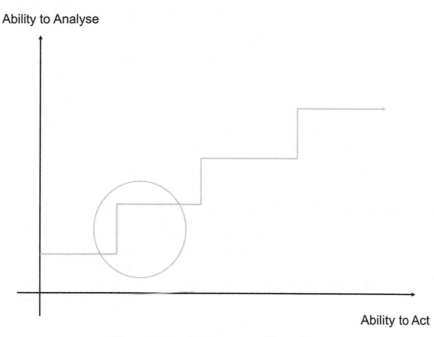

Ability to Act

Figure 13.22   Ability to act and/or analyse.

The research verified that most businesses were working in the mid-zone of the model and that they were facing difficulties about analysing, acting and not least implementing product development strategies according to the characteristics of the product development projects. Businesses had difficulties in choosing from success criteria related to the characteristics of the PD project.

The research verified that the businesses focus narrowly on the same success criteria – most often costs – on all PD projects.

13.5.2 Short- and Long-Term HS PD Success Criteria

The research verified that both the short-term and the long-term success criteria existed in the involved businesses. Yet, it was highly significant that very few of the involved businesses focused on long-term success criteria. Additionally, the formulation of success criteria together with the prioritisation of success criteria were most often made through coincidence and randomness. Very few of the businesses had a strategic view on success criteria and they were often not particularly focused on or prioritising the success criteria – with the exception of cost.

Furthermore, the businesses featuring in the empirical research seldom formulated the success criteria as measurable. This is quite different from what was observed and seen in some of the secondary cases.

Finally, it was significant that businesses seldom overview or control their objectives and success criteria. This issue was a very surprising result of the research as businesses today are so focused on the short-term criteria cost and time.

In this connection, it was also very interesting to see that businesses hardly ever paid any attention to the collection and transforming of learning from one product development project to another.

13.5.3 Central Success Criteria in Short-Term Perspective

My hypothesis that the short-term success criteria time, cost and performance were central success criteria in businesses in the time up to 2003 could be verified. Although the short-term success criteria were said to be in focus, businesses hardly ever knew why they were in focus. Furthermore, the businesses did not know how and why the success criteria were prioritised and measured.

13.5.4 Central Success Criteria in Long-Term Perspective

It could also be verified that continuous improvement, continuous innovation, and learning were not major success criteria in businesses and certainly not when seen in a long-term perspective.

From this verification it could be concluded that businesses were very focused on short-term success criteria in their product development activities and furthermore, that businesses focused more on cost than on value.

13.5.5 Product Development Management and Leadership

My suggestion was to focus on product development leadership (PUL). PUL affected and was closely related to learning and knowledge management of product development. Learning and PUL of NB HS NPD created knowledge about the interaction and development of the market, technology, network, and competence component in the product development field. PUL choose the right product development model and process and thereby the right mix of main components to "the product development field" to gain right speed.

PUL should be elaborated along with product development management (PUM) to gain right time product development. Focus had been on PUM until now where PUM is in between the four main components influencing NB HS NPD. PUM tends to be much involved in day-to-day product development management. Seen from the point of view of the product development management level, PUM has difficulties in developing and maintaining the objective view of "the product development game". It was also difficult to elaborate a flexible view and design in the "product development game" where all components are endogenous and exogenous variables played out and into the "product development field".

PUL is the ability and the know-how of the "game of product development". PUL is to know **why** the "game of product development" is, and **what** the "game of product development " is about to decide and perform how the right speed in businesses product development activity should be. As was verified in the research project, the product development managers of the businesses focused mostly on short-term success criteria and on PUM.

At the PUM level focus is on short term success criteria such as time, cost, and performance. On the PUL level focus is on CIM, CI, and to learning create long term success criteria such as right cost, right performance, and right time. CIM is in focus because the product development process has to be improved continuously both initially within the product development process, across product development projects, and on the market place. CI is in focus because businesses have to innovate new products and seek innovation possibilities at the start of a product development project, along the product development process, and when the product has been introduced to the market.

My research also showed that PUL has proved to be difficult to practice without linking it to a strategic focus on product development learning and knowledge creation. My observation showed that many businesses were "stuck" in a predefined product development model. The product development model did not "fit" the relevant product development situation, and the businesses were not able to carry out or chose a flexible product development design because of e.g. ISO standards, business policy, competences lack of PUL.

The employees and managers therefore tried to or needed to break the formal product development model and to introduce informal models and processes to reach "the want of speed" either from customers, network partners, employees, or management. It seemed as if there was a strong demand for a more flexible design in the product development of the businesses. It also appeared that there was knowledge in informal product development models and processes of the businesses which could be valuable to progress the speed and diminish the costs of NPD.

This knowledge has to be made available in an open form and transferred before it can be used to reach right speed in the businesses' product development. When PUL is implemented in businesses and learning interacts with CIM and CI, my hypothesis was that long-term success criteria such as right performance, right cost and right speed can be gained in product development.

The research project aimed at proposing a descriptive framework model for network based high speed product development. The research project suggests managers to redesign businesses' management of NPD models and processes to reach better performance, lower cost, and product development within right time.

The framework advocates that management in product development increase their focuses on product development leadership (PUL). The framework suggests that PUL is strongly related to learning and knowledge creation on how "the product development game" is played both in the specific situation and across the product development project of the businesses. The framework proposes that learning and knowledge transfer is concentrated on understanding the main components in "the product development field". In this way, the businesses will be able to chose the structure of the product development models and processes. The choice of speed in product development includes a decision on product development model and process based on a thorough understanding of "the game of product development". Moreover, a decision on speed in product development must be made continuously and overviewed and decided upon as the product moves along the product development process from idea to market introduction but also when the product has already been introduced to the market.

The framework finally recommends that businesses focus more on long term success criteria to gain right performance, right cost, and – essentially – right speed.

13.6 Summary of Impacts on Businesses Trying to Gain NB HS PD

Through the research some significant results and observations could be verified in NB HS NPD. These were:

- Pressure on time in NPD created informal processes and models within the business but also within the network
- Pressure on time in NPD created both first mover advantage and first mover bad advantage
- Pressure on speed and time in NPD would often raise total cost in businesses who focus only on costs
- Pressure on speed and time often diminished quality when focusing on costs
- Focus on right time could raise quality and diminish costs
- Focus on speed and time was often miscalculated in relation to radical and incremental product development projects
- Focus on right speed and right time in NPD could result in increased liquidity
- Pressure on speed and time could often be impossible because of the businesses' failure to choose the right PD model for the PD project – businesses got stuck in the chosen product development
- In the future NB RS NPD will be a very strong, competitive weapon
- First mover advantage and first mover bad advantage can both be the result of NB HS NPD
- Costs were primarily in focus for SMEs in 2003 when they tried to do HS in NPD
- SMEs focused primarily on direct costs and very seldom on alternative costs
- SME focused very seldom on perceived value in product development
- Quality could be increased by NB RS NPD
- Quality decreased when focus was on high speed

- Quality increased when focus was on right speed
- Many SMEs forced incremental PD into radical PD even if the PD task was incremental
- NB RS NPD may result in major liquidity possibilities because money was paid by the customers before the supplier require payment
- NB RS NPD was a major competitive weapon for the future global market
- The impact on market, technology and networks by network based right speed product development was expected to be enormous

The above-mentioned findings resulted in the hypothesis that another product development model may exist and must be elaborated to explain the high speed of product development.

I therefore proposed the following normative guideline to businesses who want to work with NB HS NPD:

1. Define "the field of product development"
2. Define the task of product development – Incremental or radical
3. Define the success criteria – time and speed – in relation to relative time and speed and in relation to right speed and right time
4. Define costs in relation to both direct and alternative costs (offer costs)
5. Define speed in relation to right speed and value instead of high speed and direct costs
6. Define the product development model – stage-gate and flexible model but choose the right model according to the task of the product development project
7. Choose to focus on long-term success criteria and not short-term criteria
8. Choose to relate the long-term criteria to value and not to costs
9. Choose to focus on product development leadership and not product development management
10. Formulate the core of the product development task in a narrow focus on CIM, CI and Learning
11. Choose the contact limits to network partners by value and advantage and improve the use of network partners to optimize and gain right time and right speed
12. Choose to involve all functions and actors in the product development activities of the business to help improve the product development within right time and right speed
13. Choose to use the high speed enablers with an outside in focus and choose to use more of the HS enablers but in an optimal way

14

Learning Perspectives of NB HS NPD
in PhD Project

This chapter presents the learning perspectives of the project and offers suggestions for further research. The presentation will be divided into three main parts. Firstly, parts referring to the learning which can be drawn from the project on NB HS NPD will be presented. This part will also include the contribution to theory which the research project claimed to make. Secondly, a suggestion for further research was made stressing the further research which had already been established together with the research which should be established for the future after 2003. Finally, a part dealing with the learning perspectives for me as a researcher will be presented.

14.1 Introduction

On the basis of the research project this chapter will present the learning perspectives of the project and will offer suggestions for further research. The learning perspectives will be divided into learning perspectives:

1. on NB HS NPD
2. for the author in particular

The proposal for further research will be given both with a general focus on the Centre of Industrial Production and with a specific focus on my own work as a future researcher.

14.1.1 Learning Perspectives on NB HS NPD

The research project verified that NB HS NPD was just at its beginning in 2003 and that there were more potentials to NB HS NPD than the researchers and the industry use and know about today. From my research on NB HS NPD the following picture of the present situation of speed related to cost in product development could be drawn:

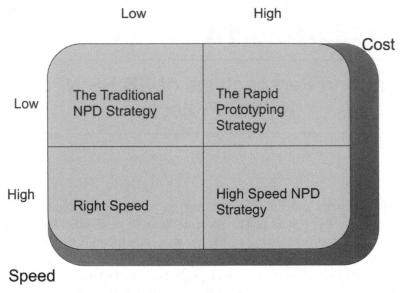

Figure 14.1   Speed and NPD strategy.

The research project verified that businesses could be placed in four generic areas of NB NPD strategies related to speed and cost in product development as shown in Figure 14.1. The characteristics of these generic types can be seen in Table 14.1.

The Traditional Business

The traditional business used same process and models to develop all their products. They did not pay much attention to speed in product development. They decided from an inside out perspective when it was appropriate to introduce new products.

The Rapid Prototyper

The research project verified that some businesses in the research proved to be placed in "the rapid prototype" area. They did not wish to develop products at a higher speed because they felt that they were developing as fast they could and would like to. However, the decision on speed was not related to strategic decisions on which speed was optimal to the product development activity or to the amount of costs and values of developing new products with rapid prototyping.

Table 14.1 Types of NB HS NPD

Types of NB HS NPD	Characteristics	Success Criteria	Product Development Models	Characteristics Related to Speed	Case Examples
The traditional	Tries to gain HS by parallel and simultaneous product development activities. Has a lot of informal processes going on in the business	Cost, speed, performance and short term success criteria	Stage-gate model or Flexible models	Feels its impossible to speed the process further	Ansager NEG Micon
The rapid prototyper	Tries to gain HS by jumping over some stage- and gates	Speed, direct cost.	Stage-gate models or Rapid prototyping or no formal proto development model	Feels they are developing at the maximum speed possible	Scanio, AKV
The high speeder	Tries to gain HS by using a somewhat alternative very flexible way of developing new products. Very much team oriented and plays with the HRM and Management enabler.	Speed, performance, continuous improvement and to some extent learning	Stage-gate model or alternative model	Do not see any reason to why they cant speed product development more.	GSI Lumonics
The right speed developer	The business focus on the product as an process	Focus on Perceived value direct and alternative cost and value. Focus on long term success criteria. Focus on knowledge of NB RS NPD.	All types of product development models, a optimal mix of HS enablers	Focus on right speed. This means that when necessary then high speed and when not slow speed. The main focus is to enter the market at the right time and harvest maximum of the market	Zara, Ryan Air, Nike

The Alternative Business

The research project verified that very few businesses in the research were placed in the alternative product developer area. Only one business was verified to be situated in this area. However, I advance the hypothesis that there were more businesses in this area. The businesses develop fast but with some alternative methods to product development. The businesses saw physical and physiological "walls" to speeding product development further. However, the alternative methods helped them to gain maximum speed. Still, whether maximum speed was the optimal right speed of the product development activities, was not being deliberated. Learning and the long-term success criteria in general were not strategically put into focus.

The Right Speeder

The empirical part of my research could not verify that some businesses were placed in "the right speed product developer" area. In the secondary case area we could verify that a few businesses were placed in the area or were moving into the area of "the right speed developer". These businesses develop products fast but always at a speed that matches the demands on the field of product development. The businesses had a strong focus on long-term success criteria and a strong focus on product development leadership. The businesses did not see a physical or a physiological "wall" to speeding product development further on. They continuously seek to develop the right speed in product development. This involved the continuous search for new products, new improvements to the product development process, and particularly new knowledge and learning from network partners. The businesses had or were moving into an area of strong competitive advantage or what could be defined as a core competence of right speed product development.

The businesses were focused on both direct, perceived and alterative, perceived value. The businesses were focusing on the way in which their customers defined and perceived value.

These businesses were focusing almost exclusively on perceived value based on exhaustive market analyses and on an "outside in" perspective. The businesses were also focused on costs, both direct and alternative costs. The research verified that the alternative costs were often worth more than the direct costs. However, only a few businesses had the tools to manage and measure these alternative costs.

This prevented many businesses from gaining real competitive advantage in NB HS PD. Figure 14.2 illustrates the four types of NB HS NPD businesses.

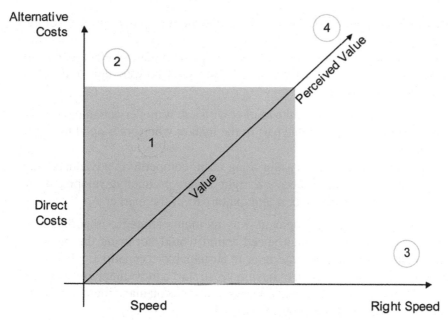

Figure 14.2 Measurement of success criteria in NPD.

The research verified that most businesses were placed in the traditional product developer position where businesses focused on direct costs, value and high speed. It was also verified that there was still a long way to go for the businesses to reach level 4. Also researchers had to do much research to find ways to reach level 4.

14.1.2 Strategic Importance and NB RS NPD

The research showed that a product development process which was vital or had major strategic importance to the business would normally compel the business to use the stage-gate model. The result of the research project verified that the stage-gate model was not always the most effective product development model when focusing on speed and time. However, the stage-gate model was the most used product development model in SMEs.

Therefore, both in general and specific terms it was verified that the product development activities of the businesses did not always follow the optimum line for speed – and especially for right speed – in product development.

Furthermore, it was verified that the stage-gate models generate informal product development models and processes which influence the alternative costs, value and perceived value of the product development projects of the businesses.

The strategic importance of right speed in product development seemed to increase in these years. Firstly, because businesses who were able to develop products at right speed can:

- harvest the major part of the market with a first mover action or activity
- harvest the most profitable part of the market when entering at the right time
- gain continuously increasing long-term competitive advantage and develop core competences on right speed product development by continuously taking the best and major part of the market

When a new product development was of major strategic importance to the business, a short-term high speed action would not bring the business nearer to long-term success criteria or to sustainable competitive advantage. Therefore it seemed important to businesses to start a movement towards new product development models and towards new focus on managing the product development activities of the businesses.

14.1.3 Speed and Time Related to PD

The PhD project showed that a definition of speed and time in relation to NB HS PD was far more complicated than it seemed at the beginning of this thesis. Original and traditional definitions had related speed and time to the physical time. Until now, businesses and researchers had related speed and time in product development to the physical time it toke for a product to be developed from the idea was born to the product was developed to the market $(t0-t1)$.

In the research I verified that time and speed must be considered in terms of relative time. Time and speed must be related to the task of the product development activity and to the field of product development. It was verified that the view of time and speed differs quite dramatically depending on whether it is seen from the point of view of the market, the technology, the network, or the business.

The PhD project showed that the classic view in SMEs on time was a much too narrow view on time and speed in PD. Firstly, it had to be stated that there was a time before and a time after $(t-1-t+1)$ the product development project.

The time before and the time after the beginning and end of the product development project was very difficult to determine according to the empirical data of my research. When this was the case, time and speed became floating and fuzzy concepts and it was definitely a question whether it was relevant to use the term of physical time as a measurement of NB HS NPD. I claimed in my research that it was not relevant to measure product development within physical time. The PhD project verified that it was relevant to define time and speed within terms of relative time and speed and within the business optimal point of entry to the business.

The business optimal point of entry is relative to each business dependent on the product development task and the characteristics of the field of product development.

Different Types of Speed

During my case and survey research I had observed different types of speed in NB HS NPD. The different types of speed are shown in Table 14.2.

Table 14.2 Types of speed

Types of Speed	Characteristics
Idea – speed	The ability to speed new ideas coming to be absorbed by the product development process of the business
Idea to market introduction speed – "time to market speed"	The ability to speed the NPD project from idea to market introduction
Stage- and gates speed	The ability to speed the single stages and gates within the product development project
Transfer speed	Speed from one stage to another gate
Complex speed	The ability to speed complex NPD projects
Concurrent speed	The ability to speed several NPD projects at the same time
Market speed	The ability to speed incremental NPD on the market

Until 2003 there were only fragmented knowledge and research on the speed types appropriate in different situations of product development and on the businesses' use of the speed types. Learning had to be established in all areas of product development to investigate the types of speed and to find normative models of right speed in NB NPD. This became of major interest in my future research after 2003.

Right Speed in Network Based Product Development

The research project verified that right speed in product development has to be learned. The critical issue before talking about speed in product development is the ability of the management to:

1. Define the task of product development (radical or incremental)
2. Analyse "the game of product development" or "the field of product development"
3. Learn from one product development project to another
4. Learn across networks
5. Transfer knowledge both vertically and horizontally within the network based product development organisation

Even more critical is the ability of the product development managers to learn through the product development process continuously both vertically and horizontally. The last learning area concerns the continuous learning process in all stages of the product development process – before the formal product development process – from idea to market introduction but also the time after market introduction.

This became a focus for a EU research project at the CIP centre after 2003.

In my research project I claimed that high speed and diminishing physical time in product development was not the issue because of a number of factors:

- High speed focus only on costs and direct costs
- High speed cannot be determined because the end and beginning of a product development project cannot be determined, is not registered in the business or, is very difficult to determine.

Instead, I propose to look at speed and time in product development as right speed and right time. In addition, high speed had a very narrow focus on cost and forgets to focus on value as well. My research verified very clearly that the value both in net profit and in competitive advantage of waiting or speeding for the right time to introduce and launch the product were very high.

Speed Related to Incremental and Radical Product Development

At a previous point in this research I discussed whether a new product can be said to have a beginning and an end. This is very much related to the discussion of incremental and radical product development. The PhD project showed that most product development was on incremental physical products and very little development is radical product development. At the same time the research verified that the life cycle of products and product development is diminishing dramatically. This meant that the pressure on time increased. The pressure on time was related to the characteristics of the product development task. Many businesses up to 2003 saw major difficulties in speeding the product development process further because the products were physical and because they found themselves stuck in too complicated radical types of product development models and processes.

I therefore claimed that businesses had to change the way they thought about product development to move towards a more agile type of strategic management of product development and of product development models and processes. Businesses had to use more types of product development models to match the need of speed in product development.

I claimed that much more incremental product development could be carried out "on the market" as illustrated in Figure 14.3 and did not

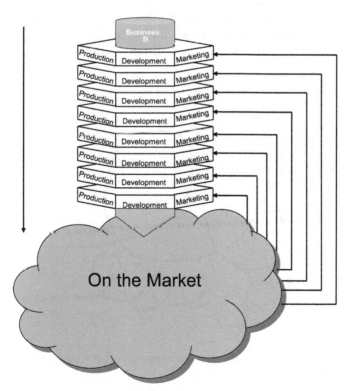

Figure 14.3 On the market product development.

needed to enter a formal product development model or process. Only the radical product development projects should enter a full-scale product development model and process. However, even some radical product development projects needed not to enter a full scale product development model.

The discussion of time becomes interesting but also more complicated when the discussion is extended to encompass a process view turning the definition of the product from a view of understanding the product to a product viewed as a mixture of product and process with no beginning and no end.

I claimed that the process view was necessary to fathom for future discussions on product development in order to allow businesses to gain competitive advantage. The process view was new to the product development theory in 2003 and at the same time very difficult and very poorly introduced to and implemented in the industry. This was particularly stressed in the research empirical part of the research project.

The process view as illustrated in Figure 14.4 puts into perspective especially the focus on the right time to introduce and implement a product on the market. This turned the discussion into a discussion of a dimension of relative time or, more specifically, to find the strategically right time to

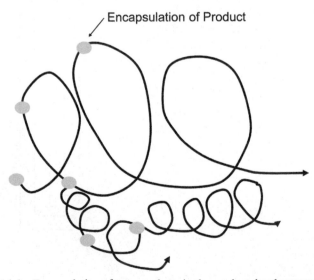

Figure 14.4   Encapsulation of new products in the product development process.

implement a product be it radical or incremental. It stresses the relationship approach where

> *"product development is based on a continuously repeated market transaction and mutual gain where product development strategy and process integrate customers, suppliers, and other network partners into the businesses design, development, manufacturing and sales and marketing process".*

In this case, the concept of right time had to be defined by the network management level of the product development project before a final statement of speed and time to develop a new product could be discussed as illustrated in Figure 14.5. The research project verified that it was impossible to define whether the product development of a business should go fast or slow before the right time of a product development introduction could be defined. The question is if this can even be defined and a process view can be used as a definition of products. I claimed that the right time is "eternal" and that the right time is moving all the time. This became subject for further research after 2003.

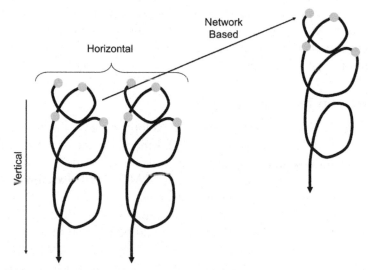

Figure 14.5 Encapsulation of new product across the vertical, horizontal and network based product development process.

Source: Inspired by Svend Hollensen.

14.1.4 When Are Time and Speed Important?

The discussion of time and speed is important and can be discussed on the basis of the following perspectives.

Time, Speed and Characteristics of Markets

The research showed that in case of stable markets and dynamic markets, it could be a preferable to move the product development at a somewhat slower speed. When markets were stable, the net profit of moving product development at high speed would not give the business major competitive advantage although minor, continuously incremental product development was necessary. In markets with dynamic characteristics it was also preferable to move product development at a lower speed because market demands had not yet stabilised or were even not yet present. The pressure on time was therefore not really existent.

On the other hand, markets that were evolving and begin to show signs of standard product market architecture needed to be handled with speed in product development in order for the business to harvest the market before the competitors gain first mover advantage.

Evolving markets should be handled with care and much attention should be paid to the degree of speed of the product development project. A wrong decision on speed was verified to be extremely critical to the business. Speed in an evolving market situation was maybe the most difficult feature for development managers to handle.

Time, Speed and Technology

The research showed that the technological development gave the businesses up to 2003 significant, exciting possibilities. Furthermore, the technology put major pressure on the field of product development and on the businesses to adapt new technological features. Many businesses had fallen victim to this pressure and had experienced first mover bad advantage.

In future after 2003, the businesses had to improve their ability to analyse and decipher when the technology should be adapted to their product. Again, it was a question of right speed in adapting the technological possibilities to the products. I claimed that this decision had to be related to an analysis of the product development task and of the field of product development.

Time, Speed and Network Component

As mentioned above I claimed that the network enabler would be of major importance to the product development of the businesses in the future after 2003. More network activities in known and unknown networks were seen up to 2003. Such networks had a high degree of mixing different network types – physical, digital and virtual. This ability would be necessary to future product development in order to keep up with competition on speed in product development on the global market.

The businesses of 2003 therefore had to put a strategic view on their network activities and involvement and start a strategic process of developing their NB HS NPD. The research verified that the businesses' involvement in network and their use of the network enabler were much too narrow in 2003. The reason was also shown in the fact that the businesses' overview of their network competences was far too restricted. Maybe this was due to the very limited use of the HRM enabler and HRM function in product development.

Time, Speed and PD Competence Component

The product development competences of a business in a stable market needed to be handled with much attention paid to the stabilisation and rationalisation of the competences of the business. It was essential to find specific and necessary competences that matched the particular needs for speed in product development.

Competences in an evolving market had to be handled with care because some competences could be of vital and core competence importance whereas others could be of minor interest seen from a product development perspective.

Therefore it was vital to the business to find the right product development competence architecture. In this way the business could develop and secure future right speed product development. Whether it should be slow or fast speed should be decided in accordance with the evolvement on the field of product development and with the business optimal point of entry. This meant that in some cases it could be preferable to out-source some competences and in others to in-source competences.

When product development competences were dynamic and very fussy and when no core competences were established in a market, it was of major importance to the business – according to the research – that businesses move at a slower speed in product development. In such cases the businesses should try to focus on finding competences of major importance to the PD project on the global market.

The HRM function could be of use when it came to spotting such competences.

14.1.5 Importance of NB RS NPD Related to Field of PD

The research showed both specific and general examples of the importance of NB RS NPD to SMEs. In general, NB RS NPD was important to all businesses operating in the global markets. We expected in 2003 to see more right speed to product development when products turned to processes, when supply chains were changed to match the new order of right speed in product development and when businesses really began to use and develop the network output in product development. I claimed that 2003's network cooperation in SMEs was only the first generation of NB HS NPD. A lot of businesses would have to change their mission, goals and strategies of their product development activities and even to change their strategic business areas. Furthermore, we would see new types of industry and business models which had never been seen before. Yet, we only knew very little in 2003 about the business possibilities of the new virtual process business areas.

The importance of NB RS NPD to industry seemed to be tremendous. In some cases NB RS NPD could change the whole industry structure. Some cases showed that NB RS NPD could change old, stable, competitive situations. Additionally, it could change market leaders to market followers or could even send the market leaders out of the market. Therefore, NB RS NPD should be handled with care.

In particular the effects of NB RS NPD show directly on the turnover and net profit of the businesses. When businesses succeed in harvesting the actual market continuously because of right speed NB PD, the effects could be tremendous and could be dangerous to businesses if the process was not managed and assessed properly.

It was obvious in 2003 that NB RS NPD could also have a major influence on the production of the businesses because production had to perform a new type of agility in production. Many production lines were designed for mass production today. Such design will fail in situations where NB RS NPD is based on a strong customer enabler – and had to reach out for customisation.

14.1.6 Strength and Weaknesses of NB HS NPD

NB HS NPD seemed not to have any weakness because nearly everybody could agree that if businesses could develop new products in networks continuously and at high speed, this would result in competitive advantage, increased net profit and significant market share. Consequently, this seemed to be the business optimal scenario for a business. However, NB HS NPD also has and had weaknesses.

If NB HS NPD turned out to be made into a routine and when businesses continuously "win the market", then the market will stabilise and products will become standard or very similar. The business will try to focus on costs and will move away from value and perceived value because the market characteristics will change from rather radical innovation, rather dynamic and rather many competitors. The market will begin to show more and more incremental product development at fewer intervals, and a rivalry on price will emerge. I stated this earlier by saying that markets float between stable and dynamic characteristics but NB RS NPD can result in a stable market situation where someone will try to prevent continuous innovation.

This change in characteristics will influence the motivation and the product development work both inside the business and in the network. This situation is very critical to businesses because they can be attacked by competitors who suddenly change the speed of product development and the conditions of market competition. Therefore, the difficult part for managers of product development who want to focus on NB RS NPD is to keep focusing continuously on NB RS NPD. This was, however in 2003, defined as a major scientific issue and a challenge for the future.

Furthermore, if a business dominates NB RS NPD on the market, there will be a tendency to "kill" the good ideas and the entrepreneurs. This can turn out to be critical and can be a weakness both to the business, to the network and to the global market.

NB HS NPD Kills Network

I claimed that NB RFS NPD had to be based on network and network partners. Dependent on this network it is possible to gain the advantage of NB RS NPD. However, as we had mentioned before the network must be based on trust. When the network stabilises or suddenly does not have a product development task, then the network partners will seek other network partners and other tasks. The motivation of the network partners to join other networks and other product development task will increase.

Core Competences Can Slip Out of Network

In the NB RS NPD situation businesses must realise that competences, knowledge and other critical resources are at a potential risk of "slipping" out of the network and being transferred to new networks. It is therefore essential to the business to make a long-time planning on how to join the NB RS NPD. It is also essential to find solutions to what competences and which knowledge the business wants to give as open source, and to which they do not want to give network partners access.

Of course there were and are many other weakness of NB RS NPD but my claim in 2003 was that the major weaknesses were found in the network component. Also the fact that working in networks and specifically in networks that include unknown partners increases the potential of the above-mentioned. Therefore, there was and still is much to learn on how businesses should join NB RS NPD.

14.1.7 Opportunities of NB RS NPD

Through the research I discovered three important competitive advantages when businesses follow a NB RS NPD line of business.

Firstly, as has already been commented on, the businesses can gain the opportunity to "harvest the market at high speed". This will leave the competitors with no or only small parts of the market.

The second important discovery is the possibility for businesses to "play with the money of suppliers and customers". This means that "right speed businesses" barely have to finance their business activities because they are playing with the money of suppliers and customers. In some cases they may even earn money on the financial part of the business transaction which can turn out to be worth more than the primary business – the multi business model approach (Lindgren 2013).

The third important discovery was the observation of the way in which the businesses "played the right HS enablers at the right time". The observation showed that the businesses could save much on costs if they played the right HS enablers. At the same time, the businesses could improve much on value and perceived value.

The PhD project showed that the use of HS enablers to NB HS NPD was very fragmented and different among businesses today. Many businesses focus very single-mindedly on one or two HS enablers shown in this research. Particularly enablers such as 2, 3 and 9 – the modularisation enabler which where the most used HS – were verified in the empirical data. The secondary cases showed, however, that the businesses used a much higher degree of a mixture of HS enablers along with a much more agile use of HS enablers. Furthermore, the research discovered two new HS enablers – the management enabler and the informal product development model enabler.

The research showed that the use of HS enablers was not founded on a strategic product development plan in the SMEs. This is probably the reason for the above-mentioned fragmented and unstructured use of HS enablers. Furthermore, the SMEs seemed to use the same HS enablers for all product development projects which were verified not to be effective in a right time and right speed perspective. The research project learned that the use of HS enablers should be much more flexible and agile initially, during and after the product development project.

From the research project the flexibility and agility of using the enablers to HS product development were claimed to be very much related to product development leadership (PUL) because it was at the PUL level that the strategic use of HS enablers should be decided. The PUM level should secure the tactical implementation of the HS enablers and the "front-end right speed product development performance".

It was therefore important that the HS enablers were considered together with PUL and PUM in a ongoing process where knowledge was absorbed at the front-end of the product development organisation and transferred to the PUM and PUL level.

14.1.8 High Speed – A Competitive Weapon

The high speed tool in product development was shown to be a strong competitive weapon in 2003. Many businesses used this weapon to stay in business when others used the weapon to press their competitors out of the market by continuous high speed product development. However, it was also discovered that businesses could use the NB HS product development strategy to change the conditions on a market. This meant that the businesses could change their position from market follower to market leader. However, this had to be seen in relation to the characteristics of the "field of product development".

14.1.9 Threats of NB HS NPD

The threats of NB HS NPD had already been stressed in the previous chapters. The main threats could be divided to the following areas:

1. The threat of coming to early to the market
2. The threat of coming to late to the market

However, there was another threat which was more industry related and where many businesses are threatened by new entrants (Porter, 1985) using NB HS NPD.

14.1.10 Demands to NB HS NPD

The research showed major demands to NB HS NPD. The research did not have particular focus on these demands to NB high speed in NPD but some demands were very significant to the researcher. These can be seen in Table 14.3.

14.1.11 Barriers to HS NPD

During the research I observed various barriers to NB HS NPD. The research did not have particular focus on barriers to high speed in NPD but some barriers were very significant to the researcher. These can be seen in Table 14.4.

14.1.12 High Speed and Right Speed

The PhD project verified that businesses today have a strong focus on high speed which is related to costs as illustrated in Figure 14.6.

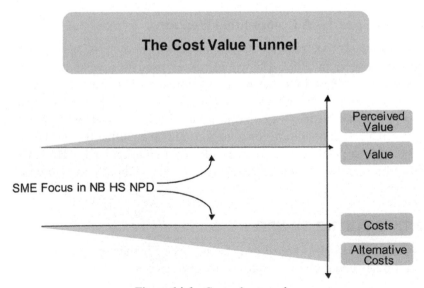

Figure 14.6   Cost value tunnel.

Table 14.3 Demands to NB HS NPD

Demands to NB HS NPD	Literature Search	Case Research	Focus Group Interviews	Survey	Other	Verified
Trust	Yes but in few businesses	Yes but in few businesses and very little	Yes but in few businesses	Yes but in few businesses	Yes	Yes
Motivation	Yes	Yes	Yes	Yes	Yes	Yes
PU model	Yes	Yes – but mainly stage-gate	Yes – mainly stage-gate	Yes – mainly stage-gate	Yes	Yes
The Network partners must be strong	Yes	Yes – but mainly customer and to some extent suppliers Limit networks	Yes – but mainly customers and to some extent suppliers Limit networks	Yes – but mainly customers and to some extent suppliers Limit networks	Yes – but mainly customers and to some extent suppliers Limit networks	Yes
A network culture must be establish in the business	Yes	Yes	Yes	Yes	Yes	Verified
HRM has to play a more active part in PD	No	Yes – but few businesses	Yes – but few businesses and limit efforts	No	Yes – but few businesses and limit efforts	Partly verified
The business must focus on long term relationships	Yes	No	No	No	To some extent	–
The business must focus on long term success criteria	Yes	To some extent – few businesses and limit efforts	No	To some extent	To some extent	Partly verified

Table 14.4 Barriers to NB HS NPD

Barriers to NB HS NPD	Literature Search	Case Research	Focus Group Interviews	Survey	Other	Verified
No Trust	Yes in few businesses	Yes in few businesses and very little	Yes but in few businesses	Yes but in few businesses	Yes	Yes
Motivation	Yes	Yes	Yes	Yes	Yes	Yes
PU – model	Yes	Yes – but mainly stage-gate	Yes – mainly stage-gate	Yes – mainly stage-gate	Yes	Yes
The Network partners are too weak or not used to NB HS NPD	Yes	Yes – but mainly customer and to some extent suppliers. Limit networks	Yes – but mainly customers and to some extent suppliers. Limit networks	Yes – but mainly customers and to some extent suppliers. Limit networks	Yes – but mainly customers and to some extent suppliers. Limit networks	Yes
Culture	Yes	Yes	Yes	Yes	Yes	Verified
HRM is not playing an active part in PD	No	Yes – but few businesses	Yes – but few businesses and limit efforts	No	Yes – but few businesses and limit efforts	Partly verified
The process enabler	Yes	No	No	No		–
The businesses think product instead of process	Yes	To some extent – few businesses and limit efforts	No	To some extent	To some extent	Partly verified
The modularisation enabler	Yes	Yes – but not fulfilled	Yes – but not fulfilled	Yes – but not fulfilled		(+)
E-development is not functioning	Yes – in a few businesses	Yes in very few businesses	Yes but in very few businesses	Yes – but in very few businesses	Yes – but in very few businesses	Partly verified

The research verified that some businesses were reaching in 2003 a limit to HS PD when focusing solely on costs. Businesses had to change their focus and management of NB HS NPD to focus both on costs and value to increase profitability and to achieve right speed.

14.1.13 Short-Term and Long-Term Success Criteria

The research project verified the businesses' strong focus on short-term success criteria and their practically non-existing attention to long-term success criteria. This was very much related to the discussion of product development leadership (PUL) and product development management (PUM). In spite of the encouragement of management literature for business leaders to focus more on leadership, my research project verified that according to product development leaders have far to go in implementing product development leadership (PUL). This lack of focus on PUL seemed to be one of the answers to the strong difficulties of businesses to implement NB HS NPD and also to implement right speed, right cost and right performance product development.

The overall fragmented focus on short term success criteria in the product development of the businesses along with a lack of a strong core in product development projects made it difficult for the businesses to "harvest" the opportunities of NB HS NPD. Furthermore, the above-mentioned did not help the businesses to strengthen their competence to perform optimal NB HS NPD.

14.1.14 NB HS NPD and the Management of PD

PUL and PUM

The research project stressed the difference between PUL and PUM. During the research project I learned that there are significant possibilities of moving the product development activities of a number of the businesses from the top to the bottom of the product development model and process or even outside the formal product development model. When 85% of the general product development activities are incremental, there is a need to analyse the possibility of "pressing" the incremental product development activities out of the top of the product development model and nearer to the market. This is shown in Figure 14.7.

The optimal carrying out of PUL and PUM presupposes that strategic and tactical limits are set for the business's product development activities. This would give the businesses a possibility to carry out "on the market product

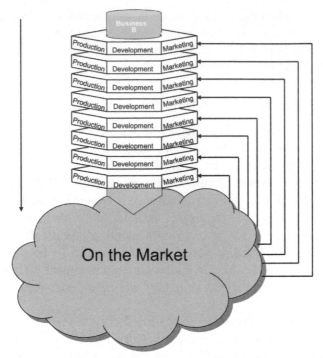

Figure 14.7   Pressure on incremental PD.

development activities". It also puts pressure on changing the management of product development.

The PUL level should focus on short term product development success criteria. Additionally, the PUM level should support them but should focus on the short term success criteria formulated according to and in line with the long term success criteria.

14.1.15 NB HS NPD Model

The research project showed that the NB HS NPD model was very much focused on the stage-gate model up 20 2003. It seems as if there were possibilities for SMEs in 2003 to use more flexible product development models when product developments were into radical, uncertain and risky areas. It was also verified in the research that businesses in the research today had the possibility of increasing the speed of PD by changing their models of PD and choosing among product development models in a more agile way. By doing so, the businesses could gain advantage of cost, speed time and performance.

Furthermore, the research showed that businesses in 2003 had the opportunity to simplify their product development models and processes, to speed their product development projects and to reduce the time to market. This observation was based on the fairly large percentage of product development projects that were placed in the incremental product development area. Many product development projects seemed in 2003 to be better off in terms of time and speed with a reduced or revised number of product development stages and gates. Some product development projects seem to be better off when not entering a formal product development process. This meant that they could easily be developing as "on the market product development".

Summing up on this issue verifies that businesses in 2003 were developing their new products at a too slow speed. It was possible to increase speed in PD much further.

However, the business must always ask the question – how much speed should the product development attend?

"On the market product development" demanded strong product development leadership and strong product development management as illustrated in Figure 14.8. This was and is still important to prevent the business from developing products in strategic and tactical bad areas.

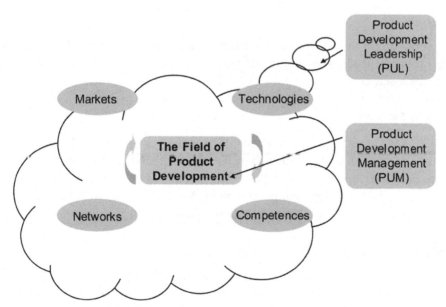

Figure 14.8 PD leadership and PD management.

In the cases of small businesses which did not have a tradition for strategic planning the product development activity was verified to use less formal product development models. Nevertheless, these businesses use product development models the structure of which look very much like a stage-gate model in structure. The time from idea phase to prototype phase was often quicker in small businesses as they did more direct prototyping. However, this activity had to be carefully steered and planned so that the product development process did not turn out to be a strategic failure to the business. However, this is the schism between innovation and motivation to idea generation. The business has to be continuously innovative and at the same time to plan and lead product development. This was proved to be a very difficult challenge to SMEs in 2003.

14.1.16 HS Enablers in Overall Perspective

The use of HS enablers to speed product development seemed to be important up to 2003. Although the knowledge of the use of HS enablers seemed to be more important. A further research should help to understand and give guidelines on how to use the HS enablers. However, some learning had been observed during the research especially on the individual HS enablers.

ICT Enabler

Quite surprisingly the ICT HS enabler turned out to be a rather poorly used enabler to speed product development in the businesses tested in the empirical part of the research.

Customer Enabler

In nearly all businesses the customer enabler was used as a high speed enabler. However, the research shows that there is more potential to the use of the customer enabler. Most businesses had their customers joining the very initial part of the product development process but it seemed as if many businesses still have an inside out perspective to product development.

PU Model Enabler

In nearly all businesses the PU model enabler was claimed to be used very much as an enabler to high speed. The use of this enabler was highly focused on the optimization of existing stage-gate models both overall and fragmented in one or two particular stages or gates.

A more agile and flexible use of the product development model enabler could in some cases speed the product development more than was the case up to 2003 especially in the case of radical product development.

Network Enabler

The research showed that the network enabler left major opportunities for improvement in future. The SMEs did not use the enabler very much today; they are left with major opportunities to increase network cooperation in product development. This could result in higher speed, lower costs and better performance in product development.

Innovation Enabler

Innovation was often seen as a "stop and go process". The innovation enabler was therefore considered and used very little when reflecting on high speed product development. By including innovation and continuous innovation in all phases of the product development process, the ability to speed product development will increase. The businesses which could absorb, develop and improve new products in this way were expected continuously to be able to increase their competitive advantage.

HRM Enabler

The research showed that the HRM enabler was very little or not at all in focus in the businesses. The expectation was that this enabler will become more in focus in the future after 2003. The HRM enabler will be a major tool to increase, develop and overview the network competences both internally in and externally of the business.

Process Enabler

The process enabler was used in very few businesses. It was my prediction that the process enabler would be used much more in future after 2003.

Product to Process Enabler

The product to process enabler was very much concentrated on a completely new way of understanding the product and the product development process. When businesses begin to work with the product as a process with no beginning and no end, then product development was continuously going on, and the product was continuously developing.

Modularisation Enabler

The modularisation enabler was very important and very much used in businesses to speed product development. However, together with other research carried out at the CIP Centre in 2003 my research showed that the effort and results of modularisation seen in a high speed product development perspective have until 2003 been very poor.

The reason for this must be seen as a result of:

1. fragmented use of the modularisation enabler
2. lack of PUL when using the modularisation enabler
3. the misunderstanding that modularisation is always the answer to high speed product development
4. businesses' failure to use the modularisation enabler together with the product architecture
5. the delayed introduction of the modularisation enabler in the product development process

E-development Enabler

When the product or the process can be kept "floating" until the very last minute before the product is introduced to the market, then the business has a competence which is very important seen in a competitive way.

Extra Enablers Found in Research

During the research I found two new HS enablers that were not registered in the secondary case research. The management enabler seemed to be of very big importance to high speed product development; especially in the cases where top-managers place them self in the product development group or network. By doing this, top managers showed that the product development project had significant, strategic importance which normally motivated the network partners to do their utmost. At the same time it was possible to speed the product development process because the manager was able to make decisions right away in the project and when needed.

The second extra enabler which was verified in the research was the informal product development model enabler. This enabler turned out to exist in almost every business. The enabler was used in many different ways to speed product development. This was subject for further research after 2003.

14.1.17 Learning Perspectives for Me as PhD Student

The learning perspectives of a PhD study are enormous. I am very grateful to those who gave me the opportunity to join the PhD study and to those – both friends and colleges – who supported me throughout the process.

In this paragraph I will stress the main learning issues of my PhD study. Such learning issues can be classified as follows:

1. Doing research
2. Writing articles
3. Working with research partners
4. Working close to the industry

14.1.18 Proposals for Future Research in General at the Centre of Industrial Production

On the basis of this research project the following further research activities have been initiated or are under preparation:

TOM Project

During my stay at Polytecnico di Milano I agreed to do a joint research together with researchers in the TOM project. The objectives of the project were to do research on knowledge management and knowledge transfer in NB HS NPD. At this point in time, the research plan has been agreed upon and we are now trying to formulated the questionnaire. The research toke place in Italy, Germany and Denmark.

E-PUIN – E-development Project – a National PD Research Project

One of the interesting HS enablers to product development in networks is the e-development enabler. At the CIP Centre we carried out a research project focusing on e-development in network based product development. Four different Danish counties were invited to join the project. Each of these counties came up with 6–8 SMEs and carry out a joint research together with the SMEs.

Research on New NB PD Models

At the CIP Centre proposed to complete a research on new network based product development models. This was carried out with major European researchers and SMEs forming a European network under the EU 6th Framework Programme. Danish Industry (DI) was the contractor of this new research project which formed the basis for future competitive advantage to European SMEs.

The project was a joint research cooperation with highly esteemed universities in Europe, industrial organisations and first class SMEs who all focus on this subject.

New Journal Article on E-development

One article for a journal on E-development was developed in Spring 2003 in cooperation with research network partners who I had met during the research project.

New Journal Article on Right Speed, Right Performance and Right Costs in NB HS NPD

One article for a journal on right speed in network based product development was drawn up in Spring 2003 in cooperation with research network partners who had met during the research project.

PART VI

Conclusion

The conclusion defines the contents of the research project and gives a statement of the problem – network based high speed product development. In the problem statement a discussion of network based high speed product development is carried out, and a systematic and precise description of the problems and issues pertaining to network based high speed product development is generated. The problem statement also delimitates the focus of this research. This includes the main questions and the main hypotheses for network based high speed product development.

This part is completed with an overview of the structure of the research project.

15

Conclusion

The conclusion will comprise three major areas:

1. The results and findings of the research on NB HS NPD
2. The comments on and analyses of the research methodological performing
3. Proposal for further research

15.1 Results and Findings of Research on NB HS NPD

Through the research significant results and findings could be verified in NB HS NPD.

Firstly, the research identified that time and speed in NPD were measured in businesses up to 2003 as physical time and subsequently transferred into cost. The businesses measured time of product development as the time from idea generation to market introduction. The businesses did not measure time and speed in accordance with time before the idea was generated and after the implementation point of entry for the new product.

However, the research verified that the businesses had difficulties in placing the beginning and the end of a product – the market introduction. This caused major problems when trying to clarify the exact physical time of a product development project and also when comparing two product development projects – the physical speed of the product development projects. The research therefore concluded that time and speed of product development measured in physical time and speed was not relevant if the businesses could not define the beginning of a new product – the exact time that a product idea was born – or the end of a product – the time when a product was implemented to the market. Secondly, through the cases and PUIN focus group meeting the research verified that "the field of product development" never had the same characteristics from one product development project to another. This meant that in terms of speed it was not theoretically possible to compare two product

495

development projects or the work of two product development teams to a new product because the conditions were not the same.

Furthermore, the views on time and speed in product development both by industry and by theorists up to 2003 did not pay attention to speed and time before and after the product development process. The research verified that the product development process runs both before and after the product development process and that product development should not be seen as a project with a beginning and an end but as a continuous product development process.

I therefore firstly proposed another view on the speed and time of product development. This view had a more relative approach to speed and time in product development. This view defined the time of product development as relative time depending on which view you take of the product development process – whether it was the macro view, the business view, the product view, the customer view, the competitor view, the technician's view or the network view. The research verified that these views were very different and present very different pictures of time and speed in NB PD. A competitor could e.g., regard the product development process of one business as running at high speed whereas the customer could at the same time regard the same product development project as running at slow speed.

Thus, in my definition of time and speed was achieved when "a business continuously *hits* the optimal point of entry to the market". This meant the point of entry where it was most business optimal to enter the market with the new product. This was called *right time*, and businesses who continuously achieve the above-mentioned, move at the right speed in product development – which was called the optimal speed and time of network based product development.

The research verified that businesses did not only prioritise HS in the second phase of the PD process – the PD phase. Businesses also prioritised time and speed in the first general phase and even in the last product development phase. When the characteristics of the product development task and of the field of product development were of a certain quality, the businesses also focused on time and speed in these phases.

The research project fully verified how important it was for businesses to focus on right speed and not high speed. However, it was verified that most businesses participating in the empirical part of the research did not focus on right speed but instead narrowly on high speed.

The research verified that the framework of the idea and concept stage/gate of high speed product development based on networks could be preferably measured with a generic model as shown in Figure 15.1.

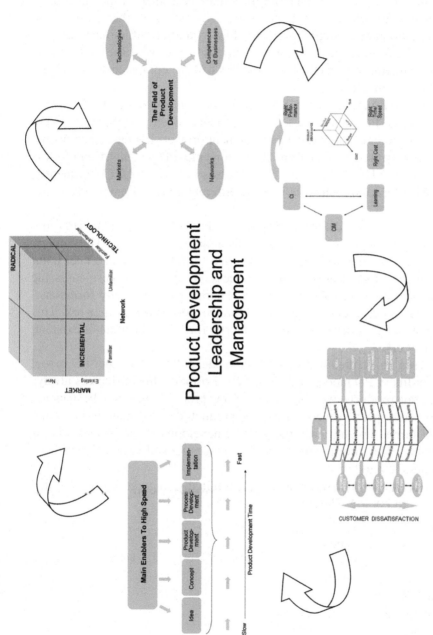

Figure 15.1   Generic model for measuring NB HS PD.

The measurement of wether the innovation task had an incremental or radical innovation character was proposed to come out of the classification related to market (existing/new), technology (familiar/unfamiliar) and network (Familiar/unfamiliar). The innovation task had further to be related to the business competences, whether the business had the competences or needed to adapt new competences.

The research verified that HS PD projects could be divided into radical and incremental PD projects/tasks. Most product development projects were classified as incremental product development projects but the research verified very clearly that the radical and the incremental PD projects do not follow different generic HS PD models and processes in businesses. Consequently, they could not be empirically verified by different generic frameworks in my research.

The research showed that HS PD projects always had a formulated PD core where mission and goals were generally formulated beforehand. The strategy of the product development project was often formulated but at a later point of time in the product development process. In all the business cases the organisational resources and network boundaries were normally formulated initially when the product development project toke place in network. This formulation of network boundaries was mostly formulated within the existent, narrow network.

The research verified that the HS PD model followed another PD model than the ordinary PD model of the business. Formally, the businesses always used the same product development model and process, but when the management stressed speed, several informal product development models and processes turned up and make the product development model and process very "fuzzy". The informal model and processes looked very different from one business to another.

The research verified 8 enablers of the originally, hypothetically identified 10 enablers to HS PD. These were

1. The ICT enabler
2. The customer enabler
3. PU model enabler
4. The Network enabler
5. The HRM enabler
6. The product to process enabler
7. The modularisation enabler
8. The E-development enabler

The innovation enabler and the process enabler could not be indisputably verified. Instead, two new enablers were verified in the case and PUIN research activities together with the other research activities. This was the Management Enabler and the Informal Process and Model Enabler. The informal process and model enabler could also be verified in the survey.

The research verified that businesses used different HS enablers which were very much identical to the HS enablers mentioned above. The use of HS enabler was narrowed down considerably in the individual businesses to one of very few HS enablers. However, the businesses showed that they did not let the HS enablers play a different role according to the PD task or to the characteristics of "the field of product development". The businesses' knowledge of the different roles and possibilities of the HS enablers was insufficient and therefore they did not "play" with the HS enablers in a strategic business economically optimal way. The research could not verify that the businesses mix the HS enablers in an business economically optimal and strategic way as seen in some of the secondary cases.

The customer enabler and the PD model enabler were verified to play the most important role in the upper phase of the HS PD phase. The network enabler could not be verified in the same significant way. It was verified that the businesses use the network enabler very narrowly and with very little use of alternatives to the physical network – e.g., digital and virtual networks.

The research verified that the success criteria for measuring high speed product development based on networks were mainly measured by businesses as short-term success criteria. These were verified as time, cost and performance. The long-term success criteria as CIM, CI and Learning were hardly existent in the businesses. Businesses do not focus particularly on these success criteria.

The success criteria for HS PD were also verified as not significantly dependent on the specific PD project – radical or incremental. This meant that the businesses did not change success criteria dependent on the characteristics of the product development task.

As mentioned above, the right speed PD success criteria could be formulated as short-term and long-term success criteria. However, as verified by my research the businesses only used the short term success criterias. As mentioned above, time, cost, and performance were central success criteria in a short-term perspective.

Finally, the research verified the importance of distinguishing between PD leadership (PUL) and PD management (PUM) when operating NB HS NPD.

The research showed that businesses focusing on product development leadership gain significant business economical result and competitive advantage.

The following impact on NB HS NPD could be verified through the research.

Firstly, my research verified that the pressure on time in NPD created informal processes and models within the business but also within the network. These informal processes existed in nearly all businesses and were most often known by the management level. The management even "relied" on the informal product development processes. Unfortunately, most businesses did not know about the value and cost of these informal product development processes and especially on the alternative costs of these informal product development processes. Furthermore, the learning from these informal models and processes were not transferred to the formal product development system to create value in the formal product development system.

The pressure on time in NPD created both first mover advantage and first mover bad advantage. Both were verified in the empirical research. Most often, the pressure on speed and time in NPD would result in an increase of total costs because pressure on speed and time often diminishes quality and – more essentially – increases alternative costs. However, the secondary literature claimed that costs diminish when speeding time in NPD. Yet, the secondary literature seemed to lack a calculation of alternative cost. The research verified that high speed pressure on product development was seldom related to actual demands in the market and it therefore resulted in unnecessary, high, alternative cost.

The research showed that a focus on right time increases performance and diminishes costs. Right speed in product development also offered the businesses competitive advantage, e.g., first mover advantage, attack on competitors when most inconvenient or keeping the competitors always out of the market. Right speed also offered internal advantages as e.g., liquidity advantage when the business played for the supplier's money.

However, the research also showed that the major part of the businesses tried to obtain high speed and time focused on cost and not value. I claimed that this was a dangerous way of working with high speed in product development because the businesses used an inside out perspective which

- was not most favourable to the optimal point of entry to the market
- was too cost consuming in relation to both direct costs and particularly alternative costs

- forgot to calculate and use the value and perceived value which market opportunities offered the business

The research verified that high speed in network based product development was often miscalculated in relation to radical and incremental product development projects. The research showed how the business used the same focus on speed regardless of whether it was a radical or an incremental product development project. Furthermore, the research showed how many businesses run the same product development model and process on all product development projects independent of incrementalness and radicalness.

Pressure on speed and time would therefore often be impossible to optimise or achieve because of the businesses' failure or lack of possibility to choose the right PD model for the particular product development project. In this respect, my research also showed how a wrong choice of product development model could be fatal to subsequent changes in the demands on the PD project as a result of changes in the field of product development. A stage-gate model had considerable difficulties in relation to cost and time when it came to changing direction and to radically changing the product in the middle of the product development process. Initially, a flexible model would result in higher costs. However, the flexible model would more easily, at a higher speed and at lower costs be able to change direction later in the product development process. None of the two types of product development models were optimal and definitely not to all PD task. We therefore had in 2003 to look into new models of NB RS NPD (Network Based Right Speed New product Development).

My research verified that NB RS NPD could be a very strong competitive weapon. However, it could give both first mover advantage and first mover bad advantage as shown in the research. The result depends on careful leadership by product development management. Another peculiar thing was that NB RS NPD may increase quality when focus was on right speed. If customers, suppliers and other internal and external network partners were involved in the product development process in the right "spots", then the product would perform at a higher level and would enter the market at a business optimal time. This was verified by my research.

The research showed examples of businesses who were pushed into radical PD because of pressure on high speed in product development. This was the case even when the product development task initially seemed to be very incremental. This case showed one of the bad effects of high speed product development when the business did not carefully interpret the product development task or the characteristics of "the field of product development".

The impacts on market, technology and networks by network based high speed product development were verified to be numerous. The research showed evidence of diminishing product lifecycle, continuously lower prices on technology and continuous introduction of new technological features, products and process possibilities. Furthermore, high speed product development proved to push the businesses on the global market into more and often unknown network activities both in the physical network area and in the digital area. However, according to my research the virtual network area had not yet been particularly developed.

The above-mentioned resulted in the hypothesis that another product development model and process had to be developed to achieve right speed of network based product development in the future. I therefore propose the following normative guideline to businesses who want to work with NB RS NPD:

1. Define the task of product development – Incremental or radical
2. Define "the field of product development"
3. Define the success criteria – time and speed – related to relative time and speed – related to right speed and right time
4. Define costs related both to direct and alternative costs
5. Define speed related to right speed and value instead of high speed and direct costs
6. Focus on both value and perceived value
7. Define the product development model – stage-gate and flexible model – but choose the right model according to the task of the product development project
8. Try to be more agile and flexible in the choice of product development models and processes
9. Choose to focus on long-term success criteria and not only on short-term criteria
10. Choose to relate the long-term criteria to both value, perceived value and direct costs and alternative costs
11. Choose to focus both on product development leadership (PUL) and product development management (PUM)
12. Formulate the core of the product development task with a focus on CIM, CI, and Learning
13. Choose the contact limits to network partners by value and advantage and improve the use of network partners to optimize and gain right time and right speed

14. Choose to involve all functions and actors in the product development activities of the business to help improve the product development within right time and right speed
15. Choose to use the high speed enablers with an outside in focus and choose to mix and use more of the HS enablers but in an optimal, right speed way
16. Change the product development focus from a focus on reaching an "encapsulated product" to a focus on a process with many product and process encapsulations

With these proposals for normative guidelines to NB RS NPD I hereby finish my comments on the results and findings of the research project on Network Based High Speed Product Development models and processes.

15.2 Comments and Analysis of Research Methodological Performance

The methodological performance of the research can be commented on in different ways. I have chosen to comment on the following areas:

1. To what extent did I fulfil my ambition in terms of a generalisation
2. Comments on the validity of the research
3. Comments on the reliability of the research
4. Comments on the strength and weakness of the research method
5. Summarising the generalisation, validity and reliability of the research

Ambition of Generalisation

The ambitions of the research were very high, primarily in the generalisation area and the realism area. Through the research project I tried to reach this ambition via a method of triangulation where I focused on the object from different angels and with a multitude of caseresearch, focusgroup, survey and other research methods. Although none of the research methods could be claimed to give a representative picture, I still claim that the collective research methods enabled me "to climb" high on the generalisation axis. This is verified by the fact that each research method produced many identical results and findings to the initially put research hyptheses and questions.

However, as already indicated by Jick's dilemma (McGrath et al. 1982) presented in chapter two, my research would be at risk of not being able to reach as high on the validity arrow. In the following, I will comment on the results of this challenge and will explain how the research covered this dilemma.

Validity Parameter

The challenge on the validity parameter turned out to be just as difficult as predicted. However, a validity discussion in relation to the research project is not considered relevant for all four validity aspects:

Predictive validity, which evaluates the ability of a research instrument or method to predict the future, was not considered relevant in 2003, since the purpose of the research on NB HS NPD was to grasp the present status of businesses involved and draw a picture of NB HS NPD in SMEs up to 2003.

Concurrent validity refers to the extent to which the results of an analysis correspond to results of similar analyses made at approximately the same time. For NB HS NPD and NB RS NPD, no standard of reference exists. Some researchers had elaborated important results on parts of product development and to some extent on product development in network. However, no one had until 2003 focused on NB HS NPD and NB RS NPD.

Another argument for the irrelevance of concurrent validity was that NB HS NPD was dynamic, and results from later research. Observations, interviews, focus group discussions etc. could necessarily be somewhat differently carried out in e.g., another business focus group, another TIP student group etc. according to the task of product development and the characteristics of the field of product development. However, I claimed in 2003 that the generic results and research findings would be the same. On the basis of the research architecture which sought to meet the demands of a triangulation research method I believed that the generic results would be the same if the same analysis was carried out in the same way and at the same time.

Content validity was important for the management of the research on NB HS NPD, since it referred to the extent to which the analysis was representative. An analysis which was not considered representative had a low content validity, whereas a high content validity meant a good balance between analysis and reality. With regard to the research on NB HS NPD, the content validity was only to some extent obtained. The case and focus group interview can by nature never be said to be representative. The survey can be said to be representative; however, the number of SMEs participating in the survey was too small to meet the representative demand. Still, the survey must be said to have a good response rate.

The case interviewed was carried out in five businesses with a general and a specific case analysis in each business. The case research was carried out in most case businesses with an interview with two or more persons responsible for or involved in the product development activities of the business. Of course it would had been better to have had all actors involved in the product development activity represented in the interview, but due to time limits this was not possible. Furthermore, it would also have been better to follow a case from start to end because the memory of the respondents could be influenced by time, policy and other components. However, when speaking to the respondents I had no impressions that this was the case.

Obviously the case interviews could only be representative within the single business but combined with other case, a stronger representativeness could be obtained. Of course it would have been better to carry out more than five case researches but this was not possibly within the time given.

The focus group interview was carried out with persons responsible of product development in ten different businesses. The selection of businesses was carefully made with a view to business-to-business businesses and to different lines of business. The last criterion was made because it was essential to the research in an explorative perspective to get as many aspects of NB HS NPD as possible. A major difference in lines of business proved to give the intended many aspects of NB HS NPD which surely could not have been obtained if all businesses had been in the same line of business.

The survey was distributed to a representative amount of SMEs in the business-to-business market. However, as will be verified later in this project, the amount of answering businesses was not high but still allowed me to draw an explorative picture of how NB HS NPD was carried out in SMEs. However, it was not possible to make general conclusions on specific lines of business.

Some other research activities have been carried out on the subject NB HS NPD. None of these activities could be said to match the criteria of representativeness. However, neither was this my intention. Instead it was my intention to find add-on information to NB HS NPD and additionally to support findings in some of the other research activities. Furthermore, especially the TIP project allowed me to come closer to NB HS NPD and to closely observe the process of NB HS NPD.

My visit to Italy added new dimensions to NB HS NPD although this could not be characterised as representative add-on to the research.

The Licentia case supported my earlier findings in the case, focus group and survey research.

Summarising on the content of validity it could be verified that each individual empirical research activity could not be said to be representative. Yet, some of the activities were more representative than others. However, this was not my intention considering the fact that the research was carried out in an explorative perspective.

Altogether the research activities try to match a triangulation on the research focus NB HS NPD. In this perspective, the general, generic findings must be said to have succeeded to the highest degree.

The construct validity refers to the extent to which the analysis maps the construct that it was intended to map. The construct validity was a complex matter and forms a general idea of something formed in the mind by combining a number of pieces of information. This could be defended by expert opinion. The expert opinion related to the construct validity of my research project comes firstly from my supervisors of the research project, colleagues at the CIP Centre and from discussions with Professors and seniors at the Polytecnico di Milano. Secondly, expert opinions come from those individuals accessing my research project. I claimed that this had succeeded to some extent, but could have been improved.

The explorative and semi-structured interview approach was exactly what Wind (1973) and Aaker and Day (1983) propose when researchers and businesses deal with product development and marketing research tasks that were into the radical and new to the market area. This was my immense inspiration for choosing this way of researching along with the fact that I felt that I had to try out and use the solutions and methods of the science of product development on a rather radical area of product development.

Reliability Parameter

With regard to reliability and instruments used in the research on NB HS NPD, the question was whether or not another researcher or research group would have developed a significantly different action plan for researching on NB HS NPD. The argument defending the reliability of the models and processes of the research project is found in:

1. The data collection, which includes not only a few businesses, but nearly 180 individual businesses.
2. The strong involvement of businesses, participants, students in case reasearch, focus group interviews, survey interviews, student NB HS NPD projects etc.

3. The strong involvement of management of product development in businesses involved in the research project.
4. The strong involvement of other researchers in the research project.

Collection of data included all phases of the product development process; however, with a strong focus on the upper part of the product development process. A high degree of verification was demanded of e.g., the enablers to high speed product development due to the number of participants involved in the research process.

The strong involvement of businesses was explained by the degree actuality of the project and by intense discussions in business networks, internal businesses and in research. All parties were interested in discovering how to perform high speed within network based product development.

The strong involvement of product development management was important in the discussion of relevance and priority of the suggested hypothetical models and processes to NB HS PD. Since costs, benefit and priority of each suggested model and process to NB HS NPD were discussed or considered at several meetings with the persons responsible of product development and with the management, only the significant model and processes survived and were verified.

Therefore it was not considered likely that a similar research would have given significantly more or different findings and suggestions than the findings presented in Chapter 16.

All in all, it could be concluded that the research undertaken in the research project NB HS NPD fulfils the aspects regrading validity and reliability. This supports the credibility of the research on NB HS NPD and the results obtained.

Strength and Weakness of Research Method

Previously, I had stressed the strength of my research on NB HS NPD: I will now comment on some of the weaknesses I had discovered when reflecting on my research.

Firstly, I had to some extent reflected on the question if the participants in the research had always understood the questions put to them. Some of the questions could be said to be complicated and to some extent only understandable by experts or researchers.

Another of my concerns was whether I had influenced the respondents by my cooperation with him/her. Throughout the research I had to be careful not to give answers and solutions to the respondent. This has been a difficult task for me as a researcher. Especially, when I could see businesses having major difficulties formulating a new NB HS NPD strategy. I had to be careful both in this research result and to future investigations.

In the focus group discussions held together with my colleagues, it was a major challenge for me not to influence the participants allthough the discussions were very close and intense.

I also hesitate in the evaluation of my findings because I saw the danger imbedded in the fact that most of my cases were made on persons responsible of product development management. My reserach could have been improved if I had conducted more interviews with diversified respondents as I did e.g., with GSI Lumonics. Due to time limits this was not possible in any of the other cases.

The last fault I would stress was the analysis and understanding of the answers I got in the research. I may have misjudged and misanalysed some answers. However I felt very convinced that the answers were given in an open and honest way. Additionally, when I received the same results and findings through more investigations it must be claimed that the risk was not significant.

Summarizing on Generalisation, Validity and Reliability

When summarizing on generalisation, validity and reliability of the research project NB HS NPD I claimed that the research was challenged with exactly the same problems as predicted in Chapter 2. My research was challenged on the validity dimension but reached very close to my ambition on the generalisation and realisation axis. I did not reach as highly as I wished on the generalisation axis because the survey did not produce as many answers as I would have wished. This was an additional challenge to my further research to reach a deeper understanding of NB HS NPD and NB RS NPD to reach a higher level on the generalisation axis.

Looking back on the research, I realise that my research very much relates to the area of the the UK OM tradition of research as shown in Figure 15.2.

I must say that not every part of this research did originally intend to reach this profile of research tradition. Parts of my research project turned out to develop like this by coincidence. A major reason for this was my own curiosity to know science and to a continuous understanding of different views on the understanding of NB HS NPD.

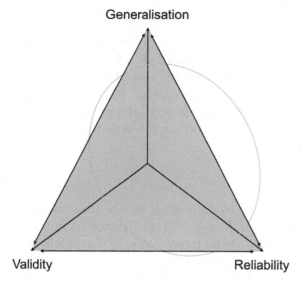

Figure 15.2 Trade-offs for scientific research.

Source: Adapted from McGrath, J. E., 1982.

With these comments I finalise my comments and considerations of the research methodology.

15.3 Proposal for Further Research

On the basis of my research project the following further research activities had come up as important in order to reach a higher and increased understanding of the research area. Some of these further research activities were already running or under establishment. Nevertheless, I will comment on some of the important ones.

Tom Project

During my stay at Polytecnico di Milano I followed the Tom Project as described in Chapter 13. In this connection I agreed to carry out joint research together with researchers in the TOM project on knowledge management and knowledge transfer in NB HS NPD. At this point in time in 2003, the research plans had been agreed upon and we began to formulate the questionnaire. The research was conducted in Italy, Germany and Denmark herafter.

E-PUIN Project

The E-PUIN E-development project – a national product development research project – had been developed. On 25 February 2003 the project was approved. One of the interesting HS enablers to product development in networks was the e-development enabler. We had therefore proposed a project focusing on e-development in network product development to four councils to collect four clusters with each 6–8 SMEs to do a joint research together with these businesses.

EU-EPUIN Project

Research on new network based product development models focused on European SME networks in Europe – EU 6th framework programme seemed to be an important focus area in the future to develop European SME businesses to stay competitive on the global market.

Together with Dansk Industri (an association of major Danish businesses) as contractor we had proposed this new research project on new network based product development models. We hope the project may form the basis for major research findings to develop competitive advantage to European SMEs.

The projects were joint research projects with highly esteemed universities in Europe, with industrial organisations and with first class SMEs who all focus on the subject.

New Journal Article on E-development

One article for a journal on E-development in the spring 2003 was intended to be prepared together with research network partners who had meet during the research project.

Learning in Entrepreneur Businesses

The RESME research group www.RESME.dk had become a new project on learning in entrepreneur businesses. This project was initiated in Spring 2003 and would be running for the next 2–3 years. I contributed with journal and conference writings in this research. I was particularly interested in learning in NB NPD by entrepreneurs.

E-Content Entrepreneurship

During my participation in the research group RESME www.RESME.dk I developed a new EU project on entrepreneurship. Because of my considerable interest to understand the first phases of a product development project, this project offered more access to how entrepreneurs performed NB HS NPD in the initial phases of product development.

One Journal Article on Right Speed, Right Performance and Right Cost in NB HS NPD

One article for a journal of right speed in network based product development was planned for Spring 2003. The article was prepared and published in cooperation with research network partners.

Book on Network Based Right Speed Product Development

Finally, I began to prepared a new book on the basis of my findings in this research project on network based right speed product development. My ambition was to finish this book in the Autumn 2003 but because of a change in my research interest and research focus into business model innovation and multi business model innovation this book is still waiting to be finished.

With these proposals for further research, I hereby finish my comments on further research on Network Based High Speed Product Development. At the same time I finalise and finish the conclusion on Network Based Product Development Models and Processes.

References

[1] Abell, D. F., 1980, 'Defining the Business: The Starting Point of Strategic Planning', Prentice-Hall, Inc., NJ.

[2] Abnor & Bjerke (2009) Methodology for Creating Business Knowledge SAGE Publications Ltd.

[3] Albaum, G., Strandskov, J., Duerr, E., Dowd, L., 1994, 'International Marketing and Export Management', 2nd edition, Addison-Wesley.

[4] Baker, M. & Hart, S., 1999, 'Product Strategy and Management', Prentice Hall, Harlow, U.K.

[5] Balachandra, R., 2000, 'An Expert System for New Product Development Projects', Industrial Management & Data Systems 100/7, pp. 317–324.

[6] Bartezzaghi, E. Spina G., Verganti R., 1996, 'Time Drivers and Systemic Process Re-Engineering', Advances in Production Management Systems, IFIP.

[7] Bessant, J., (2000) 'Challenges in Innovation Management', Centre for Research in Innovation Management, University of Brighton.

[8] Boer, H., 2001, 'Operational Effectiveness, Strategic Flexibility or Both? – A Challenging Dilemma'.

[9] Bohn, K. R. & Lindgren, P., 2002, 'Main enablers to Network Based High Speed Product Development', September 2001 (working paper).

[10] Bohn, K. R. & Lindgren, P., August 2002 'Network Based High Speed Product Development'.

[11] Bohn, K. R. & Lindgren, P., May 2002, 'Network Based High Speed Product Development – A Framework Model'.

[12] Bohn, K. & Lindgren, P., 2002, 'Right Speed in Network Based Product Development and the Relationship to Learning, CIM and CI', CINet, Helsinki.

[13] Bohn Kim R. and Peter Lindgren (2013) Produktudvikling i Netværk River Publishers ISBN: 9788792982315.

[14] Bolwijn, P. T. & Kumpe, T., 1998, 'Marktgericht ondernemen. Management van continuïteit en vernieuwing', Van Gorcum, Assen.

[15] Booz, Allen & Hamilton, 1982, 'New Products Management for the 1980s', New York.

[16] Caffyn S. J., 1998, 'The Scope for the Application of Continuous Improvement to the Process of New Product Development', University of Brighton.

[17] Child, J. & Faulkner D., 1998, 'Strategies of Co-operation – Managing Alliances, Networks, and Joint Ventures', Oxford University Press, Oxford.

[18] Christensen, J. F., 1992, 'Produktinnovation – proces og strategi', Handels-højskolens Forlag, København.

[19] Cooper, R. G., 1993, 'Winning at New Products – Accelerating the Process from Idea to Launch', 2nd edition, Perseus Books.

[20] Cooper, R. G., 1995, 'Developing new products on time, in time', Res. Tech. Management 38(5) 49–57.

[21] Corso, M. et al., 2000, 'Knowledge Management in Product Innovation: An interpretative Review', International Journal of Management review Vol. 3, No. 4, pp. 341–352.

[22] Corso, M., Martini, A., Pellegrini, L. & Paolucci, E., 2001, 'Knowledge Management in SMEs – Does Internet Make a Difference?', Italy.

[23] Crawford, C. M., 1977, 'Marketing Research and the New Product Failure Rate', Journal of Marketing, 41 (April), pp. 51–61.

[24] Crawford, C. M., 1991, 'New Product Management', 3rd edition, Richard D. Irwin, Inc.

[25] Drejer, A. & Riis, J. O., 2000, 'Competence Strategy', Børsens Forlag, ISBN 87-7553-740-0.

[26] Fine, C. H., 1998 'Clockspeed', Perseus Books.

[27] Gieskes, J., 2001, 'Learning in Product innovation Processes – Managerial Action on Improving Learning Behaviour', ISBN: 90-365-1651-x.

[28] Goldman, Nagel & Price, 1998, 'Agile Competitors and Virtual Organisations', Van Nostrand Reinhold, New York.

[29] Grunert, K. G. & Harmsen, H., 1997, 'The Interaction of R&D and Market Orientation in Improving Business performance – An Empirically Based Framework for Understanding what Drives Innovation Activity', IMDA, Hummelstown, PAS.

[30] Grunert, K. G., Harmsen, H. & Göransson, G., 1997, 'Innovation in the Food Sector: A Revised Framework', Blackie A&P, London.

[31] Hansen, P. K. & Thyssen, J., 2000, 'Continuous Improvement and Modularization in the Product Development Process, CINet, Denmark.

[32] Håkansson, H. & Johanson, J., 1992, 'A Model of Industrial Networks', in Axelsson, B. and Easton, G. (eds) Industrial Networks: A New View of Reality, pp. 28–34, Routledge, London.

[33] Hamel, G. & Prahalad, C. K., 1994, 'Competing for the Future', Harvard Business School Press.

[34] Hein, Lars & Andreasen, M. Myrup, 1985, 'Integreret produktudvikling', Jernets Arbejdsgiverforening, København.

[35] Hollensen, S., 2003, 'Marketing Management – A Relationship Approach', Pearson Education Limited, Essex, U.K.

[36] Hørlück, J., Kræmmergaard, P., Nielsen, J. S. & Lindgren, P., 2002, 'Interorganizational Project Management", CINet, Helsinki.

[37] Kotler, P. & Armstrong, G., 1996, 'Principles of Marketing', 7th edition, Prentice Hall International Editions, NJ.

[38] Kotler, P., 2000, 'Marketing Management', the Millennium Edition, Prentice Hall International, Inc.

[39] Leifer, R., December 2002, 'Critical Factors Predicting Radial Innovation Success', Rensselaer Polytechnic Institute, NY.

[40] Lindgren, P. and Ole Horn Ramussen 2013, Journal of Multi Business Model Innovation and Technology, 135–182 River Publishers.

[41] MacCormack A. & Verganti R., 2002, 'Managing Uncertainty in Software Development: How to Match Process and Context'.

[42] MacCormack A., Verganti, R. and Iansiti, M., January 2001, 'Developing Products on "Internet time": The Anatomy of a Flexible Development Process', Management Science/Vol. 47, No, 1.

[43] Bohn, K. R. & Lindgren, P., 2000, 'Begreber i netværksbaseret produktudvikling under høj hastighed', Working Paper, Denmark.

[44] MacCormack, A. & Iansiti, M., 1997, 'Product Development Flexibility', 4th International Product Development Management Conference, EIASM, Stockholm, Sweden.

[45] Nonaka, I., & Takeuchi, H., 1995, 'The Knowledge-Creating Business', New York: Oxford University Press, pp. 93–138.

[46] Pine, J. B., 1993, 'Mass Customization', Harvard Business School Press, Boston.

[47] Prahalad & Hamel, 1990, Harvard Business Review.

[48] Riis, J. O. & Johansen, J., 2001, 'Developing a Manufacturing Vision', International Working Coneference on Strategic Manufacturing, Denmark.

[49] Rosenau, M. D., 1993, 'Managing the Development of the New Products', ITP, pp. 39–41.

[50] Sanchez, R., 1996, 'Strategic Product Creation: Managing New Interactions of Technology, Markets and Organizations', European Management Journal Vol. 14. No. 2, pp. 121–138.

[51] Sanchez, R., 2000, 'Product, Process, and Knowledge Architectures in Organizational Competence', Research Working Paper, Oxford University Press, 2000–11.

[52] Sanchez, R., 2001, 'Modularity, Strategic Flexibility, and Knowledge Management', Oxford University Press.

[53] Smeds, R., Olivari, P. & Corso, M., 2001, 'Continuous Learning in Global Product Development: A Cross-Cultural Comparison', Finland.

[54] Takeuchi, H. & Nonaka, I., 1986, 'The New New Product Development Game', Harvard Business Review, Vol. 64.

[55] Ulrich, K. T. & Eppinger, S. D., 2000, 'Product Design and Development", 2nd edition, Irwin McGraw-Hill.

[56] Verganti, R., MacCormack, A. and Iansiti, M., 1998, 'Rapid Learning and Adaptation in Product Development: An Empirical study on the Internet Software Industry', EIASM 5th International Product Development Management Conference, Como, Italy (25–26 May 1998).

[57] Verwest, P. et al., 2005, 'Smart Business Network', Von Norstrand.

[58] Wheelwright, S. C. & Clark, K. B., 1992, 'Revolutionizing Product Development. Quantum Leaps in Speed, Efficiency, and Quality', Free Press, New York, NY.

[59] Wind, Y., 1975, 'Product Policy'.

Ph.D. Review Comity 2003

Professor John Bessant, University of Exeter, United Kingdom
Professor of Innovation and Entrepreneurship

Originally a chemical engineer, Professor John Bessant has been active in research, teaching and consultancy in technology and innovation management for over 25 years. He currently holds the Chair in Innovation and Entrepreneurship at Exeter University where he is also Research Director. In 2003, he was awarded a Fellowship with the Advanced Institute for Management Research and was also elected a Fellow of the British Academy of Management. He served on the Business and Management Panel of both the 2001 and 2008 Research Assessment Exercises. He has acted as advisor to various national governments and to international bodies including the United Nations, The World Bank and the OECD.

Professor Bessant is the author of over 20 books and monographs and many articles on the topic and has lectured and consulted widely around the world. His most recent books include Managing innovation (now in its 4th edition) and High involvement innovation (both published by John Wiley and Sons).

Awards and Honours

- Elected Fellow, British Academy of Management
- Fellow, Sunningdale Institute
- Senior Fellow, Advanced Institute of Management Research

Professor Anders Drejer, Aalborg University, Denmark

Full Professor in strategy and business development at Aalborg University from 2010. Anders Drejer holds a Full Professorship in strategy and business development at Aarhus School of Business and holds a Ph.D. in business strategy and competence development from Aalborg Universitet from 1996.

Anders Drejer was head of the Strategy Lab at Aarhus School of Business (2007–2010) at Aalborg University. Member of the NEWGIBM research project (2005–2007, Ministry of Science, Technology and Innovation.

Anders Drejer has carried out several Research management functions: 1996–1999: Member of the management group for "Center for Technology Management", 1997–1999: Member of the project management of the EU project "Enhancing Competitiveness of SMEs via Innovation", 1999–2003: Member of the management group for "Centre for Industrial Production", CIP, 1999–2003: Programme manager for Intelligent manufacturing and New Product Development research programmes under "Centre for Industrial Production", CIP.

Anders Drejer has further in 2001 been Keynote speaker at "the 10th International Conference on Management of Technology IAMOT 2001", June 19–22, Lausanne, Switzerland, 2001. He was from 2002–2003 Chairman for the organizing committee for the Conference, Managing Innovative Manufacturing (MIM) 2003 and from 2003 Leader of the Strategic Management Group at Aarhus School of Business.

Anders Drejer has besides many international research publications published several books focused on management and practice. Anders Drejer is a often used speaker at conference, coach and board member in Danish businesses.

Associate Professor Svend Hollensen, University of Southern Denmark, Denmark

Svend Hollensen is Ph.D. (Dr) and Associate Professor of International Marketing at University of Southern Denmark (Sønderborg). He has practical experience from a job as International Marketing Coordinator in a large Danish multinational enterprise (Danfoss) as well as from being International Marketing Manager in a business producing agricultural machinery.

After working in industry he received his Ph.D. from Copenhagen Business School (CBS) in 1992.

He has published articles in international recognized journals and is the author of globally published textbooks, eg Global Marketing, 6th ed, which was published in September 2013. Indian and Spanish editions have been developed in co-operation with co-authors. The textbook Global Marketing has also been translated into Chinese, Russian, Spanish and Dutch.

By the end of 2014 his 'Marketing Management', 2nd edition, will come out in an Arabic translation, published in Amman, Jordan.

Currently (2014) 'Global Marketing' is no. 1 in sales outside United States, and no. 2 or 3 worldwide (in the segment 'International Marketing' textbooks). The lifetime sales of that textbook is more than 100.000 copies.

Svend Hollensen's research interests are within Relationship Marketing, Globalization, Global Branding and Internationalization of businesses (especially 'Entry mode decisions').

In February 2008 he was awarded the BHJ Fund award for 'Excellent Education and Teaching' at University of Southern Denmark.

Through his business, Hollensen ApS, Svend has also worked as a business consultant for several multinational businesses, as well as global organizations like World Bank.

CPSIA information can be obtained
at www.ICGtesting.com
Printed in the USA
LVOW02*1353250417
531591LV00013B/1/P